THE WORLD'S CLASSICS
THE SWISS FAMILY ROBINSON

JOHANN DAVID WYSS (1743–1818), who first wrote the story of a castaway Swiss minister and his family with the purpose of amusing and educating his children, was a chaplain to the Swiss army for many years before affiliating himself with the ministry at Berne, in 1794.

His son, Johann Rudolf Wyss (1781–1830) was born in Ber and educated in theology and philosophy at Tübingen, Göttingen, and Halle. At the age of 23, he was appointed professor of philosophy at the Berne Academy. An accomplished writer and poet, an educator and editor, who devoted himself to collecting and publishing the folklore of his homeland, and who in 1811 composed the Swiss national anthem, *... Thou, My Fatherland?*, to the tune of *God Save the King*, the younger is best known for having completed and seen publication his father's novel, originally entitled *Der Schweizerische Robinson* (Zurich, 1812, 1813), authorship of which is generally credited to him.

Baroness Isabel de Montolieu, the most obscure of the three contributors to *The Swiss Family Robinson*, having translated the German version into French, greatly enlarged the original by means of sequels published between 1824 and 1826.

JOHN SEELYE is a Graduate Research Professor at the University of Florida. He has written novels, literary criticism, and studies of American culture, including *The True Adventures of Huckleberry Finn* (1970, 1988) and *Prophetic Waters: The River in Early American Life and Literature* (1977). He has edited Mark Twain's *Life on the Mississippi* in World's Classics.

THE OXFORD CLASSICS

THE SWISS FAMILY ROBINSON

JOHANN DAVID WYSS (1743–1818), who first wrote the story of the Swiss Family Robinson for his family, had the purpose of amusing and educating his children, while it led him to the Swiss army (military) service while himself with the military.

JOHANN RUDOLF WYSS (1781–1830) was born in Bern and expanded on theology and philosophy at Tübingen, Göttingen and Halle. At the age of 23, he was appointed professor of philosophy at the Bern Academy. An accomplished writer and poet in German and in Italian, he devoted himself to collecting and publishing the folk-songs of his homeland, and was in 1811 prompted that the Swiss national anthem, Call's Wood ('Rufst du mein...') in the time of Carl Society. Karl Wyss the younger's brother became well acquainted and worked to publication, his dictionary of words ... eventually ... of the Schweiz ... Idioticon (Leipzig, 1812–1822), authorship of which he finally decided to him.

Barbara Kimber held the Foundation, the most important of the foundation in the early twentieth century, having translated the German version into French, then published the original by means of which publication occurred, 1954 and 1959.

John Seelye is a Graduate Research Professor at the University of Florida. He has written novels, literary criticism, and studies of American culture, including The Exclusiveness of Barbarous (1970, 1998) and Prophetic Waters: The American Land... American Literature, 1979. He has edited Mary Rowland...

Life on the Mississippi, Classics.

THE WORLD'S CLASSICS

═══

JOHANN WYSS

The Swiss Family Robinson

═══

Edited with an Introduction by
JOHN SEELYE

Oxford New York
OXFORD UNIVERSITY PRESS
1991

Oxford University Press, Walton Street. Oxford OX2 6DP

Oxford New York Toronto
Delhi Bombay Calcutta Madras Karachi
Petaling Jaya Singapore Hong Kong Tokyo
Nairobi Dar es Salaam Cape Town
Melbourne Auckland

and associated companies in
Berlin Ibadan

Oxford is a trade mark of Oxford University Press

Introduction, Note on the Text, Select Bibliography, Text © John Seelye 1991
Explanatory Notes © Robert A. Kosten 1991
This edition first published 1991 as a World's Classics Paperback

British Library Cataloguing in Publication Data
Wyss, Johann David
The Swiss family Robinson.—(The World's classics)
I. Title   II. Seelye, John D.
833'.6 PT2583.W9S313
ISBN 0-19-282724-3

Library of Congress Cataloging in Publication Data
Wyss, Johann David, 1743-1818
[Schweizerische Robinson. English]
The Swiss family Robinson. / Johann Wyss ; edited with an
introduction by John Seelye.
p. cm.–(The World's Classics)
Translation of : Schweizerische Robinson
Includes bibliographical references (p.   )
Summary : Relates the fortunes of a shipwrecked family as they
adapt to life on an island with abundant animal and plant life.
[1. Family life – Fiction. 2. Survival – Fiction.]
I. Seelye, John D. II Title. III. series.
PT2583.W9S313 1991   833'6—dc20   91-7830
ISBN 0-19-282724-3

Typeset by Pure Tech Corporation, Pondicherry, India.
Printed in Great Britain by
BPCC Hazell Books
Aylesbury, Bucks

CONTENTS

INTRODUCTION

IF it can be said that the English novel begins with *The Adventures of Robinson Crusoe*, then it can also be claimed that *The Swiss Family Robinson* occupies a similar place in the history of novels written for children. True, there were other long books written for young readers that appeared well before it was published, books of a similarly didactic design, but few of these have had the long shelf life of *Swiss Family Robinson* and none has survived deep into the twentieth century. The earnest purpose of books like J. H. Campe's *Robinson der Jüngere* (1779), or Thomas Day's *Sandford and Merton* (1783–9)—which within a century of its publication was inspiring parodists not imitators — simply could not cross the great divide which we can identify with the emergence of Robert Louis Stevenson on one side of the Atlantic and Mark Twain on the other. It is a division that demarcates a later tradition of books written largely to entertain from those of the earlier dispensation, in which entertainment was a thin coating upon heavily instructional material. It is a tribute to the authors of this story about a family marooned on a tropical island that they so cannily lightened the burden of instruction with an almost breathless series of encounters, discoveries, and adventures, that children have continued for almost two centuries to find the book exciting and readable.

I have written 'authors' advisedly, for the credit often given on the title-page to Johann Rudolf Wyss is misleading, as is the title itself. For not only is the book not about a Swiss family named 'Robinson', it was not written by Johann Rudolf Wyss (1781–1830) but by his father, Johann David (1743–1818), a former army chaplain and pastor in Berne. The younger Wyss, a poet and composer as well as a professor of philosophy, took up the manuscript that had been written for himself and his brothers as children and revised it for publication, a first volume appearing in 1812, to be followed by a second the next year. This text was then translated from German into French by Baroness Isabelle de Montolieu, who apparently made a few minor changes, which were then incorporated in the first

English translation, from the French, by either William Godwin
or his second wife, Mary, in 1814. Then in 1816 the Godwins
brought out a 'Second Edition' 'from the German of M. WISS',
which remained for many years the basis for the standard
English edition.

The Baroness had in the meantime not been idle. The story
written by the Wysses had ended two years after the shipwreck,
with the family still on their island. A rescue ship had arrived,
but before the castaways could embark, a storm forced the
rescuers out to sea, taking with them the father's manuscript
account of his family's adventures, which was carried back to
Europe and published. The possibilities of making much more
of a good thing inspired de Montolieu to continue the story in
five additional volumes, published between 1824 and 1826. In
1847 this continuation was translated into English by J. D.
Clinton Locke, and, combined with a somewhat modified
version of the Godwin translation, first published in 1826 as
the 'Sixth Edition', Locke's translation remained in print
throughout the century. Yet another sequel was provided in *Le
Pilote Willis*, by Adrien Paul, translated as *Willis the Pilot* by 1857,
which continued the adventures of the two sons who left the
island at the end of Montolieu's version but who took almost
as long as Ulysses before reaching home. By mid-century,
moreover, revisionists were already at work, reducing the bulk
of the expanded text or inserting 'improvements' of their own.
At least three major new translations had been made by the
end of the century, by editors who took considerable liberties
with the story, either reducing or expanding the pious and
moralizing elements or introducing new adventures while cutting out old.

This continuing process, by means of which editors and
translators felt free to tinker with the story, taking their cue
from the Baroness de Montolieu, is relatively unique in the
annals of literature, though certainly the textual history of
Robinson Crusoe (as adapted for children) provides a parallel. I
intend to make some distinctions between the first part of the
story, with its definably Swiss imprint, and the latter half, which
evinces considerable differences in tone and emphasis introduced by the French author, but here the point to be made is
that *The Swiss Family Robinson*, as a text, resembles the

adventures of the family themselves, in that it is a communal, even corporate, product. Moreover, like those adventures and their ideological implications, the fact of corporate authorship is virtually inseparable from the intellectual and social climate of the period in which the book was first written. We can start from the point when Wyss senior set down a series of adventures in which he, his wife, and his four sons were cast as castaways, in a continuing drama clearly designed as both a family entertainment and an educational device, and end with that critical moment in de Montolieu's continuation of the story when the oldest boy, now a young man, rescues an English girl from her own lonely island. These chronological, compositional points bracket the late period of the Enlightenment and the early stages of Romanticism, resulting in a text with an impressive contextual range.

Thus the book as first conceived was a quintessential product of the 1790s, and was born on the high tide of the Enlightenment, with its emphasis on the progressive improvements men could make to their world thanks to the power of God-given reason. Indeed, most of the objections that have been made to the book, including the unlikely assemblage of flora and fauna found by the Swiss family on their island and the rather easy time they have of it creating their arcadian colony, can be explained by the Enlightenment context. But perhaps most important in that regard is the role played by the pastor as the informing centre of his family, who as loving husband and father combines the roles of fond parent and thoughtful tutor. To find the source of this formulaic construct, we must go to Rousseau's *Émile*, and there also we will find the original seed for the Swiss family's story, with its obvious indebtedness to Defoe. For in determining the ideal formative environment with which his pupil was to be surrounded, Rousseau legislated a single companion mentor (himself), and provided Émile with only one book for his diversion, *Robinson Crusoe*.

It has been said that the Wyss story puts Defoe's hero 'in the bosom of his family', but there is very little in common between the two books apart from the castaway situation. We should perhaps think of the Wyss story as placing Rousseau, not Crusoe, in the midst of *his* family, for it is through the

French savant's theory of education that Defoe's tale of terrible hardships overcome is filtered. Young Crusoe is cast away because he disobeyed his father, breaking one of the great Commandments, and because he ignored divine warnings to abandon the seafaring life and return to his proper station in life. His is a penal as well as a penitential isolation, and he is granted a companion only after he has been stranded on his island for the number of years equal to his age when he was first cast up from the sea; a Levitical sentence, surely. Defoe's story is a marvel of articulated adversity, signalled by the cross concealed in Crusoe's name. It is a Calvin-inspired version of the fortunate fall, by means of which the narrator–hero comes to acknowledge the absolute authority of God. The God who rules over the Swiss family's island is certainly omnipotent and ever-present, but there is none of the punitive element in the Wyss story, quite the opposite.

Defoe's novel illustrates the dark side of Calvin's notion of election by grace, Crusoe being one of the chosen few who is selected for special treatment so that he will come to his senses —and salvation—a rather ponderous mechanism that involves the destruction of a ship and its crew so that the hero will see the light. But Wyss's family are hardly sinners—Jonahs— singled out for punishment. They are victims of economic recessions that followed the French Revolution, and have set sail as colonists for the Australia–New Zealand area. The ship-wreck can be laid to God's account, in that all must be attributed to him in His world, but if anyone is 'punished' it is the captain and crew who desert the family during the storm and are drowned in their attempt to escape. The family, forced to remain on the ship, are preserved, and everything that happens to them thereafter is for their good, a divinely ordered dispensation to which the minister–father constantly calls his children's attention. And yet, despite these frequent references to God's goodness, the emphasis of the story is hardly religious, for having celebrated their first Sabbath, the family (or at least the narrative) neglects that holy day thereafter. Instead, succeeding events promote the idea that with the aid of proper tools and information, there is nothing that human kind can-not accomplish—a distinctly Enlightenment notion, derived certainly from *Robinson Crusoe* but without identifiable scriptural authority otherwise.

A sharp contrast to the relative ease with which the family overcome difficulties is provided by an analogous text already mentioned, J. H. Campe's retelling of *Robinson Crusoe* for children, in which Robinson 'the Younger' is denied his shipload of supplies during the first part of his ordeal, and is forced to improvise his own crude tools. But in the Swiss version, adversity is seldom more than a temporary nuisance, quickly overcome, and the narrative is a sequence of solvable problems presumably contrived to create a positive frame of mind in the children involved, whether as castaways or as readers of the adventures of the castaways. Wyss created a bravura new world of instant accomplishments, tasks in which work and play are absolutely conflated, so that generations of children have yearned to join the family in their continuously expanding playground, little realizing that all the fun and games have a serious purpose.

Like *Émile,* and like Wyss's predecessor in the mentor genre, *Sandford and Merton* (also heavily influenced by Rousseau), *The Swiss Family Robinson* creates a schoolhouse environment without perceptible walls. Early on, the father expresses his philosophy of parenting, which is tantamount to a theory of education:

> One of the points of my system of education for my sons was, to awaken their curiosity by interesting observations, to leave time for the activity of the imagination, and then to correct any error they might fall into. I contented myself now, however [the family are struggling to launch their improvised tub-boat from the wrecked ship], with this general remark, that God sufficiently compensated the natural weakness of man by the gifts of reason, of invention, and the adroitness of the hands; and that human meditation and skill had produced a science, called mechanics, the object of which was, to teach us how to make our own natural strength act to an incredible distance, and with extraordinary force, by the intervention of instruments.

This observation is attached to the father's explanation to his curious sons concerning the operations of the lever with which they are attempting to slide their clumsy boat into the ocean, creating a moment that has struck at least one reader as highly unlikely, given the exigencies of the situation. But to object to the father's seizing upon every opportunity, however inopportune, to instruct his sons is to be ignorant of the Rousseauean context.

Rousseau especially prized Defoe's novel because it put such a stress on *doing* as a way of learning, and this principle is essential to the Swiss father's theory of education. Rousseau saw the value of Crusoe's situation as forcing him 'to judge all things . . . in relation to their own utility', reminding us that another Enlightenment savant who admired *Robinson Crusoe* was Benjamin Franklin, and at times the minister–father seems like Ben in the heart of *his* family. 'When a variety of duties present themselves for our choice,' he maximizes, 'we should always give the preference to that which can confer the most solid advantage', an opinion familiar to readers of Franklin's *Autobiography*. Late in their second year on the island, the family having claimed possession of and developed many of its salient sites, the father opines that 'We had possession of the most eligible premises: the sole business was to turn them to the best account; and how to effect this was our unceasing theme'. Turning things to 'the best account' is an ideal central not only to the utilitarian scheme of values but to the Yankee way of life, and there is a decided, even determined, American dimension to the story of the castaway Swiss family.

We must remember that Johann David Wyss was raising his children during the very period when European eyes were fixed on North America, the emerging United States presenting a virtual allegorical demonstration of certain Enlightenment ideals regarding the possibilities for human perfection within a framework of divine dispensation. Though there are few direct references to political systems, and only scattered allusions to the American continent, the very layout of the island on which the family find themselves has a distinct American cast of features. Let us first recall, however, that the family have sailed with a shipload of goods with the intention of joining a colony in the South Pacific, a situation that explains the useful tools and machines with which they are provided but which also evokes the American experience prior to 1775. Early on, the father delivers a sermon in the form of a parable in which he explains the divine plan for mankind in terms of colonization. Obviously based on Christ's parable of the talents, the pastor's little tale puts forth an eschatology in which life is a colony the returns from which will determine an individual's salvation. Now the Swiss nation has seldom been associated with the colonization process, indeed Switzerland is a virtual byword for insular integrity, not expansion. So Johann David was assuredly looking elsewhere for his model, and the emerging shape of the island on which his family are marooned does at times suggest America in the 1790s.

First of all, although it is called an 'island' throughout, the map which accompanies the early editions of the book shows something much more closely resembling a peninsula. The family never completely explore the place, nor do they circumnavigate it, although they have a boat equal to the task. Instead, having in the first part of the book made tentative forays into the interior region, from which they are separated by a line of barrier cliffs, and in the second part having found it to be a 'desert' region, centred by a fertile zone defined by a river flowing from some undiscovered source, a region moreover largely inhabited by dangerous beasts, the family erect a protective barricade across the entranceway and leave off exploration thenceforth. We can certainly read this act of barricading as an essentially Swiss action (though contrived by de Montolieu), the famous neutrality of that nation being attributed to its inaccessibility, but the division of their 'island' into two zones, the fertile, settler-friendly area which the family calls home, and the desert, inhospitable zone beyond, which stretches to unexplored limits, also suggests the map of North America in 1790, in which the region beyond the Appalachians was for the most part *terra incognita*, defined by the great Mississippi River. Notably, the first beasts discovered in the desert region of the island are buffalo, an animal early associated with the American trans–Allegheny frontier.

It is in the second part of the book, by de Montolieu, that the interior region becomes associated with even more exotic creatures from the gigantic boa constrictor that provides one of the most horrific episodes in the book—a signal departure from the tone maintained by Wyss—to elephants, hyenas, crocodiles, ostriches, and other creatures associated with Africa not North America. Still, the Baroness's use of the word

'savannah' for the interior region and her allusion to William
Bartram's account of his travels along the St Johns River in
Florida, from which she apparently obtained the word, keeps
the American context consistent. Bartram's narrative is well
known for having inspired a number of romantic writers,
including Coleridge and Wordsworth, but that it contributed
to the romantic aspects of *The Swiss Family Robinson* is not
common knowledge.

And yet many aspects of the first part of the book as well
may be aligned to comparable elements in Bartram's *Travels*,
elements likewise traceable to the Enlightenment context. For
Bartram seems to have set out with Linnaeus in tow, and his
narrative is packed with learned references to species and
genus, couched in the scientific terms of the day, including
Latin labels. In Florida and South Georgia he found an amazing abundance of exotic flora and fauna, most especially the
ferocious alligator, about which he wrote lengthy and excited
descriptions. In Bartram's view, Nature was an aspect of God's
beneficence, endowing mankind with a multitude of useful and
profitable commodities, set often in a landscape of supernal
beauty. Although there is no inter-textual proof that Pastor
Wyss had read Bartram's book, the first part of *Swiss Family
Robinson*, even more than the part written by de Montolieu,
contains descriptions of natural scenery that perfectly harmon-
ize with Bartram's ecstatic, even mystical, sense of the beauty
and abundance of divine creation.

Equally harmonious with Bartram is the common emphasis
on the element of utility, the beneficial aspects of natural
abundance which put a Horatian handle to the beautiful blade,
a combination equivalent to an axe. Again and again, Bartram
lists the good uses to which products of nature may be put, a
practice devoutly followed by the pastor–tutor as he guides his
family through the prolixity of products provided by their
island. This aspect of the book has drawn subsequent criticism
also, for the island contains an impossible plenitude of wildlife,
and if the duck-billed platypus, crocodile, and kangaroo are
native to the Australian regions, the elephant, lion, and hyena
are not. Most especially after the Baroness has opened up her
generous pen, the island begins to resemble a zoological and
botanical garden combined, but this too is part and parcel of

the Enlightenment spirit, as yet further recourse to Bartram's example suggests.

For Bartram's putative purpose in travelling through Florida was to gather seeds to be sent back to his English patron, who was not alone among wealthy Europeans with private botanical gardens, equivalents to those 'cabinets' of natural curiosities that late eighteenth-century savants set up in their homes—much like the one the Swiss family mount in their cave. And we may date the zoo as a civic establishment from late in the same century, with the menagerie established by Louis XIV at Versailles, which while hardly a public institution did serve the uses of naturalists until it was destroyed during the Revolution. But by 1793 the Paris Museum of Natural History, with an attached menagerie, had been established for the enlightenment of citizens, serving a function similar to the congeries of wildlife assembled on the island for the use and education of the Swiss castaways.

Given the Rousseau connection and the Bartram analogues, what is chiefly missing from this insular preserve is any presence of native populations, from *Robinson Crusoe* on as rigorous a convention in castaway narratives as the obligatory cave. For the notion of the noble savage, in part inspired by the example provided by Friday, is essential to Rousseau's concept of nature. It was expanded upon by an early Anglo-American version of the Robinsonade, *Mr Penrose*, by William Williams (1727–1790), first published in a truncated version in 1815. In this uxorious fiction, set in the Caribbean, Robinson Crusoe is indeed found in the bosom of his (extended) family, since Penrose takes a succession of Indian women for his wives. And Bartram likewise, though with no recorded liaisons, had a ready eye for native beauty, and found in the Seminoles of Florida real-life counterparts to Rousseau's idealized 'savage'. But although the Swiss pastor masters the elements of the Malay language in preparation for possible encounters, none such occur, and since the Malays are infamous for acts of piratical cruelty, something less than a Rousseauean context is established.

This is a definitive departure from the Enlightenment pattern, but it does merge handily with the imperialistic aspects of the castaway genre. We will recall that Friday stands out from his fellow cannibals, who appear to Crusoe as repulsive examples of bestial humanity, and Crusoe's first impulse is to exterminate them wholesale. Only his intimate experiences with Friday somewhat redeem the rest of the Caribees in Crusoe's eyes, but his companion remains exceptional, and Robinson never thinks of living amongst Friday's people. In nineteenth-century Robinsonades, moreover, starting with Captain Frederick Marryat's *Masterman Ready* in 1841 (a massive, even retaliatory, response to *Swiss Family Robinson*), English castaways are constantly threatened by hostile natives, generally depicted as cannibals, save when, as in that most famous of castaway tales for Victorian children, *The Coral Island*(1858), they have been converted by missionaries to Christianity and civilized uses.

In this regard, *Swiss Family Robinson* is a transition narrative, for though no savages ever appear, we have the marauding monkeys, one of whom becomes an adopted member of the family, while the others are slaughtered in repeated expeditions of extermination. It is not difficult to see the subtext here as sublimated imperialism, most especially since the activities of the monkeys are destructive to the extended wings of the little Swiss colony. This is the other side of the Enlightenment vision, the licensing of imperial ventures with the claim of bringing civilized order to unenlightened peoples, often in the form of missionary endeavours, efforts which predictably end in the deaths of the unconverted. Similarly, when the family had 'made the first steps towards a condition of civilization' at the end of their first two years on the island, the father 'vowed vengeance against the malicious monkeys, who would soon render our island desolate'.

But it is in the second part of the narrative that these lines appear, for the Baroness takes a much harder imperial line than that of the Wysses, who tend to emphasize a relatively restricted zone of occupation, as the map of their domain suggests. And yet the Wyss family Robinson are prototypical colonists, in that they struggle to replicate their old way of life in new, even exotic, surroundings. Notably, they early on give order to their island by giving place-names to prominent features of the landscape, and many of the structures they erect have familiar outlines, much as colonists in the new world built houses not much different from those they lived in at home.

But even in such an idiosyncratic dwelling as the family's famous tree-house the ordering impulse obtains, for no better diagram of the Swiss family's domain can be found than the stairway they build inside the great tree, rendering the former haven for bees a useful adjunct to their own busy lives.

Indeed, the complex act of taking possession of their towering fig, from measuring its height to preparing it for their first winter residence, can serve as an epitome for the book as a whole. The tree, whose branches 'grew close to each other, and in an exactly horizontal direction . . . in every respect convenient . . . for my undertaking', is type and symbol of accommodating nature, while the building of the staircase is an 'undertaking' typical of so many of the family's works, 'appearing at first beyond our powers; but intelligence, patience, time, and a firm resolution vanquished all obstacles'. One only wishes that the father had rendered in more detail the method by which he 'fixed in the centre [of the hollow] the trunk of a tree about twenty feet in length, and a foot thick . . . in order to carry my winding staircase round it'. We can only conclude that 'firm resolution adopted by all', combined with 'patient industry and constant efforts to the end' can indeed accomplish miracles.

Again, replication of the old world in a new is essential to the act of colonization: we need only remember that the first articles of furniture constructed by Robinson Crusoe, once he has dug out and fortified his cave, is a table and chair, so that he may eat his meals in civilized, English fashion, not squatting on his hams like a savage. Where the Swiss family differ from Crusoe in this regard has to do with their ornate sense of interior decoration, most especially in the second part of the book, in which the sensibilities of the Baroness de Montolieu take charge. At the start of the narrative, and holding pretty much through the entire first part, the family build relatively spartan—if snug—accommodation, but with their happy discovery of the salt cavern (yet another demonstration of accommodating nature), with its glittering crystals and stalactites, a decided shift in aesthetics takes place. Where the aesthetic element hitherto has for the most part been concentrated in descriptions of the paradisiac natural surroundings on the island (the new cave home, for example, commands an excellent view), the reader's attention during much of the second part of the book is turned to the interior, both of the island and of the Swiss family's new home.

Much as the family explores the desert region beyond the range of cliffs, then compartmentalizes the area where the wild things are by building a protective barricade sealing it off from the region they have chosen to domesticate, so the cave is broken up from one common room into many chambers. Chairs and tables are manufactured and to these are added 'all sorts of furniture' recovered from the wreck, including 'very handsome toilet-cases and bureau-tables', and 'a splendid clock, with an automaton figure'. The 'grotto' (as it is called) is a familiar adjunct to the highly artificialized landscape gardens of the eighteenth century, often a man-made structure tricked out with all manner of gimcracks and souvenir rockery, but in the hands of the Baroness the family's cave begins to take on a semblance to a dwelling furnished in *haut empire* taste. But this exuberance of interior decorating is but an extreme instance of the colonizing process throughout the book, by means of which the mechanics of domestication pairs the taming of wild animals to make them amenable to the family's use with the erection of dwellings throughout the occupied zone of the island.

Despite this domestic emphasis and despite the omnipresence of the mother, who with her magic bag and general resourcefulness plays a central role throughout the story, readers have objected to the preponderance of the male element in Wyss's novel. Given the biographical context, it could hardly have been otherwise, fate having ordained that the Wysses would have only sons. Fate also apparently ordained that the boys would fall into convenient categories, from the athletic, courageous Fritz to the studious, indolent Ernest, from the brash, boastful Jack to the rather amorphous but pleasant youngest son, Francis. I shall have more to say about Jack anon, but for now it needs only to be pointed out that for whatever cultural and biographical reasons, the family life on the island is distinctly patriarchal, with the mother providing at best a mollifying function. When Ernest decides to remain at home and help his mother this is not intended as a favourable note on his behalf. Instead, we are meant to prefer the stalwart acts of his oldest brother, who soon emerges as the father's favourite companion.

Ironically, the Baroness reinforced this programme of values. It is notable that the opening chapter of the second part is taken up with a series of competitions, athletic in nature, by means of which the boys are pitted against one another to determine their superiority. Here, as in earlier episodes, their reactions to the differing situations provides a diagram of character. But in the first part of the book, the emphasis is communal not competitive, and all are urged to work together for the common good. There are reasons for this difference, traceable to changing styles in educational theory, but the Baroness in other ways hardens the distinctions between the sons, much as she emphasizes the martial and hunting arts. 'Bang, bang, went their pistols, and two more animals bit the dust' is a frequent refrain. The mother scolds them for their profligacy—'Why should you abuse the provision nature has so liberally provided, by killing more than we require?'—but the killing goes on. In the first part of the book, notably, it was the father who counselled moderation: ' "Why", said I, "must we be always applying the means of death and annihilation to the creatures that fall in our way?" ', and the shift of this voice from father to mother is definitive. The Baroness likewise introduces whipping as a proper form of discipline—of animals, not children, let it be added—and the father begins to express notions of education associated more with Rugby than Rousseau: consoling his youngest son who is fatigued by the burden of the beast he has just slain, the father 'tried to impress upon his mind that pain always accompanies pleasure in every glory', and he opines at a later point that 'self-love' is 'the natural stimulant to action and foe of idleness'. He likewise maximizes that 'human vanity always has a share in in all the actions of life', and though such a theme is intrinsic to Enlightenment thought, we associate such sentiments with sceptics like Voltaire and Dr Johnson, not idealists like Rousseau.

Chief victim of this shift in tone is son Jack, who is portrayed in the first part as being often rash in his actions, haste which brings with it its own punishment, but who in the second part emerges as a braggart and a coward, the perpetual butt of his brothers and parents. Where Ernest at least has the virtue of learnedness, by the end of the second part having taken over the father's role of walking encyclopaedia, Jack has little to redeem him. He comes close to the stereotypical 'bad boy' in early children's literature, and though this role was sanctioned also by Enlightenment literature, as in the pioneering instance of little Tommy Merton in Thomas Day's novel, it eventually hardened into Victorian excess. In *Masterman Ready* this role is filled to repletion by another Master Tommy, whose carelessness results in the loss of the family's water supply during an attack by natives and consequently in the death of the titular hero, an old salt whose efforts have guaranteed the survival of the family in his care. Marryat's Master Tommy is made painfully aware of his guilt, a weight we are told that assisted in his reformation, assuring us that we have arrived in the Victorian era of children's literature, to which, once again, the second part of *The Swiss Family Robinson* serves as threshold.

To move from the first to the second part of the book, from the relatively benign Jacobinism of the story as written by the Wysses, father and son, to the increasingly severe behavioural criteria of the story as continued by the French author, is to trace a chronological graph from the mood of exhilaration regarding the potential of mankind that accompanied the French Revolution to the antithetical (and reactionary) emphasis on moral rigour that signalled the first half of the nineteenth century.

Control is the key word, a prevailing drive to mastery that covers most aspects of the family's life, from barricading the pass into the wild country to devising means of harnessing such unlikely steeds as an ostrich. True, much of this can be attributed to literary necessity, for the family during the first two years of their stay ensured their survival and basic comfort, leaving de Montolieu to devise episodes that would fill up a second volume, with a not surprising emphasis on embroidery work, adding more pets, more territory, further exploration, greater defences, etc. But the hardening of lines and drawing tighter reins is undeniable.

Where the book might have gone had the Baroness given her own literary impulses full exercise is suggested by the last chapters, in which the French author finally wound up her

sequel, presumably having arrived at the point where emulation, having exhausted all possible repetitions, began to veer
toward invention. This final sequence is dominated by the
adventures of Fritz, who in his kayak wanders farther and
farther away from the family at home, a symbolic journey
testifying to his manhood and consequent independence.
A further signal is given by his discovery and rescue of the
English girl, Emily Montrose, an episode that allowed de
Montolieu to give full flight to her interior decorating talents.
For the taste evinced by little Emily in furnishing out *her* cave
brings us very far indeed from Robinson Crusoe's bare-bones
utilitarianism, and suggests instead the items to be found in
the kind of Victorian parlours burlesqued by Mark Twain in
Life on the Mississippi and *Huckleberry Finn*:

Fritz had made her a box which held them all, and they really were
very curious . . . There were fish-lines made of the twisted hair of the
young girl's head, with fish-hooks attached, made of the
mother-of-pearl; some needles made from fish-bones; piercers and
bodkins, made of the beaks of birds; two beautiful needle-cases, one
made of a pelican's feather, the other of the bone of a sea-calf. The
skin of a young walrus sewed together served for a bottle, a lamp made
of a shell, with a wick of cotton drawn from her handkerchief, over the
lamp another shell served as boiler; a turtle-shell used to cook food in,
by throwing in hot stones, some fish-bladders, shells of all sizes, serving
for glasses, spoons, dishes, etc; little sacks full of seeds gathered by the
young solitary, a quantity of plants, such as the cochelaria, sorrel,
celery, and cress, which grow among the rocks.

The inventory goes on, for another two paragraphs, but
enough is sufficient to make the point. Providing Fritz with a
female partner, thus definitively marking his arrival at maturity,
may have been a fictive necessity, but the Baroness seems to
have outdone herself in matters of furnishing out Emily's
hope-chest and designing her castaway couture: 'For wearing
apparel, she had a hat made of the downy breast of the
cormorant . . .'.

To these details, we may add the account of Fritz's voyage
of rescue, which brings him into increasingly exotic surroundings, the threshold to which is a marvellous sea-cave, resembling 'one of those old Gothic cathedrals . . . a temple elevated
to the Eternal, in the midst of immensity'. It is beyond this

place that Fritz discovers a fortune in pearl-oysters and finds Emily, a crossing over that brings us definitively out of the realm of the eighteenth-century sensibility into the romantic zone. The wildlife in this region becomes increasingly exotic also, signalling the dangers Fritz must brave before he can attain his beloved, and the rewards include not only pearls and Emily but truffles, the ultimate gustatorial gift discovered by the family in their insular pantry of abundant delights.

In short, for it is time that we also should wind up our account, by the last pages of *The Swiss Family Robinson* the tenor of this corporate product has been quite entirely taken up by the romantic spirit, even to the definitive presence of the castaway girl, a fair captive to be rescued by the chivalric Fritz. The ritual journey ends with the father granting his son freedom from 'all subordination', a gesture which the bookish Ernest correctly interprets as 'the ceremony of the "toga virilis"', although as he hints Fritz may have simply traded one set of 'leading-strings' for another. Still, that matter was left for a possible third sequel, and the adventures of the family are quickly swept up into a tidy pile of manuscript, with the arrival of Emily's father and the dividing up of the family between those who wish to remain on the island and those who seek repatriation to the old world.

It is at this point that the book at last becomes a complete text, relocating the closure found in the earlier version that ended with the accidental discovery of the family by an English ship, which was driven away before the castaways could be taken on board. And it is finally as a text that we should consider two parts of *The Swiss Family Robinson* combined, a book inspired by a book that becomes increasingly bookish as the story proceeds, until in Fritz's final adventure we find ourself in the realm of high romance, of rescue and young love. As children we are innocent of the meaning of these complications, yet I must testify that as a child I was never able to finish the book, however many times I read over the familiar adventures of the first part, and though I cannot now recall the point where my interest flagged, I would put it in the general region of that mysterious inland desert with its interminable parade of exotic wildlife. Enough, even for the tastes of childhood, is enough, and my preference still holds for the

first part of the story, with the beautiful tree-house, the bountiful shipwreck, the sparkling salt cave, and that marvellous little
pinnace with its signal guns. The pinnace is a vessel that the
Baroness de Montolieu seems to have been at a loss to accommodate to her powers of invention, though most little boys of
my generation would have accepted it gladly in lieu of the prize
offered up in Emily Montrose, feather hat, sea-calf waistcoat,
'other garments' of birdskin, and all.

A NOTE ON THE TEXT

As my introductory remarks have suggested, there are few classics of children's literature with such a tangled textual history as this, a situation that presents problems when viewed in the light of modern ideals of textual purity, which put a high premium on authorial manuscript, when available, and regard subsequent meddling with the text as garbage to be purged. The original manuscript of *The Swiss Family Robinson*, along with charming illustrations by one of the Wyss sons, is still in the care of the family, but even if it were possible to do so, this is not the proper occasion for its recovery. Nor would the earliest publication in German be suitable for an English language text, without commissioning a new translation, not practicable either. It might have been worthwhile to reproduce the Godwin translation, which has an abundant charm of its own, rather than the adaptation which first appeared in 1826, but then what would we do with the Baroness de Montolieu's continuations, which were not translated into English until more than twenty years later, but which subsequently became part of the text, familiar to young readers for nearly a century and a half? What is needed, perhaps, is a complete concordance to this transmogrified text, but ours is not the occasion for that, either, nor can we even license a new translation throughout of a combined edition of the Wyss and Montolieu books.

What we have done, instead, is to use here the long-lived modified Godwin text of 1826, published by Baldwin, Cradock, and Joy, the firm to which M. J. Godwin and Company was sold after its failure in 1825. It is not known who made the substantive revisions to the Godwin translation in producing that text, but the result was reprinted many times in the nineteenth century, and was used in the first (1832) and subsequent American editions. Further standardization (and circulation) was guaranteed when the 1826 text was used by Ernest Rhys as the basis for the first part of his Everyman edition in 1910—with the usual imposition of revisions and deletions. The Rhys version also incorporated the Locke translation (1848) of the de Montolieu sequels, as have we here, thereby making up the basic story known to so many generations of readers. As the Everyman text demonstrates, any such marriage necessitates a certain amount of tinkering. We have, therefore, dropped some material from an introductory chapter added by Locke which repeats (and revises) details found in the concluding section of the 1826 edition. We have also removed the penultimate chapter of that earlier text, which accounts for the recovery of the father's journal without the rescue of the family itself, an episode rendered superfluous by the continuation of their adventures. We have retained, however, the account found in Locke's first chapter of the training of a young bull-calf and its assignment as a steed to Francis, since the animal plays a role in episodes that follow. This still leaves unexplained the calf's miraculous birth, more than a year after the family was stranded. Of a similar sort are the sudden appearance in Chapter 6 of the sequel of Jack's pet jackal, whose origin is unaccounted for, and the dog Flora, who becomes Billy in the de Montolieu continuation, only to reappear as Flora in the final chapters. But these would not be the most serious violations of the laws of nature in this marvellous book.

Perhaps the better to justify the bull-calf's birth, Locke (or de Montolieu) in Chapter 2 of the sequel staged a celebration, shortly before the second rainy season endured by the family, of the first anniversary of their arrival on the island. This, despite the father's observation, in his closing words to the original story, that 'nearly two years [had] elapsed' since the family escaped the shipwreck. Though chronological matters in Wyss's book are not precise, we can ascertain that well over 300 days elapse between the family's arrival on the island and the first rainy season, and this, along with the storm that wrecks their ship, would suggest they arrived at the end of a previous season of rain. Rhys, or some previous hand, corrected this error in the Locke translation, as have we. But this still leaves the matter of the celebration anachronistic, since it takes place prior to the second rainy season spent on the island by the family, when it should have followed that interlude. This problem was also simple to correct, and we have transposed Chapter 2 of the sequel so that it follows Chapter 10. Although de Montolieu was no Henry James, her prolixity is not dissimilar, so that a transposed chapter is hardly noticeable, as anyone who reads the story before glancing at this note will attest.

Finally, we have preserved the head matter to the 1826 edition and the 1848 translation, as being of historical interest, but for the sake of continuity have placed them both in the front of our book. The tables of contents have likewise been combined, necessitating the renumbering of the chapters in the sequel. Amusingly outdated annotations supplied by Godwin have been dropped, there being no counterpart in the Locke translation. The notes for this edition have been prepared with the help of Mr Robert A. Kosten, who likewise supplied the biographical information on Wyss and Madame de Montolieu.

Let me here enter a statement of gratitude to Rita Smith and the staff of the Baldwin Library of Children's Literature at the University of Florida for their assistance in preparing these textual remarks, and to Dr Ruth Baldwin especially, whose untimely death occurred as this edition was in its last stages of preparation. Given her own interest in the Robinsonade tradition and her sizeable collection of the many versions of *Swiss Family Robinson*, I regret that she did not live to see the results of our labour published. It seems fitting, however, that the introduction and annotations to this edition be dedicated to Dr Baldwin, in recognition of her great contribution to the study of children's literature, as a memorial to her life-long pursuit of its preservation.

SELECT BIBLIOGRAPHY

AVERY, GILLIAN, *Nineteenth Century Children: Heroes and Heroines in English Children's Stories, 1780–1900*. London: Hodder and Stoughton, 1965.

—— and BRIGGS' JULIA, (eds), *Children and their Books: A Celebration of the Work of Iona and Peter Opie*. Oxford: Clarendon Press, 1989. Contains William St Clair, 'William Godwin as Children's Bookseller', pp. 165–79.

DARTON, F. J. HARVEY, *Children's Books in England*, 2nd edn., with an Introduction by Kathleen Lines. Cambridge: Cambridge University Press, 1960.

GREEN, MARTIN, *Dreams of Adventure, Deeds of Empire*. New York: Basic Books, 1979.

——, *The Robinson Crusoe Story*. University Park: Pennsylvania State University Press 1990.

HOLDEN, PHILIP, 'A Textual History of J. R. Wyss's *The Swiss Family Robinson*', unpublished MA thesis. University of Florida, 1986.

HÜRLIMANN, BETTINA, *Three Centuries of Children's Books in Europe*, trans. and ed. Brian W. Alderson. Cleveland and New York: World Publishing Company, 1959.

TARG, WILLIAM (ed.), *Bibliophile in the Nursery: A Bookman's Treasury of Collectors' Lore on Old and Rare Children's Books*. Cleveland and New York: World Publishing Company, 1957. Contains Vincent Starrett, 'Of Castaways and Islands', pp. 347–70.

WILLIAMS, WILLIAM, *Mr Penrose: The Journal of Penrose, Seaman*, ed. with Introduction and Notes by David Howard Dickason. Bloomington and London: Indiana University Press, 1969.

The Swiss Family Robinson

or

Adventures of
a Father and Mother and Four Sons
in a Desert Island

━━━━━

The Genuine Progress of the Story Forming a
Clear Illustration of the First Principles of
Natural History, and Many Branches of
Science which Most Immediately Apply to
the Business of Life

PREFACE
TO THE SEVENTH EDITION OF
THE SWISS FAMILY ROBINSON.

THE Editor confesses that he expected the favour which has been shown by the Public to the present work, in the sale of six large editions; for the situations it exhibits of the best affections of our nature, are such as to 'come home to every bosom', to interest and gratify both parents and children of all ages and descriptions. In its pages the useful, the moral, and the entertaining, so naturally mix with or succeed each other, that every generous taste is suited. No story can be better calculated than this of the Swiss Pastor and his Family to awaken and reward curiosity, to excite amiable sympathies, to show the young inquirer after good, that the accidents of life may be repaired by the efforts of his own thought, and the constancy of his own industry; and to rouse the most inert to emulation. —What youthful reader of lively dispositions, who would not try to possess; or dream of the possession of, a saw, a hammer, and some nails, and hurry in fancy to the contrivance of a *Family Bridge*, a *staircase* to *Falcon's Nest*, or *a parlour, a bedroom*, and *a kitchen*, in a *Rock of Salt*? What lad who can see with unflushed cheek, *Ernest*, and *Jack*, and *Francis*, all together on the back of their ferocious but now subdued *buffalo*, and guiding his motions at their pleasure; or descry '*Fritz driving along our avenue like lightning*', on his disciplined *onagra*, without feeling his personal courage expand as he turns the page? What heart that will not swell with kindness for the exemplary mother of the family, who tastes not of the refreshing draught obtained by her own care and labour, till all her dear ones have drunk and are refreshed?—or what eye repress a tear when her little Francis, clinging to her side, cries, '*Welcome!*' too, '*though not well knowing whether he was to be sad or merry*'? Who does not partake the affecting sentiment of the interesting group, who, '*with their heads sinking on their bosoms, took the road to Tent House*', when the ship had disappeared forever! And can we sufficiently admire the fortitude, the self-sacrifice, the industry, the fervour, the almost unexampled skill, that guided this affectionate pair to results so happy, in circumstances of such dismay and danger!

PREFACE
TO THE TRANSLATION OF
DE MONTOLIEU'S CONTINUATION

THE first part of the Swiss Family Robinson was published many years ago, and has ever since that time remained one of our standard juvenile works: I say juvenile, but it can be read by persons of all ages with pleasure and instruction. The descriptions of the different animals, their nature and habits, the uses of different plants and other natural productions of the earth, are delineated with the utmost fidelity. The work shows us how much human means can effect, when thrown upon its own resources.

The continuation of the story, by the same author, fell into the hands of the translator about a year since; and, as he remembered the intense interest with which he and others had read the former part, he was induced to undertake the following translation, which he submits to the public.

J. C. LOCKE

SING-SING, *March* 12, 1847

ADVERTISEMENT BY THE EDITOR

A PASTOR or Clergyman of West Switzerland, having lost his fortune in the Revolution of 1798, resolved, on reflecting on the family he had to bring up, to become a voluntary exile, and to seek in other climates the means of support. He sailed, accordingly, with his wife and children, four sons, from 12 to 5 years of age, for England, where he accepted an appointment of Missionary to Otaheite; not that he had any desire to take up his abode in that Island, but that he had conceived the plan of passing from thence to Port Jackson, and domiciliating himself there as a free settler. He possessed a considerable knowledge of agriculture, and by this means hoped, with the aid of his sons, to gain an advantageous establishment, which his own country, convulsed with the horrors of war, denied him. He turned the small remnant of his fortune into money, and bought with it seeds of various sorts, and a few cattle, as a farming stock. The family took their passage accordingly, satisfied with this consolation—that they should still remain together; and they sailed with favourable winds till in sight of New Guinea. Here they were attacked by a destructive and unrelenting tempest; and it is in this crisis of their Adventures that the Swiss Pastor, or Family Robinson, begins the Journal which is now presented to the Public.

CONTENTS

Settlement of the Swiss Pastor and his Family in the Desert Island

A.   Arcadia
B.   Sugar-Canes
C.   Cabbage Palm Wood
D.   Gourd Wood
E.   Bamboos
F.   Pass—Drawbridge
G.   Acorn Wood
H.   Rice Marsh
I.   Monkey Wood
K.   The Farm
L.   Lake

M.   Cotton Wood
N.   Flamingo Marsh
O.   Cascade
P.   Falcon's Nest
Q.   Coconut Palm Wood
R.   Family Bridge
S.   Potato Plantation
T.   Tent House
U.   Grotto
V.   Marsh
W.   Shark's Island

THE SWISS FAMILY ROBINSON

1

A Shipwreck, and Preparations for Deliverance

... ALREADY the tempest had continued six days; on the seventh its fury seemed still increasing; and the morning dawned upon us without a prospect of hope, for we had wandered so far from the right track, and were so forcibly driven toward the south-east, that none on board knew where we were. The ship's company were exhausted by labour and watching, and the courage which had sustained them, was now sinking. The shivered masts had been cast into the sea; several leaks appeared, and the ship began to fill. The sailors forbore from swearing; many were at prayer on their knees, while others offered miracles of future piety and goodness, as the condition of their release from danger. 'My beloved children,' said I to my four boys, who clung to me in their fright, 'God can save us, for nothing is impossible to him. We must however hold ourselves resigned, and instead of murmuring at his decree, rely that what he sees fit to do is best, and that should he call us from this earthly scene, we shall be near him in heaven, and united through eternity. Death may be well supported when it does not separate those who love.'

My excellent wife wiped the tears which were falling on her cheeks, and from this moment became more tranquil: she encouraged the youngest children, who were leaning on her knees; while I, who owed them an example of firmness, was scarcely able to resist my grief at the thought of what would most likely be the fate of beings so tenderly beloved. We all fell on our knees, and supplicated the God of Mercy to protect us; and the emotion and fervour of the innocent creatures, are a convincing proof that, even in childhood, devotion may be felt and understood, and that tranquillity and consolation, its natural effects, may at that season be no less certainly experienced. Fritz, my eldest son, implored in a loud voice, that God would deign to save his dear parents and his brothers, generously unmindful of himself: the boys rose from their posture with a state of mind so improved, that they seemed forgetful of the impending danger. I myself began to feel my hopes increase as I beheld the affecting group. Heaven will surely have pity on them, thought I, and will save their parents to guard their tender years!

At this moment a cry of 'Land, Land!' was heard through the roaring of the waves, and instantly the vessel struck against a rock with so violent a motion as to drive every one from his place; a tremendous cracking succeeded as if the ship was going to pieces; the sea rushed in, in all directions; we perceived that the vessel had grounded, and could not long hold together. The captain called out that all was lost, and bade the men lose not a moment in putting out the boats. The sounds fell on my heart like a thrust from a dagger: 'We are lost!' I exclaimed; and the children broke out into piercing cries. I then recollected myself, and addressing them again, exhorted them to courage, by observing that the water had not yet reached us, that the ship was near land, and that Providence would assist the brave. 'Keep where you are,' added I, 'while I go and examine what is best to be done.'

I now went on the deck. A wave instantly threw me down, and wetted me to the skin; another followed, and then another. I sustained myself as steadily as I could; and looking around, a scene of terrific and complete disaster met my eyes: the ship was shattered in all directions, and on one side there was a complete breach. The ship's company crowded into the boats till they could contain not one man more, and the last who entered, were now cutting the ropes to move off. I called to them with almost frantic entreaties to stop and receive us also, but in vain; for the roaring of the sea prevented my being heard, and the waves, which rose to the height of mountains, would have made it impossible to return. All hope from this source was over, for while I spoke, the boats, and all they contained, were driving out of sight. My best consolation now was to observe, that the slanting position the ship had taken, would afford us present protection from the water; and that the stern, under which was the cabin that inclosed all that was dear to me on earth, had been driven upwards between two rocks, and seemed immovably fixed. At the same time, in the distance southward, I descried through clouds and rain several nooks of land, which, though rude and savage in appearance, were the objects of every hope I could form in this distressing moment.

Sunk and desolate from the loss of all chance of human aid, it was yet my duty to appear serene before my family: 'Courage, dear ones,' cried I, on entering their cabin, 'let us not desert ourselves: I will not conceal from you that the ship is aground; but we are at least in greater safety than if she were beating upon the rocks: our cabin is above water; and should the sea be more calm tomorrow, we may yet find means to reach the land in safety.'

What I had just said, appeased their fears; for my family had the habit of confiding in my assurances. They now began to feel the advantage of the ship's remaining still; for its motion had been most distressing, by jostling them one against another, or whatever happened to be nearest. My wife, however, more accustomed than the children to read my inmost thoughts, perceived the anxiety which devoured me. I made her a sign which conveyed an idea of the hopelessness of our situation, and I had the consolation to see that she was resolved to support the trial with resignation: 'Let us take some nourishment,' said she, 'our courage will strengthen with our bodies; we shall perhaps need this comfort to support a long and melancholy night.'

Soon after, night set in: the fury of the tempest had not abated; the planks and beams of the vessel separated in many parts with a horrible crash. We thought of the boats, and feared that all they contained must have sunk under the foaming surge.

My wife had prepared a slender meal, and the four boys partook of it with an appetite to which their parents were strangers. They went to bed, and exhausted by fatigue, soon were snoring soundly. Fritz, the eldest, sat up with us: 'I have been thinking', said he after a long silence, 'how it may be possible to save ourselves. If we had some bladders or cork-jackets for my mother and my brothers, you and I, father, would soon contrive to swim to land.'

'That is a good thought,' said I: 'we will see what can be done.'

Fritz and I looked about for some small empty firkins:* these we tied two and two together with handkerchiefs or towels, leaving about a foot distance between them, and fastened them as swimming-jackets under the arms of each child, my wife at the same time preparing one for herself. We provided ourselves with knives, some string, some turfs, and other necessaries which could be put into the pocket, proceeding upon the hope, that if the ship went to pieces in the night, we should either be able to swim to land, or be driven thither by the waves.

Fritz, who had been up all night, and was fatigued with his laborious occupations, now lay down near his brothers, and was soon asleep; but their mother and I, too anxious to close our eyes, kept watch, listening to every sound that seemed to threaten a further change in our situation. We passed this awful night in prayer, in agonizing apprehensions, and in forming various resolutions as to what we should next attempt. We hailed with joy the first gleam of light which shot through a small opening of the window. The raging of the winds had begun to abate, the sky was become serene, and hope throbbed in my bosom, as I beheld the sun already tinging the horizon. Thus revived, I summoned my wife and the boys to the deck, to partake of the scene. The youngest children, half forgetful of the past, asked with surprise why we were there alone, and what had become of the ship's company? I led them to the recollection of our misfortune, and then added: 'Dearest children, a Being more powerful than man has helped us, and will, no doubt, continue to help us, if we do not abandon ourselves to a fruitless despair. Observe, our companions, in whom we had so much confidence, have deserted us, and that Divine Providence, in its goodness, has given us protection! But, my dear ones, let us show ourselves willing in our exertions, and thus deserve support from Heaven. Let us not forget this useful maxim, and let each labour according to his strength.'

Fritz advised that we should all throw ourselves into the sea, while it was calm, and swim to land.—'Ah! that may be well enough for you,' said Ernest, 'for you can swim; but we others should soon be drowned. Would it not be better to make a float of rafts, and get to land altogether upon it?'

'Vastly well,' answered I, 'if we had the means for contriving such a float, and if, after all, it were not a dangerous sort of conveyance. But come, my boys, look each of you about the ship, and see what can be done to enable us to reach the land.'

They now all sprang from me with eager looks, to do as I desired. I, on my part, lost no time in examining what we had to depend upon as to provisions and fresh water. My wife and the youngest boy visited the animals, whom they found in a pitiable condition, nearly perishing with hunger and thirst. Fritz repaired to the ammunition room; Ernest to the carpenter's cabin, and Jack to the apartment of the captain; but scarcely had he opened the door, when two large dogs sprang upon him, and saluted him with such rude affection, that he roared for assistance, as if they had been killing him. Hunger, however, had rendered the poor creatures so gentle, that they licked his hands and face, uttering all the time a low sort of moan, and continuing their caresses till he was almost suffocated. Poor Jack exerted all his strength in blows to drive them away: at last he began to understand, and to sympathize in their joyful movements, and put himself upon another footing: he got upon his legs; and gently taking the largest dog by the ears, sprang upon his back, and with great gravity presented himself thus mounted before me, as I came out of the ship's hold. I could not refrain from laughing, and I praised his courage: but I added a little exhortation to be cautious, and not go too far with animals of this species, who, in a state of hunger, might be dangerous.

By and by my little company were again assembled round me, and each boasted of what he had to contribute. Fritz had two fowling-pieces,* some powder, and small shot, contained in horn flasks, and some bullets in bags.

Ernest produced his hat filled with nails, and held in his hands a hatchet and a hammer; in addition, a pair of pincers, a pair of large scissors, and an auger,* peeped out at his pocket-hole.

Even the little Francis carried under his arm a box of no very small size, from which he eagerly produced what he called some little sharp-pointed hooks. His brothers smiled scornfully. 'Vastly well, gentlemen,' said I; 'but let me tell you that the youngest has brought the most valuable prize: and this is often the case in the world; the person who least courts the smiles of Fortune, and in the calm of his heart is scarcely conscious of her existence, is often he to whom she most readily presents herself. These little sharp-pointed hooks, as Francis calls them, are fishing-hooks, and will probably be of more use in preserving our lives, than all we may find besides in the ship. In justice, however, I must confess, that what Fritz and Ernest have contributed, will also afford essential service.'

'I, for my part,' said my wife, 'have brought nothing; but I have some tidings to communicate which I hope will secure my welcome: I have found on board, a cow and an ass, two goats, six sheep, and a sow big with young: I have just supplied them with food and water, and I reckon on being able to preserve their lives.'

'All this is admirable,' said I to my young labourers; 'and there is only master Jack, who, instead of thinking of something useful, has done us the favour to present us two personages, who, no doubt, will be principally distinguished by being willing to eat more than we shall have to give them.'

'Ah!' replied Jack, 'but if we can once get to land, you will see that they will assist us in hunting and shooting.'

'True enough,' said I, 'but be so good as to tell us how we are to get to land, and whether you have contrived the means?'

'I am sure it cannot be very difficult,' said Jack, with an arch motion of his head. 'Look here at these large tubs. Why cannot each of us get into one of them, and float to the land? I remember I succeeded very well in this manner on the water, when I was visiting my godfather at S——.'

'Every one's thought is good for something,' cried I, 'and I begin to believe that what Jack has suggested is worth a trial: quick! then, boy, give me the saw, the auger, and some nails; we will see what is to be done.' I recollected having seen some empty casks in the ship's hold we went down, and found them floating in the water which had got into the vessel; it cost us but little trouble to hoist them up, and place them on the lower deck, which was at this time scarcely above water. We saw with joy, that they were all sound, well guarded by iron hoops, and in every respect in good condition; they were exactly suited for the object; and, with the assistance of my sons, I instantly began to saw them in two. In a short time I had produced eight tubs,

of equal size, and of the proper height. We now allowed ourselves some refreshment of wine and biscuit. I viewed with
delight my eight little tubs, ranged in a line. I was surprised to
see that my wife did not partake our eagerness; she sighed
deeply as she looked at them: 'Never, never,' cried she, 'can I
venture to get into one of these.'

'Do not decide so hastily, my dear,' said I: 'my plan is not
yet complete; and you will see presently, that it is more worthy
of our confidence than this shattered vessel, which cannot
move from its place.'

I then sought for a long pliant plank, and placed my eight
tubs upon it, leaving a piece at each end, reaching beyond the
tubs; which, bent upward, would present an outline like the
keel of a vessel: we next nailed all the tubs to the plank, and
then the tubs to each other as they stood, side by side, to make
them the firmer, and afterwards two other planks, of the same
length as the first, on each side of the tubs. When all this was
finished, we found we had produced a kind of narrow boat,
divided into eight compartments, which I had no doubt would
be able to perform a short course, in calm water.

But now we discovered that the machine we had contrived
was so heavy, that with the strength of all united, we were not
able to move it an inch from its place. I bade Fritz fetch me a
crow,* who soon returned with it: in the meanwhile, I sawed a
thick round pole into several pieces, to make some rollers. I
then, with the crow, easily raised the foremost part of my
machine, while Fritz placed one of the rollers under it.

'How astonishing,' cried Ernest, 'that this engine, which is
smaller than any of us, can do more than our united strength
was able to effect! I wish I could know how it is constructed.'

I explained to him as well as I could, the power of
Archimedes' lever,* with which he said he could move the
world, if you would give him a point from which his mechanism
might act, and promised to explain the nature of the operation
of the crow when we should be safe on land.

One of the points of my system of education for my sons
was, to awaken their curiosity by interesting observations, to
leave time for the activity of the imagination, and then to
correct any error they might fall into. I contented myself now,
however, with this general remark, that God sufficiently

compensated the natural weakness of man by the gifts of reason, of invention, and the adroitness of the hands; and that human meditation and skill had produced a science, called mechanics, the object of which was, to teach us how to make our own natural strength act to an incredible distance, and with extraordinary force, by the intervention of instruments.

Jack here remarked, that the action of the crow was very slow.

'Better slow than never, Jack,' replied I. 'Experience has ever taught, and mechanical observations have established as a principle, that what is gained in speed, is lost in strength: the purpose of the crow is not to enable us to raise any thing rapidly, but to raise what is exceedingly heavy; and the heavier the thing we would move, the slower is the mechanical operation. But are you aware what we have at our command, to compensate this slowness?'

'Yes, it is turning the handle quicker.'

'Your guess is wrong; that would be no compensation: the true remedy, my boy, is to call in the assistance of patience and reason: with the aid of these two fairy powers, I am in hopes to set my machine afloat.' As I said this, I tied a long cord to its stern, and the other end of it to one of the timbers of the ship, which appeared to be still firm, so that the cord being left loose, would serve to guide and restrain it when launched. We now put a second and a third roller under, and applying the crow, to our great joy our machine descended into the water with such a velocity, that if the rope had not been well fastened, it would have gone far out to sea. But now a new difficulty presented itself: the boat leaned so much on one side, that the boys all exclaimed they could not venture to get into it. I was for some moments in the most painful perplexity; but it suddenly occurred to me, that ballast only was wanting to set it straight. I drew it near, and threw all the useless things I could find into the tubs, so as to make weight on the light side: by degrees the machine became quite straight and firm in the water, seeming to invite us to take refuge in its protection. All now would get into the tubs, and the boys began to dispute which should be first. I drew them back, and seeking a remedy for this kind of obstacle, I recollected that savage nations make use of a paddle for preventing their canoes from upsetting. I once more set to work, to make one of these.

NATURAL HISTORY OF SELBOURNE

Date Ordered: 28/10/93

White,G
Penguin
PENGUIN PAPERBACK BOOKS LTD
PBAC   AN

HEATHCOTE
BOOKS

TATTERSALLS BOOKSHOP
Sales Order No: F511183

0140431128

Order Quantity:

Price: 4.99   *

Pub. Date: 00/00/00

Classics

A/C No: 93041B
Customer Ref: FAX 28/10/93

* ACTUAL PRICE MAY DIFFER

I took two poles of equal length, upon which the sails of the vessel had been stretched, and having descended into the machine, fixed one of them at the head, and the other at the stern, in such a manner as to enable us to turn them at pleasure to right or left, as should best answer the purpose of guiding and putting it out to sea. I stuck the end of each pole, or paddle, into the bung-hole of an empty brandy-keg, which served to keep the paddles steady, and to prevent any interruption in the management of our future enterprise.

There remained nothing more to do, but to find in what way I could clear out from the encumbrance of the wreck. I got into the first tub, and steered the head of the machine, so as to make it enter the cleft in the ship's side, where it could remain quiet. I then remounted the vessel, and sometimes with the saw, and sometimes with the hatchet, I cleared away to right and left, every thing that could obstruct our passage; and that being effected, we next secured some oars for the voyage we resolved on attempting.

We had spent the day in laborious exertions; it was already late; and as it would not have been possible to reach the land that evening, we were obliged to pass a second night in the wrecked vessel, which at every instant threatened to fall to pieces. We next refreshed ourselves by a regular meal; for, during the day's work, we had scarcely allowed ourselves to take a bit of bread, or a glass of wine. Being now in a more tranquil and unapprehensive state of mind than the day before, we all abandoned ourselves to sleep; not, however, till I had used the precaution of tying the swimming apparatus round my three youngest boys and my wife, in case the storm should again come on. I also advised my wife to dress herself in the clothes of one of the sailors, which were so much more convenient for swimming, or any other exertions she might be compelled to engage in. She consented, but not without reluctance, and left us to look for some that might best suit her size. In a quarter of an hour she returned, dressed in the clothes of a young man who had served as volunteer on board the ship. She could not conceal the timid awkwardness so natural to her sex in such a situation: but I soon found means to reconcile her to the change, by representing the many advantages it gave her, till at length she joined in the merriment her dress occasioned, and one and all crept into our separate hammocks, where a delicious repose prepared us for the renewal of our labours.

2
A Landing, and Consequent Occupations

BY break of day we were all awake and alert, for hope as well as grief is unfriendly to lengthened slumbers. When we had finished our morning prayer, I said, 'We now, my best beloved, with the assistance of Heaven, must enter upon the work of our deliverance. The first thing to be done, is to give to each poor animal on board a hearty meal; we will then put food enough before them for several days; we cannot take them with us; but we will hope it may be possible, if our voyage succeeds, to return and fetch them. Are you now all ready? Bring together whatever is absolutely necessary for our wants. It is my wish that our first cargo should consist of a barrel of gunpowder, three fowling-pieces, and three carbines,* with as much small shot and lead, and as many bullets as our boat will carry; two pair of pocket-pistols, and one of large ones, not forgetting a mould to cast balls in; each of the boys, and their mother also, should have a bag to carry game in; you will find plenty of these in the cabins of the officers.'—We added a chest containing cakes of portable soup, another full of hard biscuits, an iron pot, a fishing-rod, a chest of nails, and another of different utensils, such as hammers, saws, pincers, hatchets, augers, etc., and lastly, some sailcloth to make a tent. Indeed, the boys brought so many things, that we were obliged to reject some of them, though I had already exchanged the worthless ballast for articles of use in the question of our subsistence.

When all was ready, we stepped bravely each into a tub. At the moment of our departure the cocks and hens began to cluck, as if conscious that we had deserted them, yet were willing to bid us a sorrowful adieu. This suggested to me the idea of taking the geese, ducks, fowls, and pigeons with us; observing to my wife, that if we could not find means to feed them, at least they would feed us.

We accordingly executed this plan. We put ten hens and an old and a young cock into one of the tubs, and covered it with planks; we set the rest of the poultry at liberty, in the hope that instinct would direct them towards the land, the geese and the ducks by water, and the pigeons by the air.

We were waiting for my wife, who had the care of this last part of our embarkation, when she joined us loaded with a large bag, which she threw into the tub that already contained her youngest son. I imagined that she intended it for him to sit upon, or perhaps to confine him so as to prevent his being tossed from side to side. I therefore asked no questions concerning it. The order of our departure was as follows:

In the first tub, at the boat's head, my wife, the most tender and exemplary of her sex, placed herself.

In the second, our little Francis, a lovely boy 6 years old, remarkable for the sweetest and happiest temper, and for his affection to his parents.

In the third, Fritz, our eldest boy, between 14 and 15 years of age, a handsome curl-pated youth, full of intelligence and vivacity.

In the fourth was the barrel of gunpowder, with the cocks and hens and the sailcloth.

In the fifth, the provisions of every kind.

In the sixth, our third son Jack, a light-hearted, enterprising, audacious, generous lad, about 10 years old.

In the seventh, our second son Ernest, a boy of 12 years old, of a rational, reflecting temper, well informed for his age, but somewhat disposed to indolence and the pleasures of the senses.

In the eighth, a father, to whose paternal care the task of guiding the machine for the safety of his beloved family was entrusted. Each of us had useful implements within reach; the hand of each held an oar, and near each was a swimming apparatus in readiness for what might happen. The tide was already at half its height when we left the ship, and I had counted on this circumstance as favourable to our want of strength. We held the two paddles longways, and thus we passed without accident through the cleft of the vessel into the sea. The boys devoured with their eyes the blue land they saw at a distance. We rowed with all our strength, but long in vain, to reach it: the boat only turned round and round: at length I had the good fortune to steer in such a way that it proceeded in a straight line. The two dogs perceiving we had abandoned them, plunged into the sea and swam to the boat; they were too large for us to think of giving them admittance, and I dreaded lest they should jump in and upset us. Turk was an English dog, and Flora a bitch of the Danish breed. I was in great uneasiness on their account, for I feared it would not be possible for them to swim so far. The dogs, however, managed the affair with perfect intelligence. When fatigued, they rested their fore-paws on one of the paddles, and thus with little effort proceeded.

Jack was disposed to refuse them this accommodation, but he soon yielded to my argument, that it was cruel and unwise to neglect creatures thrown on our protection, and who indeed might hereafter protect us in their turn, by guarding us from harm, and assisting in our pursuit of animals for food. 'Besides,' added I, 'God has given the dog to man to be his faithful companion and friend.'

Our voyage proceeded securely, though slowly; but the nearer we approached the land, the more gloomy and unpromising its aspect appeared. The coast was clothed with barren rocks, which seemed to offer nothing but hunger and distress. The sea was calm; the waves, gently agitated, washed the shore, and the sky was serene; in every direction we perceived casks, bales, chests, and other vestiges of shipwrecks, floating round us. In the hope of obtaining some good provisions, I determined on endeavouring to secure some of the casks. I bade Fritz have a rope, a hammer, and some nails ready, and to try to seize them as we passed. He succeeded in laying hold of two, and in such a way that we could draw them after us to the shore. Now that we were close on land, its rude outline was much softened; the rocks no longer appeared one undivided chain; Fritz with his hawk's eye already described some trees, and exclaimed that they were palm trees. Ernest expressed his joy that he should now get much larger and better coconuts* than those of Europe. I for my part was venting audibly my regret, that I had not thought of bringing a telescope that I knew was in the captain's cabin, when Jack drew a small one from his pocket, and with a look of triumph presented it to me.

The acquisition of the telescope was of great importance; for with its aid I was able to make the necessary observations,

and was more sure of the route I ought to take. On applying
it to my eye, I remarked that the shore before us had a desert
and savage aspect, but that towards the left, the scene was more
agreeable: but when I attempted to steer in that direction, a
current carried me irresistibly towards the coast that was rocky
and barren. By and by we perceived a little opening between
the rocks, near the mouth of a creek, towards which all our
geese and ducks betook themselves; and I, relying on their
sagacity, followed in the same course. This opening formed a
little bay; the water was tranquil, and neither too deep nor too
shallow to receive our boat. I entered it, and cautiously put on
shore on a spot where the coast was about the same height
above the water as our tubs, and where, at the same time, there
was a quantity sufficient to keep us afloat. The shore extended
inland in something of the form of an isosceles triangle,* the
upper angle of which terminated among the rocks, while the
margin of the sea formed the basis.

All that had life in the boat jumped eagerly on land. Even
little Francis, who had been wedged in his tub like a potted
herring, now got up and sprang forward; but, with all his
efforts, he could not succeed without his mother's help. The
dogs, who had swam on shore, received us as if appointed to
do the honours of the place, jumping round us with every
demonstration of joy: the geese kept up a loud cackling, to
which the ducks, from their broad yellow beaks, contributed a
perpetual thorough-bass: the cocks and hens, which we had
already set at liberty, clucked: the boys chattering all at once,
produced altogether an overpowering confusion of sounds: to
this was added the disagreeable scream of some penguins and
flamingos,* which we now perceived, some flying over our
heads, others sitting on the points of the rocks at the entrance
of the bay. By and by the notes of the latter had the ascendant,
from their numbers; and our annoyance was increased by a
comparison we could not avoid making, between the sounds
they uttered; and the harmony of the feathered musicians of
our own country. I had however one advantage in perspective;—it was that, should we hereafter be short of food, these
very birds might serve for our subsistence.

The first thing we did on finding ourselves safe on *terra
firma*, was to fall on our knees, and return thanks to the

Supreme Being who had preserved our lives, and to recommend ourselves with entire resignation to the care of his paternal kindness.

We next employed our whole attention in unloading the boat. Oh! how rich we thought ourselves in the little we had been able to rescue from the merciless abyss of waters! We looked about for a convenient place to set up a tent under the shade of the rocks; and having all consulted and agreed upon a place, we set to work. We drove one of our poles firmly into a fissure of the rock; this rested upon another pole, which was driven perpendicularly into the ground, and formed the ridge of our tent. A frame for a dwelling was thus made secure. We next threw some sailcloth over the ridge, and stretching it to a convenient distance on each side, fastened its extremities to the ground with stakes. Lastly, I fixed some tenter-hooks* along the edge of one side of the sailcloth in front, that we might be able to enclose the entrance during night, by hooking in the opposite edge. The chest of provisions and other heavy matters we had left on the shore. The next thing was to desire my sons to look about for grass and moss, to be spread and dried in the sun, to serve us for beds. During this occupation, in which even the little Francis could take a share, I erected near the tent a kind of little kitchen. A few flat stones, I found in the bed of a fresh-water river, served for a hearth. I got a quantity of dry branches: with the largest I made a small enclosure round it; and with the little twigs, added to some of our turf, I made a brisk cheering fire. We put some of the soup-cakes, with water, into our iron pot, and placed it over the flame; and my wife, with her little Francis for a scullion, took charge of preparing the dinner.

In the meanwhile Fritz had been reloading the guns, with one of which he had wandered along the side of the river. He had proposed to Ernest to accompany him; but Ernest replied, that he did not like a rough, stony walk, and that he should go to the sea-shore. Jack took the road towards a chain of rocks which jutted out into the sea, with the intention of gathering some of the mussels which grew upon them.

My own occupation was now an endeavour to draw the two floating casks on shore, but in which I could not succeed; for our place of landing, though convenient enough for our machine, was too steep for the casks. While I was looking about to find a more favourable spot, I heard loud cries proceeding from a short distance, and recognized the voice of my son Jack. I snatched my hatchet, and ran anxiously to his assistance. I soon perceived him up to his knees in water in a shallow, and that a large sea lobster had fastened its claws in his leg. The poor boy screamed pitiably, and made useless efforts to disengage himself. I jumped instantly into the water; and the enemy was no sooner sensible of my approach, than he let go his hold, and would have scampered out to sea, but that I indulged the fancy of a little malice against him for the alarm he had caused us. I turned quickly upon him, and took him up by the body and carried him off, followed by Jack, who shouted our triumph all the way. He begged me at last to let him hold the animal in his own hand, that he might himself present so fine a booty to his mother. Accordingly, having observed how I held it to avoid the gripe, he laid his own hand upon it in exactly the same manner; but scarcely had he grasped it, than he received a violent blow on the face from the lobster's tail, which made him loose his hold, and the animal fell to the ground. Jack again began to bawl out, while I could not refrain from laughing heartily. In his rage he took up a stone and killed the lobster with a single blow. I was a little vexed at this conclusion to the scene.—'This is what we call killing an enemy when he is unable to defend himself, Jack; it is wrong to revenge an injury while we are in a state of anger: the lobster, it is true, had given you a bite; but then you, on your part, would have eaten the lobster. So the game was at least equal. Another time, I advise you to be both more prudent and more merciful.'—'But pray, father, let me carry it to my mother,' said Jack, fearless now of further warfare; and accordingly he carried it to the kitchen, triumphantly exclaiming, 'Mother, mother, a sea lobster!—Ernest, a sea lobster! Where is Fritz? Where is Fritz? Take care, Francis, he will bite you.' In a moment all were round him to examine the wonderful creature, and all proclaimed their astonishment at his enormous size, while they observed that its form was precisely that of the common lobster so much in use in Europe.

'Yes, yes,' said Jack, holding up one of the claws; 'you may well wonder at his size: this was the frightful claw which seized my leg, and if I had not had on my thick sea pantaloons, he would have bit it through and through; but I have taught him what it is to attack *me:* I have paid him well.'

'Oh, oh! Mr Boaster,' cried I, 'you give a pretty account of the matter. Now *mine* would be, that if I had not been near, the lobster would have shown you another sort of game; for the slap he gave you in the face compelled you, I think, to let go your hold. And it is well it should be thus; for he fought with the arms with which nature had supplied him, but you had recourse to a great stone for your defence. Believe me, Jack, you have no great reason to boast of the adventure.'

Ernest, ever prompted by his savoury tooth, bawled out that the lobster had better be put into the soup, which would give it an excellent flavour: but this his mother opposed, observing, that we must be more economical of our provisions than that, for the lobster of itself would furnish a dinner for the whole family. I now left them and walked again to the scene of this adventure, and examined the shallow: I then made another attempt upon my two casks, and at length succeeded in getting them into it, and in fixing them there securely on their bottoms.

On my return, I complimented Jack on his being the first to procure an animal that might serve for subsistence, and promised him, for his own share, the famous claw, which had furnished us with so lively a discussion.

'Ah! but *I* have seen something too, that is good to eat,' said Ernest; 'and I should have got it if it had not been in the water, so that I must have wetted my feet——'

'Oh, that is a famous story,' cried Jack: 'I can tell you what he saw,—some nasty mussels: why, I would not eat one of them for the world.—Think of my lobster!'

'That is not true, Jack; for they were oysters, and not mussels, that I saw: I am sure of it, for they stuck to the rock, and I know they must be oysters.'

'Fortunate enough, my dainty gentleman,' interrupted I, addressing myself to Ernest; 'since you are so well acquainted with the place where such food, can be found, you will be so obliging as to return and procure us some. In such a situation as ours, every member of the family must be actively employed for the common good; and, above all, none must be afraid of so trifling an inconvenience as wet feet.'

'I will do my best, with all my heart,' answered Ernest; 'and at the same time I will bring home some salt, of which I have seen immense quantities in the holes of the rocks, where I have reason to suppose it is dried by the sun. I tasted some of it, and it was excellent. Pray, father, be so good as to inform me whether this salt was not left there by the sea?'

'—No doubt it was, Mr Reasoner, for where else do you think it could come from? You would have done more wisely if you had brought us a bag of it, instead of spending your time in profound reflections upon operations so simple and obvious; and if you do not wish to dine upon a soup without flavour, you had better run and fetch a little quickly.'

He set off, and soon returned: what he brought had the appearance of sea-salt, but was so mixed with earth and sand, that I was on the point of throwing it away; but my wife prevented me, and by dissolving, and afterwards filtering some of it through a piece of muslin, we found it admirably fit for use.

'Why could we not have used some seawater,' asked Jack, 'instead of having all this trouble?'

'Seawater', answered I, 'is more bitter than salt, and has, besides, a sickly taste.' While I was speaking, my wife tasted the soup with a little stick with which she had been stirring it, and pronounced that it was all the better for the salt, and now quite ready. 'But', said she, 'Fritz is not come in. And then how shall we manage to eat our soup without spoons or dishes? Why did we not remember to bring some from the ship?'—'Because, my dear, one cannot think of every thing at once. We shall be lucky if we have not forgotten even more important things.'—'But, indeed,' said she, 'this is a matter which cannot easily be set to rights. How will it be possible for each of us to raise this large boiling pot to his lips?'

I soon saw that my wife was right. We all cast our eyes upon the pot with a sort of stupid perplexity, and looked a little like the fox in the fable, when the stork* desires him to help himself from a vessel with a long neck. Silence was at length broken, by all bursting into a hearty laugh at our want of every kind of utensil, and at the thought of our own folly, in not recollecting that spoons and forks were things of absolute necessity.

Ernest observed, that if we could but get some of the nice coconuts he often thought about, we might empty them, and use the pieces of the shells for spoons.

'Yes, yes,' replied I; '*if we could but get*,—but we have them not; and if wishing were to any purpose, I had as soon wish at once for a dozen silver spoons; but alas! of what use is wishing?'

'But at least,' said the boy, 'we can use some oyster-shells for spoons.'

'Why, this is well, Ernest,' said I, 'and is what I call a useful thought. Run then quickly for some of them. But, gentlemen, I give you notice, that no one of you must give himself airs because his spoon is without a handle, or though he chance to grease his fingers in the soup.'

Jack ran first, and was up to his knees in the water before Ernest could reach the place. Jack tore off the fish with eagerness, and threw them to slothful Ernest, who put them into his handkerchief, having first secured in his pocket one shell he had met with of a large size. The boys came back together with their booty.

Fritz not having yet returned, his mother was beginning to be uneasy, when we heard him shouting to us from a small distance, to which we answered by similar sounds. In a few minutes he was among us, his two hands behind him, and with a sort of would-be-melancholy air, which none of us could well understand.—'What have you brought?' asked his brothers; 'let us see your booty, and you shall see ours.'—'Ah! I have unfortunately nothing.'—'What! nothing at all?' said I.— 'Nothing at all,' answered he. But now, on fixing my eye upon him, I perceived a smile of proud success through his assumed dissatisfaction. At the same instant Jack, having stolen behind him, exclaimed, 'A sucking pig! a sucking pig!' Fritz, finding his trick discovered, now proudly displayed his prize, which I immediately perceived, from the description I had read in different books of travels, was an agouti,* an animal common in that country, and not a sucking pig, as the boys had supposed. 'The agouti', says M. de Courtills,* in his voyage to St Domingo, 'is of the size of a hare, and runs with the same swiftness; but its form is more like the pig, and he makes the same grunting noise. He is not a voracious animal, but is nice in the choice of his food. When his appetite is satiated, he buries what remains, and keeps it for another time. He is naturally of a gentle temper; but if provoked, his hair becomes erect, he bites, and strikes the ground with his hind feet like the rabbit, which he also resembles in digging himself a burrow under ground: but this burrow has but one entrance; he conceals himself in it during the hottest part of the day, taking care to provide himself with a store of patates* and bananas. He is usually taken by coursing, and sometimes by dogs, or with nets. When it is found difficult to seize him, the sportsman has only to whistle. As soon as the agouti hears the sound, he is instantly still, remains resting on his hind feet, and suffers himself to be taken. His flesh is white, like that of the rabbit; but it is dry, has no fat, and never entirely loses a certain wild flavour, which is disagreeable to Europeans. He is held in great esteem by the natives, particularly when the animal has been feeding near the sea on plants impregnated with salt. They are therefore caught in great numbers, and for this reason the species is much diminished.'—'Where did you find him? How did you get at him? Did he make you run a great way?' asked all at once the young brothers. 'Tell me, tell us all . . .' etc. I, for my part, assumed a somewhat serious tone—'I should have preferred,' observed I, 'that you had in reality brought us nothing, to your asserting a falsehood. Never allow yourself, even in jest, my dear boy, to assert what you know to be an untruth. By such trifles as these, a habit of lying, the most disgusting of vices, may be induced. Now then that I have given you this caution, let us look at the animal. Where did you find it?'

Fritz related, that he had passed over to the other side of the river. 'Ah!' continued he, 'it is quite another thing from this place; the shore is low, and you can have no notion of the quantity of casks, chests, and planks, and different sorts of things washed there by the sea. Ought we not to go and try to obtain some of these treasures?'—'We will consider of it soon,' answered I, 'but first we have to make our voyage to the vessel, and fetch away the animals; at least you will all agree, that of the cow we are pretty much in want.'—'If our biscuit were soaked in milk, it would not be so hard,' observed our dainty Ernest.—'I must tell you too,' continued Fritz, 'that over on the other side there is as much grass for pasturage as we can desire; and, besides, a pretty wood, in the shade of which we could repose. Why then should we remain on this barren desert side?'—'Patience,' replied I, 'there is a time for every thing, friend Fritz; we shall not be without something to undertake tomorrow, and even after tomorrow. But, above all, I am eager to know if you discovered, in your excursion, any traces of our ship companions?'—'Not the smallest trace of man, dead or alive, on land or water; but I have seen some other animals, that more resembled pigs than the one I have brought you, but with feet more like those of the hare; the animal I am speaking of leaps from place to place; now sitting on his hind legs, rubbing his face with his front feet, and then seeking for roots, and gnawing them like the squirrel. If I had not been afraid of his escaping me, I should have tried to catch him with my hands, for he appeared almost tame.'

We had now notice that our soup was ready, and each hastened to dip his shell into the pot, to get out a little; but, as I had foreseen, each drew out a scalded finger, and it was who could scream the loudest. Ernest was the only one who had been too cautious to expose himself to this misfortune: he quietly took his mussel-shell, as large and deep as a small saucer, from his pocket, and carefully dipping it into the pot, drew it out filled with as much soup as was his fair share, and casting a look of exultation on his brothers, he set it down till it should be cold enough to eat.

'You have taken good care of yourself, I perceive', said I. 'But now answer me, dear boy, is the advantage worth the pains you take to be better off than your companions? Yet this is the constant failing of your character. As your best friend, I feel it my duty to balk you of the expected prize; I therefore adjudge your dish of delicious soup to our faithful followers, Turk and Flora. For ourselves, we will all fare alike; we will simply dip our shells into the pot till hunger is appeased; but the picked dish for the dogs, Ernest; and *all the rest alike!*'

This gentle reproach sunk, I perceived, into his heart; he placed the shell, filled with soup, upon the ground, and in an instant the dogs had licked up every drop. We on our parts were as sharp set as they, and every eye was fixed on the pot, watching for the steam to subside a little, that we might begin dipping; when, on looking round, we saw Turk and Flora standing over the agouti, gnawing and tearing him fiercely with their teeth and paws. The boys all screamed together: Fritz seized his gun, and struck them with it; called them the unkindest names, threw stones at them; and was so furious, that if I had not interfered, it is probable he would have killed them. He had already bent his gun with the blows he had given them, and his voice was raised so high as to be re-echoed from the rocks.

When he had grown a little cool, I seriously remonstrated with him on his violence of temper. I represented to him what distress he had occasioned his mother and myself for the event of a rage so alarming: that his gun, which might have been so useful, was now spoiled; and that the poor animals, upon whose assistance we should probably so much depend, he had, no doubt, greatly injured: 'Anger', continued I, 'is always a bad counsellor, and may even lead the way to crimes: you are not ignorant of the history of Cain,* who, in a moment of violent anger, killed his brother.'—'Say no more, my dearest father', interrupted Fritz in a tone of horror.—'Happy am I to recollect on this occasion,' resumed I, 'that it was not human creatures you treated thus. But an angry person never reasons; he scarcely knows whom he attacks. The most convincing proof of this is, that *you* just now fell upon two dumb animals, incapable of judgement, and who most likely thought that your agouti was placed there, as the soup had been before, for them to eat. Confess, too, that it was vanity which excited the furious temper you exhibited. If another than yourself had killed the agouti, you would have been more patient under the accident.' Fritz agreed that I was right, and, half drowned in tears, entreated my forgiveness.

Soon after we had taken our meal, the sun began to sink into the west. Our little flock of fowls assembled round us, pecking here and there what morsels of our biscuit had fallen on the ground. Just at this moment my wife produced the bag she had so mysteriously huddled into the tub. Its mouth was now opened; it contained the various sorts of grain for feeding poultry— barley, pease, oats, etc., and also different kinds of seeds and roots of vegetables for the table. In the fullness of her kind heart she scattered several handfuls at once upon the ground, which the fowls began eagerly to seize.

I complimented her on the benefits her foresight had secured for us; but I recommended a more sparing use of so valuable an acquisition, observing, that the grain, if kept for sowing, would produce a harvest, and that we could fetch from the ship spoiled biscuit enough to feed the fowls. Our pigeons sought a roosting place among the rocks; the hens, with the two cocks at their head, ranged themselves in a line along the ridge of the tent; and the geese and ducks betook themselves in a body, cackling and quacking as they proceeded, to a marshy bit of ground near the sea, where some thick bushes afforded them shelter.

A little later, we began to follow the example of our winged companions, by beginning our preparations for repose. First, we loaded our guns and pistols, and laid them carefully in the tent; next, we assembled together and joined in offering up our thanks to the Almighty for the succour afforded us, and supplicating his watchful care for our preservation. With the last ray of the sun we entered our tent, and after drawing the sailcloth over the hooks to close the entrance, we laid ourselves down close to each other on the grass and moss we had collected in the morning.

The children observed, with surprise, that darkness came upon us all at once; that night succeeded to day without an intermediate twilight.—'This', replied I, 'makes me suspect that we are not far from the equator, or at least between the tropics, where this is of ordinary occurrence; for the twilight is occasioned by the rays of the sun being broken in the atmosphere; the more obliquely they fall, the more their feeble light is extended and prolonged; while, on the other hand, the more perpendicular the rays, the less their declination: consequently the change from day to night is much more sudden when the sun is under the horizon.'

I looked once more out of the tent to see if all was quiet around us. The old cock, awaking at the rising of the moon, chanted our vespers,* and then I lay down to sleep. In proportion as we had been during the day oppressed with heat, we were now in the night inconvenienced by the cold, so that we clung to each other for warmth. A sweet sleep began to close the eyes of my beloved family; I endeavoured to keep awake till I was sure my wife's solicitude had yielded to the same happy state, and then I closed my own. Thanks to the fatigue we had undergone, our first night in the desert island was very tolerably comfortable.

3
Voyage of Discovery

I WAS roused at the dawn of day by the crowing of the cocks. I awoke my wife, and we consulted together as to the occupations we should engage in. We agreed, that we would seek for traces of our late ship companions, and at the same time examine the nature of the soil on the other side of the river, before we determined on a fixed place of abode. My wife easily perceived that such an excursion could not be undertaken by all the members of the family; and full of confidence in the protection of Heaven, she courageously consented to my proposal of leaving her with the three youngest boys, and proceeding myself with Fritz on a journey of discovery. I entreated her not to lose a moment in giving us our breakfast. She gave us notice that the share of each would be but small, there being no more soup prepared.—'What then,' I asked, 'is to become of Jack's lobster?'—'That he can best tell you himself,' answered his mother. 'But now pray step and awake the boys, while I make a fire and put on some water.'

The children were soon roused; even our slothful Ernest submitted to the hard fate of rising so early in the morning. When I asked Jack for his lobster, he ran and fetched it from a cleft in the rock, in which he had concealed it: 'I was determined,' said he, 'that the dogs should not treat my lobster as they did the agouti, for I knew them for a sort of gentlemen to whom nothing comes amiss.'—'I am glad to see, son Jack,' said I, 'that that giddy head upon your shoulders can be prevailed upon to reflect. "Happy is he who knows how to profit by the misfortunes of others," says the proverb. But will you not kindly give Fritz the great claw, which bit your leg (though I promised it to you), to carry with him for his dinner in our journey?'

'What journey?' asked all the boys at once. 'Ah! we will go too: a journey!' repeated they, clapping their hands, and jumping round me like little kids.*—'For this time,' said I, 'it is impossible for all of you to go; we know not yet what we are to set about, nor whither we are going. Your eldest brother and myself shall be better able to defend ourselves in any danger, without you; besides that with so many persons we could proceed but slowly. You will then all three remain with your mother in this place, which appears to be one of perfect safety, and you shall keep Flora to be your guard, while we will take Turk with us. With such a protector, and a gun well loaded, who shall dare treat us with disrespect? Make haste, Fritz, and tie up Flora, that she may not follow us; and have your eye on Turk, that he may be at hand to accompany us; and see the guns are ready.'

At the word guns, the colour rose in the cheeks of my poor boy. His gun was so bent as to be of no use; he took it up and tried in vain to straighten it; I let him alone for a short time; but at length I gave him leave to take another, perceiving with pleasure that the vexation had produced a proper feeling in his mind. A moment after, he attempted to lay hold of Flora to tie her up; but the dog recollecting the blows she had so lately received, began to snarl, and would not go near him. Turk behaved the same, and I found it necessary to call with my own voice, to induce them to approach us. Fritz then in tears entreated for some biscuit of his mother, declaring that he would willingly go without his breakfast to make his peace with the dogs; he accordingly carried them some biscuit, stroked and caressed them, and in every motion seemed to ask their pardon. As of all animals, without excepting man, the dog is least addicted to revenge, and at the same time is the most sensible of kind usage, Flora instantly relented, and began to lick the hands which fed her; but Turk, who was of a more fierce and independent temper, still held off, and seemed to feel a want of confidence in Fritz's advances.—'Give him a claw of my lobster,' cried Jack, 'for I mean to give it all to you for your journey.'

'I cannot think why you should give it all,' interrupted Ernest, 'for you need not be uneasy about their journey. Like Robinson Crusoe,* they will be sure enough to find some coconuts, which they will like much better than your miserable lobster: only think, a fine round nut, Jack, as big as my head, and with at least a teacup full of delicious sweet milk in it!'

'Oh! brother Fritz, pray do bring me some,' cried little Francis.

We now prepared for our departure: we took each a bag for game, and a hatchet: I put a pair of pistols in the leather band round Fritz's waist, in addition to the gun, and provided myself with the same articles, not forgetting a stock of biscuit and a flask of fresh river water. My wife now called us to breakfast when all attacked the lobster; but its flesh proved so hard, that there was a great deal left when our meal was finished, and we packed it for our journey without further regret from any one. The sea lobster is an animal of considerable size, and its flesh is much more nutritious, but less delicate, than the common lobster.

Fritz urged me to set out before the excessive heat came on.—'With all my heart,' said I, 'but we have forgot one thing.'—'What is that?' asked Fritz, looking round him; 'I see nothing to do but to take leave of my mother and my brothers.' —'I know what it is,' cried Ernest; 'we have not said our prayers this morning.'—'That is the very thing, my dear boy,' said I. 'We are too apt to forget God, the giver of all, for the affairs of this world; and yet never had we so much need of his care, particularly at the moment of undertaking a journey in an unknown soil.'

Upon this our pickle Jack began to imitate the sound of church-bells, and to call 'Bome! bome! bidi bome, bidiman, bome. To prayers, to prayers, bome, bome!'—'Thoughtless boy!' cried I, with a look of displeasure, 'when, oh! when will you be sensible of that sacredness in devotion that banishes for the time every thought of levity or amusement? Recollect yourself, and let me not have again to reprove you on a subject of so grave a nature.'

In about an hour we had completed the preparations for our departure. I had loaded the guns we left behind, and I now enjoined my wife to keep by day as near the boat as possible, which in case of danger was the best and most speedy means of escape. My next concern was to shorten the moment of separation, judging by my own feelings those of my dear wife; for neither could be without painful apprehensions of what new misfortune might occur on either side during the interval. We all melted into tears;—I seized this instant for drawing Fritz away, and in a few moments the sobs and often repeated adieus of those we left behind, died away in the noise of the waves which we now approached, and which turned our thoughts upon ourselves and the immediate object of our journey.

The banks of the river were everywhere steep and difficult, excepting at one narrow slip near the mouth on our side, where we had drawn our fresh water. The other side presented an unbroken line of sharp, high, perpendicular rocks. We therefore followed the course of the river till we arrived at a cluster of rocks at which the stream formed a cascade: a few paces beyond, we found some large fragments of rock which had fallen into the bed of the river: by stepping upon these, and making now and then some hazardous leaps, we contrived to reach the other side. We proceeded a short way along the rock we ascended in landing, forcing ourselves a passage through tall grass, which twined with other plants, and were rendered more capable of resistance by being half dried by the sun. Perceiving, however, that walking on this kind of surface in so hot a sun would exhaust our strength, we looked for a path to descend and proceed along the river, where we hoped to meet with fewer obstacles, and perhaps to discover traces of our ship companions.

When we had walked about a hundred paces we heard a loud noise behind us, as if we were pursued, and perceived a rustling motion in the grass, which was almost as tall as ourselves. I was a good deal alarmed, thinking that it might be occasioned by some frightful serpent, a tiger, or other ferocious animal. But I was well satisfied with Fritz, who, instead of being frightened, and running away, stood still and firm to face the danger, the only motion he made being to see that his piece was ready, and turning himself to front the spot from whence the noise proceeded. Our alarm was, however, short; for what was our joy on seeing rush out, not an enemy, but our faithful Turk, whom in the distress of the parting scene we had forgotten, and whom no doubt our anxious relatives had sent on to us! I received the poor creature with lively joy, and did not fail to commend both the bravery and discretion of my son, in not yielding to even a rational alarm, and for waiting till he was sure of the object before he resolved to fire: had he done otherwise, he might have destroyed an animal likely to afford us various kinds of aid, and to contribute by the kindness of his temper to the pleasures of our domestic scene.—'Observe, my dear boy,' said I, 'to what dangers the tumult of the passions exposes us: the anger which overpowered you yesterday, and the error natural to the occasion we have this moment witnessed, if you had unfortunately given way to it, might either of them have produced an irretrievable misfortune.'

Fritz assured me he was sensible of the truth and importance of my remarks; that he would watch constantly over the defects of his temper: and then he fell to caressing the faithful and interesting animal.

Conversing on such subjects as these, we pursued our way. On our left was the sea, and on our right the continuation of the ridge of rocks which began at the place of our landing, and ran along the shore, the summit everywhere adorned with fresh verdure and a great variety of trees. We were careful to proceed in a course as near the shore as possible, casting our eyes alternately upon its smooth expanse and upon the land in all directions, to discover our ship companions, or the boats which had conveyed them from us; but our endeavours were in vain.

Fritz proposed to fire his gun from time to time, that, should they be anywhere concealed near us, they might thus be led to know of our pursuit.

'This would be vastly well,' I observed, 'if you could contrive that the savages, who are most likely not far distant, should not hear the sound, and come in numbers upon us.'—'I am thinking, father,' interrupted Fritz, 'that there is no good reason why we should give ourselves so much trouble and uneasiness about persons who abandoned us so cruelly, and thought only of their own safety.—'

'There is not only one good reason, but many,' replied I: 'first, we should not return evil for evil; next, it may be in their power to assist us; and lastly, they are perhaps at this moment in the greatest want of assistance. It was their lot to escape with nothing but life from the ship, if indeed they are still alive, while we had the good fortune to secure provisions enough for present subsistence, to a share of which they are as fully entitled as ourselves.'

'But, father, while we are wandering here, and losing our time almost without a hope of benefit to them, might we not be better employed in returning to the vessel, and saving the animals on board?'

'—When a variety of duties present themselves for our choice, we should always give the preference to that which can confer the most solid advantage. The saving of the life of a man is a more exalted action than the contributing to the comfort of a few quadrupeds, whom we have already supplied with food for several days; particularly as the sea is in so calm a state, that we need entertain no apprehension that the ship will sink or go entirely to pieces just at present.'

My son made no reply to what I said, and we seemed by mutual silent consent to take a few moments for reflection.

When we had gone about two leagues, we entered a wood situated a little further from the sea: here we threw ourselves on the ground, under the shade of a tree, by the side of a clear running stream, and took out some provisions and refreshed ourselves. We heard the chirping, singing, and motion of birds in the trees, and observed, as they now and then came out to view, that they were more attractive by their splendid plumage than by any charm of note. Fritz assured me that he had caught a glimpse of some animals like apes among the bushes, and this was confirmed by the restless movements of Turk, who began to smell about him, and to bark so loud that the wood resounded with the noise. Fritz stole softly about to be sure, and presently stumbled on a small round body which lay on the ground: he brought it to me, observing that it must be the nest of some bird.—'What makes you of that opinion?' said I. 'It is, I think, much more like a coconut.'*

'But I have read that there are some kinds of birds, which build their nests quite round; and look, father, how the outside is crossed and twined.'

'But do you not perceive that what you take for straws crossed and twined by the beak of a bird, is in fact a coat of fibres formed by the hand of Nature? Do you not remember to have read, that the nut of a coco shell is inclosed within a round, fibrous covering, which again is surrounded by a skin of a thin and fragile texture? I see that in the one you hold in your hand, this skin has been destroyed by time, which is the reason that the twisted fibres (or inner covering) are so apparent; but now let us break the shell, and you will see the nut inside.'

We soon accomplished this; but the nut, alas! from lying on the ground, had perished, and appeared but little different from a bit of dried skin, and not the least inviting to the palate.

Fritz was much amused at this adventure. 'How I wish Ernest could have been here!' cried he. 'How he envied me the fine large coconuts I was to find, and the whole teacup full of sweet delicious milk which was to spring out upon me from the inside!—But, father, I myself believed that the coconut contained a sweet refreshing liquid, a little like the juice of almonds: travellers surely tell untruths!'

'Travellers certainly do sometimes tell untruths, but not, I believe, on the subject of the coconut, which is well known to contain the liquid you describe, just before they are in a state of ripeness. It is the same with our European nuts, with the difference of quantity; and one property is common to both, that as the nut ripens, the milk diminishes, by thickening, and becoming the same substance as the nut. If you put a ripe nut a little way under the earth, in a good soil, the kernel will shoot and burst the shell; but if it remain above ground, or in a place that does not suit its nature, the principle of vegetation is extinguished by internal fermentation, and the nut perishes as you have seen.'

'I am now surprised that this principle is not extinguished in every nut; for the shell is so hard, it seems impossible for a softer substance to break it.'

'The peach-stone is no less hard; the kernel, notwithstanding, never fails to break it, if it is placed in a well-nurtured soil.'

'Now I begin to understand. The peach-stone is divided into two parts, like a mussel-shell; it has a kind of seam round it, which separates of itself when the kernel is swelled by moisture: but the coconut in my hand is not so divided, and I cannot conceive of its separating.'

'I grant that the coconut is differently formed; but you may see by the fragments you have just thrown on the ground, that Nature has in another manner stepped in to its assistance. Look near the stalk, and you will discover three round holes, which are not, like the rest of its surface, covered with a hard impenetrable shell, but are stopped by a spongy kind of matter; it is through these that the kernel shoots.'

'Now, father, I have the fancy of gathering all the pieces together and giving them to Ernest, and telling him these particulars: I wonder what he will say about it, and how he will like the withered nut.'

'Now the fancy of your father, my dear boy, would be to find you without so keen a relish for a bit of mischief. Joke with Ernest, if you will, about the withered nut; but I should like to see you heal the disappointment he will feel, by presenting him at last with a sound and perfect nut, provided we should have one to spare.'

After looking for some time, we had the good luck to meet with one single nut. We opened it, and finding it sound, we sat down and ate it for our dinner, by which means we were enabled to husband the provisions we had brought. The nut, it is true, was a little oily and rancid; yet, as this was not a time to be nice, we made a hearty meal, and then continued our route. We did not quit the wood, but pushed our way across it, being often obliged to cut a path through the bushes overrun by creeping plants, with our hatchet. At length we reached a plain, which afforded a more extensive prospect and a path less perplexed and intricate.

We next entered a forest to the right, and soon observed that some of the trees were of a singular kind. Fritz, whose sharp eye was continually on a journey of discovery, went up to examine them closely. 'O heavens! father, what odd trees, with wens growing all about their trunks!' I had soon the surprise and satisfaction of assuring him that they were of the gourd-tree kind,* the trunks of which bear fruit. Fritz, who had never heard of such a tree, could not conceive the meaning of what he saw, and asked me if the fruit was a sponge or a wen.—'We will see', I replied, 'if we cannot unravel the mystery. Try to get down one of them, and we will examine it minutely.'

'I have got one,' cried Fritz, 'and it is exactly like a gourd, only the rind is thicker and harder.'

'It then, like the rind of that fruit, can be used for making various utensils,' observed I; 'plates, dishes, basins, flasks. We will give it the name of the gourd-tree.'

Fritz jumped for joy.—'How happy my mother will be!' cried he in ecstasy; 'she will no longer have the vexation of thinking when she makes soup, that we shall all scald our fingers!'

'What, my boy, do you think is the reason that this tree bears its fruit only on the trunk and on its topmost branches?'

'I think it must be because the middle branches are too feeble to support such a weight.'

'You have guessed exactly right.'

'But are these gourds good to eat?'

'At worst they are, I believe, harmless; but they have not a very tempting flavour. The negro savages set as much value on the rind of this fruit as on gold, for its use to them is indispensable. These rinds serve them to keep their food and drink in, and sometimes they even cook their victuals in them.'

'Oh father! it must be impossible to cook their victuals in them; for the heat of fire would soon consume such a substance.'

'I did not say the rind was put upon the fire.'

'How droll! pray how are victuals to be cooked without fire?'

'Nor did I say that victuals could be cooked without a fire; but there is no need to put the vessel that contains the food upon the fire.'

'I have no idea what you mean; there seems to be a miracle.'

'So be it, my son. A little tincture of enchantment is the lot of man. When he finds himself deficient in intelligence, or is too indolent to give himself the trouble to reflect, he is driven by his weakness to ascribe to a miracle, or to witchcraft, what is, most likely, nothing but the most ordinary operation of Art or Nature.'

'Well, father, I will then believe in what you tell me of these rinds.'

'That is, you will cut the matter short, by resolving to be sure on the word of another: this is a good way to let your own reason lie fallow. Come, come, no such idleness; let me help you to understand this amazing phenomenon. When it is intended to dress food in one of these rinds, the process is, to cut the fruit into two equal parts, and scoop out the inside; some water is put into one of the halves, and into the water some fish, a crab, or whatever else is to be dressed; then some stones red hot, beginning with one at a time, are thrown in, which impart sufficient heat to the water to dress the food, without the smallest injury to the pot.'

'But is not the food spoiled by ashes falling in, or by pieces of the heated stones separating in the water?'

'Certainly it is not easy to make fine sauces or ragouts in such a vessel; but a dressing of the meat is actually accomplished, and the negroes and savages, who are the persons to make use of what is thus cooked, are not very delicate: but I can imagine a tolerable remedy for even the objection you have found. The food might be inclosed in a vessel small enough to be contained in our capacious half of a gourd, and thus be cooked upon the principle so much used in chemistry; the application of a milder heat than fire. And this method of cooking has also another advantage, that the thing contained cannot adhere to the sides or bottom of the vessel.'

We next proceeded to the manufacture of our plates and dishes. I taught my son how to divide the gourd with a bit of string, which would cut more equally than a knife; I tied the string round the middle of the gourd as tight as possible, striking it pretty hard with the handle of my knife, and I drew tighter and tighter till the gourd fell apart, forming two regular shaped bowls or vessels; while Fritz, who had used a knife for the same operation, had entirely spoiled his gourd by the irregular pressure of his instrument. I recommended his making some spoons with the spoiled rind, as it was good for no other purpose. I, on my part, had soon completed two dishes of convenient size, and some smaller ones to serve as plates.

Fritz was in the utmost astonishment at my success. — 'I cannot imagine, father,' said he, 'how this way of cutting the gourd could occur to you!'

'I have read the description of such a process', replied I, 'in books of travels; and also that such of the savages as have no knives, and who make a sort of twine from the bark of trees are accustomed to use it for this kind of purpose. So you see what benefit may be derived from reading, and from afterwards reflecting on what we read.'

'And the flasks, father; in what manner are they made?'

'For this branch of their ingenuity they make preparation a long time beforehand. If a negro wishes to have a flask or bottle with a neck, he binds a piece of string, linen, bark of a tree, or any thing he can get, round the part nearest the stalk of a very young gourd; he draws this bandage so tight, that the part at liberty soon forms itself to a round shape, while the part which is confined contracts, and remains ever after narrow. By this method it is that they obtain flasks or bottles of a perfect form.'

'Are then the bottle-shaped gourds I have seen in Europe trained by a similar preparation?'

'No, they are of another species, and what you have seen is their natural shape.'

Our conversation and our labour thus went on together. Fritz had completed some plates, and was not a little proud of the achievement. 'Ah, how delighted my mother will be to eat upon them!' cried he. 'But how shall we convey them to her? They will not, I fear, bear travelling well.'

'We must leave them here on the sand for the sun to dry them thoroughly; this will be accomplished by the time of our return this way, and we can then carry them with us; but care must be taken to fill them with sand, that they may not shrink or warp in so ardent a heat.' My boy did not dislike this task; for he had no great fancy to the idea of carrying such a load on our journey of further discovery. Our sumptuous service of porcelain was accordingly spread upon the ground, and for the present abandoned to its fate.

We amused ourselves as we proceeded, in endeavouring to fashion some spoons from the fragments of the gourd-rinds. I had the fancy to try my skill upon a piece of coconut; but I must needs confess that what we produced had not the least resemblance to those I had seen in the Museum at London, and which were shown there as the work of some of the islanders of the Southern Seas. A European without instruments must always find himself excelled in such attempts by the superior adroitness and patience of savages; in this instance too of ourselves, we had the assistance of knives, while the savages have only flat stones with a sharp edge to work with.

'My attempt has been scarcely more successful than your own,' I cried; 'and to eat soup with either your spoon or mine, we ought to have mouths extending from ear to ear.'

'True enough, father,' answered Fritz; 'but it is not my fault. In making mine, I took the curve of my bit of rind for a guide; if I had made it smaller, it would have been too flat, and it is still more difficult to eat with a shovel than with an oyster-shell. But I am thinking that they may serve till I have improved upon my first attempt, and I am quite sure of the pleasure they will afford my mother. I imagine it pleases God sometimes to visit his creatures with difficulties, that they may learn to be satisfied with a little.'

'That is an excellent remark, my boy,' said I, 'and gives me more pleasure than a hundred crowns* would do.' Fritz burst into a fit of laughter.—'You do not rate my remark very high when you say this, father,' cried he, 'for of what use would a hundred crowns be to you at present? If you had said a good soup, or a hundred coconuts, I should be much prouder for having made it.'

'But as it is, my son, you have a right to be proud. I am well pleased to find you are beginning to estimate things according to their real value and usefulness, instead of considering them as good or bad, like children, without understanding the true reason. Money is only a means of exchange in human society; but here, on this solitary coast, Nature is more generous than man, and asks no payment for the benefits she bestows.'

While these conversations and our labours had been going on, we had not neglected the great object of our pursuit,—the making every practicable search for our ship companions. But our endeavours, alas! were all in vain.

After a walk of about four leagues in all, we arrived at a spot where a slip of land reached far out into the sea, on which we observed a rising piece of ground or hill. On a moment's reflection we determined to ascend it, concluding we should obtain a clear view of all adjacent parts, which would save us the fatigue of further rambles. We accordingly accomplished the design.

We did not reach the top of the hill without many efforts and a plentiful perspiration: but when there, we beheld a scene of wild and solitary beauty, comprehending a vast extent of land and water. It was, however, in vain that we used our telescope in all directions; no trace of man appeared. A truly embellished nature presented herself; and we were in the highest degree sensible of her thousand charms. The shore, rounded by a bay of some extent, the bank of which ended in a promontory on the further side; the agreeable blue tint of its surface; the sea, gently agitated by waves in which the rays of the sun were reflected; the woods of variegated hues and verdure, formed altogether a picture of such magnificence, of such new and exquisite delight, that, if the recollection of our unfortunate companions, engulfed perhaps in this very ocean, had not intruded to depress our spirits, we should have yielded to the ecstasy the scene was calculated to inspire. In reality, from this moment we began to lose even the feeble hope we had entertained, and sadness stole involuntarily into our hearts. We however, became but the more sensible of the goodness of the Divine Being, in the special protection afforded to ourselves, in conducting us to a home where there was no present cause for fear of danger from without, where we had not experienced the want of food, and where there was a prospect of future safety for us all. We had encountered no venomous or ferocious animals; and, as far as our sight could yet reach, we were not threatened by the approach of savages. I remarked to Fritz, that we seemed destined to a solitary life, and that it was a rich country which appeared to be allotted us for a habitation;—'at least, my son, our habitation it must be, unless some vessel should happen to put on shore on the same coast, and be in a condition to take us back to our native land. And God's will be done!' added I, 'for he knows what is best for us. Having left our native country, fixed in the intention of inhabiting some propitious soil, it was natural at first to encounter difficult adventures. Let us therefore consider our situation as no disappointment in any essential respect. We can pursue our scheme for agriculture. We shall learn to invent arts. Our only want is numbers.'

'As for me,' answered Fritz, 'I care but little about being so few of us. If I have the happiness of seeing you and my mother well in health and easy, I shall not give myself much uneasiness about those wicked unkind ship companions of ours.'

'No, my boy; they were not all bad people; and they would have become better men here, because not exposed to the temptations of the world. Common interest, united exertions, mutual services and counsels, together with the reflections which would have grown in such a state as this, tend to the improvement of the heart's affections.'

'We however of ourselves', observed Fritz, 'form a larger society than was the lot of Adam before he had children; and, as we grow older, we will perform all the necessary labour, while you and my mother enjoy ease and quiet.'

'Your assurances are as kind as I can desire, and they encourage me to struggle with what hardships may present themselves. Who can foresee in what manner it may be the will of Heaven to dispose of us? In times of old, God said to one of his chosen, "I will cause a great nation to descend from thy loins." '*

'And why may not we too become patriarchs, father?'

'Why not? you ask;—and I have not now time to answer. But come, my young patriarch,* let us find a shady spot, that we may not be consumed with the fierce heat of the sun before the patriarchal condition can be conferred upon us. Look yonder at that inviting wood: let us hasten thither to take a little rest, then eat our dinner, and return to our dear expecting family.'

We descended the hill, and made our way to a wood of palms, which I had just pointed out to Fritz: our path was clothed with reeds, entwined with other plants, which greatly obstructed our march. We advanced slowly and cautiously, fearing at every step to receive a mortal bite from some serpent that might be concealed among them. We made Turk go before, to give us timely notice of anything dangerous. I also cut a reed-stalk of uncommon length and thickness, for my defence against any enemy. It was not without surprise that I perceived a glutinous sap proceed from the divided end of the stalk. Prompted by curiosity, I tasted this liquid, and found it sweet and of a pleasant flavour, so that not a doubt remained that we were passing through a plantation of sugar-canes. I again applied the cane to my lips, and sucked it for some moments, and felt singularly refreshed and strengthened. I determined not to tell Fritz immediately of the fortunate discovery I had made, preferring that he should find it out for himself. As he was at some distance before me, I called out to him to cut a reed for his defence. This he did, and, without any remark, used it simply for a stick, striking lustily with it on all sides to clear a passage. The motion occasioned the sap to run out abundantly upon his hand, and he stopped to examine so strange a circumstance. He lifted it up, and still a larger quantity escaped. He now tasted what was on his fingers. Oh! then for the exclamations—'Father, father, I have found some sugar!—some syrup! I have a sugar-cane in my hand! Run quickly, father!'—We were soon together, jointly partaking of the pleasure we had in store for his dear mother and the younger brothers. In the meantime Fritz kept sucking the juice of the single cane he had cut, till his relish for it was appeased. I thought this a profitable moment to say a word about excesses; of the wisdom of husbanding even our lawful pleasures; of the advantages of moderation in our most rational enjoyments.

'But, father, we will take home a good provision of sugar-canes, however. I shall only just taste of them once or twice as I walk along. But it will be so delightful to regale my mother and my little brothers with them!'

'I have no objection; but do not take too heavy a load, for you have other things to carry, and we have yet far to go.'

Counsel was given in vain. He persisted in cutting at least a dozen of the largest canes, tore off their leaves, tied them together, and, putting them under his arm, dragged them, as well as he was able, through thick and thin to the end of the plantation. We regained the wood of palms without accident; here we stretched our limbs in the shade, and finished our repast. We were scarcely settled, when a great number of large monkeys, terrified by the sight of us and the barking of Turk, stole so nimbly, and yet so quietly up the trees, that we scarcely perceived them till they had reached the topmost parts. From this height they fixed their eyes upon us, grinding their teeth, making horrible grimaces, and saluting us with screams of hostile import. Being now satisfied that the trees were palms, bearing coconuts, I conceived the hope of obtaining some of this fruit in a milky state, through the monkeys. Fritz, on his part, prepared to shoot at them instantly. He threw his burdens on the ground, and it was with difficulty I, by pulling his arm, could prevent him from firing.

'Ah, father, why did you not let me fire? Monkeys are such malicious, mischievous animals! Look how they raise their backs in derision of us!'

'And is it possible that this can excite your vengeance, my most reasonable Mr Fritz? To say the truth, I have myself no predilection for monkeys, who, as you say, are naturally prone to be malicious. But as long as an animal does us no injury, or that his death can in no shape be useful in preserving our own lives, we have no right to destroy it, and still less to torment it for our amusement, or from an insensate desire of revenge. But what will you say if I show you that we may find means to make living monkeys contribute to our service? See what I am going to do;—but step aside, for fear of your head. If I succeed, the monkeys will furnish us with plenty of our much desired coconuts.'

I now began to throw some stones at the monkeys; and though I could not make them reach to half the height at which they had taken refuge, they showed every mark of excessive anger. With their accustomed trick of imitation, they furiously tore off, nut by nut, all that grew upon the branches near them, to hurl them down upon us; so that it was with difficulty we avoided the blows; and in a short time a great number of coconuts lay on the ground round us. Fritz laughed heartily at the excellent success of our stratagem; and as the shower of coconuts began to subside, we set about collecting them. We chose a place where we could repose at our ease, to feast on this rich harvest. We opened the shells with a hatchet, but first enjoyed the sucking of some of the milk through the three small holes, where we found it easy to insert the point of a knife. The milk of the coconut has not a pleasant flavour; but it is excellent for quenching thirst. What we liked best was a kind of solid cream which adheres to the shell, and which we scraped off with our spoons. We mixed with it a little of the sap of our sugar-canes, and it made a delicious repast.

Our meal being finished, we prepared to leave the wood of palms. I tied all the coconuts which had stalks together, and threw them across my shoulder. Fritz resumed his bundle of sugar-canes. We divided the rest of the things between us, and continued our way towards home.

4

Return from the Voyage of Discovery; a Nocturnal Alarm

MY poor boy now began to complain of fatigue; the sugar-canes galled his shoulders, and he was obliged to shift them often. At last, he stopped to take breath.—'No,' cried he, 'I never could have thought that a few sugar-canes could be so heavy. How sincerely I pity the poor negroes who carry heavy loads of them! Yet how glad I shall be when my mother and Ernest are tasting them!'

While we were conversing and proceeding onwards, Fritz perceived that from time to time I sucked the end of a sugar-cane, and he would needs do the same. It was in vain, however, that he tried; scarcely a drop of the sap reached his eager lips.—'What can be the reason,' said he, 'that though the cane is full of juice, I cannot get out a drop?'

'The reason is,' answered I, 'that you make use neither of reflection nor of your imagination.'

'Ah! I recollect now; is it not a question about air? Unless there were a particular opening in the cane, I may suck in vain; no juice will come.'

'You have explained the nature of the difficulty; but how will you manage to set it right?'

'Father, lend me your cane an instant.'

'No, no, that will not do; what I wish is, that you should yourself invent the remedy.'

'Let me see: I imagine that I have only to make a little opening just above the first knot, and then the air can enter.'

'Exactly right. But tell me what you think would be the operation of this opening near the first knot; and in what manner can it make the juice get into your mouth?'

'The pith of the cane being completely interrupted in its growth by each knot, the opening made below could have no effect upon the part above: in sucking the juice, I draw in my breath, and thus exhaust the air in my mouth; the external air presses at the same time through the hole I have made, and fills this void: the juice of the cane forms an obstacle to this effort, and is accordingly driven into my mouth. But how shall I manage when I have sucked this part dry, to get at the part above?'

'Oh, oh, Mr Philosopher, what should prevent you, who have been reasoning so well about the force and fluidity of the air, from immediately conceiving so simple a process as that of cutting away the part of the cane you have already sucked dry, and making a second perforation in the part above, so that——'

'Oh, I have it, I have it, I understand;—but if we should become too expert in the art of drawing out the juice, I fear but few of the canes will reach our good friends in the tent.'

'I also am not without my apprehensions, that of our acquisition we shall carry them only a few sticks for firewood; for I must bring another circumstance to your recollection: the juice of the sugar-cane is apt to turn sour soon after cutting, and the more certainly in such heat as we now experience; we may suck them, therefore, without compunction at the diminution of their numbers.'

'Well, then, if we can do no better with the sugar-canes, at least I will take them a good provision of the milk of coconuts, which I have here in a tin bottle; we shall sit round on the grass and drink it so deliciously!'

'In this too, my generous boy, I fear you will be disappointed. You talk of milk; but the milk of the coconut, no less than the juice of the sugar-cane, when exposed to the air and heat, turns soon to vinegar. I would almost wager that it is already sour; for the tin bottle which contains it is particularly liable to become hot in the sun.'

'O heavens, how provoking! I must taste it this very minute.'—The tin bottle was lowered from his shoulder in the twinkling of an eye, and he began to pull the cork; as soon as it was loose, the liquid flew upwards, hissing and frothing like champagne.

'Bravo, Mr Fritz! you have manufactured there a wine of some mettle. I must now caution you not to let it make you tipsy.'

'Oh, taste it, father, pray taste it, it is quite delicious; not the least like vinegar; it is rather like excellent new wine; its taste is sweet, and it is so sparkling! do take a little, father. Is it not good? If all the milk remains in this state, the treat will be better even than I thought.'

'I wish it may prove so, but I have my fears; its present state is what is called the first degree of fermentation; the same thing happens to honey, dissolved in water, of which hydromel* is made. When this first fermentation is past, and the liquid is clear, it is become a sort of wine or other fermented liquor, the quality of which depends on the materials used. By the application of heat, there next results a second and more gradual fermentation, which turns the fluid into vinegar. But this may be prevented by extraordinary care, and by keeping the vessel that contains it in a cool place. Lastly, a third fermentation takes place in the vinegar itself, which entirely changes its character, and deprives it of its taste, its strength, and its transparency. In the intense temperature of this climate, this triple fermentation comes on very rapidly, so that it is not improbable that, on entering our tent, you might find your liquids turned to vinegar, or even to a thick liquid of ill odour: we may therefore venture to refresh ourselves with a portion of our booty, that it may not all be spoiled. Come, then, I drink your health, and that of our dear family. I find the liquor at present both refreshing and agreeable; but I am pretty sure that, if we would arrive sober, we must not venture on frequent libations.'

Our regale imparted to our exhausted frames an increase of strength and cheerfulness. We reached the place where we had left our gourd utensils upon the sands; we found them perfectly dry, as hard as bone, and not the least misshapen. We now, therefore, could put them into our game bags conveniently enough, and this done, we continued our way. Scarcely had we passed through the little wood in which we breakfasted, when Turk sprang away to seize upon a troop of monkeys, who were skipping about and amusing themselves without observing our approach. They were thus taken by surprise; and before we could get to the spot, our ferocious Turk had already seized one of them; it was a female who held a young one in her arms, which she was caressing almost to suffocation, and which encumbrance deprived her of the power of escaping. The poor creature was killed, and afterwards devoured; the young one hid himself in the grass, and looked on, grinding his teeth all the time that this horrible feat was performing. Fritz flew like lightning to make Turk let go his hold. He lost his hat, threw down his tin bottle, canes, etc. but all in vain; he was too late to prevent the murder of the interesting mother.

The next scene that presented itself was of a different nature, and comical enough. The young monkey sprang nimbly on Fritz's shoulders, and fastened his feet in the stiff curls of his hair; nor could the squalls of Fritz, nor all the shaking he gave him, make him let go his hold. I ran to them, laughing heartily, for I saw that the animal was too young to do him any injury, while the panic visible in the features of the boy made a ludicrous contrast with the grimaces of the monkey, whom I in vain endeavoured to disengage. 'There is no remedy, Fritz,' said I, 'but to submit quietly and carry him; he will furnish an addition to our stock of provisions, though less alluring, I must needs confess, than for your mother's sake we could wish. The conduct of the little creature displays a surprising intelligence; he has lost his mother, and he adopts you for his father; perhaps he discovered in you something of the air of a father of a family.'

'Or rather the little rogue found out that he had to do with a chicken-heart, who shrinks from the idea of ill-treating an animal which has thrown itself on his protection.—But I assure you, father, he is giving me some terrible twitches, and I shall be obliged to you to try once more to get him off.'

With a little gentleness and management I succeeded. I took the creature in my arms as one would an infant, nor could I help pitying and caressing him. He was not larger than a kitten, and quite unable to help himself: its mother was at least as tall as Fritz.

'What shall I do with thee, poor orphan?' cried I; 'and how, in our condition, shall I be able to maintain thee? We have already more mouths to fill than food to put into them, and our workmen are too young to afford us much hope from their exertions.'

'Father,' cried Fritz, 'do let me have this little animal to myself. I will take the greatest care of him: I will give him all my share of the milk of the coconuts till we get our cows and goats; and who knows? his monkey instinct may one day assist us in discovering some wholesome fruits.'

'I have not the least objection,' answered I. 'You have conducted yourself throughout this tragi-comic adventure like a lad of courage and sensibility, and I am well satisfied with every circumstance of your behaviour. It is therefore but just that the little protégé should be given up to your management and discretion; much will depend on your manner of educating him; by and by we shall see whether he will be fittest to aid us with his intelligence, or to injure us by his malice; in this last case we shall have nothing to do but to get rid of him.'

While Fritz and I were talking about the young monkey, Turk was taking his fill of the remains of its unfortunate mother. Fritz would have driven him away, but besides the difficulty of restraining him, we had to consider, that we might, ourselves, be in danger from the pressing hunger of so powerful an animal; all the food we had before given him in the day seemed too little for the appeasing his unbounded appetite.

We now thought of resuming our journey. The little orphan jumped again on the shoulder of his protector, while I on my part relieved my boy of the bundle of canes. Scarcely had we proceeded a quarter of a league when Turk overtook us full gallop. Fritz and I received him without the usual marks of kindness, and reproached him with the cruel action he had committed, as if he could feel and understand us; but he showed no sign of concern about the matter, following quietly behind Fritz with an air of cool and perfect satisfaction. The young monkey appeared uneasy from seeing him so near, and passed round and fixed himself on his protector's bosom, who did not long bear so great an inconvenience without having recourse to his invention for a remedy. He tied some string round Turk's body in such a way, as to admit of the monkey's being fastened on his back with it, and then in a tone of genuine pity, he said, 'Now, Mr Turk, since you had the cruelty to destroy the mother, it is for you to take care of her child.' At first the dog was restive, and resisted; but by degrees, partly by menaces, and partly by caresses, we succeeded in gaining his good will, and he quietly consented to carry the little burden; and the young monkey, who also had made some difficulties, at length found himself perfectly accommodated. Fritz put another string round Turk's neck, by which he might lead him, a precaution he used to prevent him from going out of sight. I must confess, we had not the sin of too great haste to answer for, so that I had leisure for amusing myself with the idea, that we should arrive at our home with something of the appearance of keepers of rare animals for show. I enjoyed in foresight the jubilations of our young ones when they should see the figure we made.—'Ah!' cried Fritz, 'I promise you, brother Jack will draw materials enough from the occasion for future malicious jokes.'—'Do you then, my son,' said I, 'like your admirable mother, who never fails to make allowance for the buoyant spirits of youth, and is ever ready to find a charitable motive in every thing. As for the question of Turk, let me observe that it would in our situation be dangerous to teach our dogs not to attack and kill, if they can, what unknown animals they meet with. You will see that he will soon regard your little monkey as a member of our family; already he is content to carry him on his back. But we must not discourage him in his fancy for attacking wild beasts: Heaven bestowed the dog on man to be his safeguard and ally, and the horse the same. How conspicuous is the goodness of the Almighty, in the natural dispositions he has bestowed on these useful creatures, who discover so much affection for man, and so easily submit to the slavery of serving him! A man on horseback, and accompanied by a troop of well-conditioned dogs, need not fear any species of wild beasts, not even the lion, nor the hyena;* he may even baffle the voracious rapacity of the tiger.'

'I feel how fortunate we are in the possession of two such creatures: but what a pity that the horses we had on board died during our voyage, and leave us with only an ass!'

'Let us take care how we treat even our ass with disdain. I wish we had him safe on land. Fortunately he is large, and strong, and not of the common kind. We may train him to do us the same services as are performed by the horse; and it is not improbable that he will even improve under our care, and from the excellent pasture he will find in this climate.'

In such conversation as this, on subjects equally interesting to both, we forgot the length of our journey, and soon found ourselves on the bank of the river, and near our family, before we were aware. Flora from the other side announced our approach by a violent barking, and Turk replied so heartily, that his motions unseated his little burden, who in his fright jumped the length of his string from his back to Fritz's shoulder, which he could not afterwards be prevailed upon to leave. Turk, who began to be acquainted with the country, ran off to meet his companion, and shortly after, our much-loved family appeared in sight, with demonstrations of unbounded joy at our safe return. They advanced along by the course of the river, till they on one side, and we on the other, had reached the place we crossed in the morning. We repassed it again in safety, and threw ourselves into each other's arms. Scarcely had the young ones joined their brother, than they again began their joyful exclamations: 'A monkey, a live monkey! Papa, mamma, a live monkey! Oh, how delightful! How happy shall we be! How did you catch him? What a droll face he has!'—'He is very ugly,' said little Francis, half afraid to touch him.—'He is much prettier than you,' retorted Jack; 'only see, he is laughing: I wish I could see him eat.'—'Ah! if we had but some coconut!' cried Ernest; 'could you not find any? Are they nice?'—'Have you brought me any milk of almonds?'* asked Francis.—'Have you met with any unfortunate adventure?' interrupted my wife. In this manner, questions and exclamations succeeded to each other with such rapidity as not to leave us time to answer them.

At length, when all became a little tranquil, I answered them thus: 'Most happy am I to return to you again, my best beloved, and God be praised! without any new misfortune. We have even the pleasure of presenting you with many valuable acquisitions; but in the object nearest my heart, the discovery of our ship companions, we have entirely failed.'

'Since it pleases God that it should be so,' said my wife, 'let us endeavour to be content, and let us be grateful to him for having saved us from their unhappy fate, and for having once more brought us all together: I have had much uneasiness about your safety, and imagined a thousand evils that might beset you. The day appeared an age. But now I see you once more safe and well! But put down your burdens; we will all help you; for though we have not spent the day in idleness, we are less fatigued than you. Quick then, my boys, and take the loads from your father and your brother. Now then sit down, and tell us your adventures.'

Jack received my gun, Ernest the coconuts, Francis the gourd-rinds, and my wife my game-bag. Fritz distributed the sugar-canes, and put his monkey on the back of Turk, to the great amusement of the children, at the same time begging Ernest to relieve him of his gun. But Ernest, ever careful of his ease, assured him, that the large heavy bowls with which he was loaded were the most he had strength to carry. His mother, a little too indulgent to his lazy humour, relieved him of these; and thus we proceeded all together to our tent.

Fritz whispered to me, that if Ernest had known what the large heavy bowls were, he would not so readily have parted with them. Then turning to his brother, 'Why, Ernest,' cried he, 'do you know that these bowls are coconuts, your dear coconuts, and full of the sweet nice milk you have so much wished to taste?'

'What, really and truly coconuts, brother? Pray give them to me, mother; I will carry them, if you please, and I can carry the gun too.'

'No, no, Ernest,' answered his mother, 'you shall not tease us with more of your long-drawn sighs about fatigue: a hundred paces, and you would begin again.' Ernest would willingly have asked his mother to give him the coconuts, and take the gun herself, but this he was ashamed to do: 'I have only', said he, 'to get rid of these sticks, and carry the gun in my hand.'

'I would advise you not to find the sticks heavy, either,' said Fritz drily; 'I know you will be sorry if you do; and for this good reason—the sticks are sugar-canes!'

'Sugar-canes! Sugar-canes!' exclaimed they all; and, surrounding Fritz, made him give them full instructions on the sublime art of sucking sugar-canes.

My wife also, who had always entertained a high respect for the article of sugar in her household management, was quite astonished, and earnestly entreated we would inform her of all particulars. I gave her an account of our journey and our new acquisitions, which I exhibited one after the other for her inspection. No one of them afforded her more pleasure than the plates and dishes, because, to persons of decent habits, they were articles of indispensable necessity. We now adjourned to our kitchen, and observed with pleasure the preparations for an excellent repast. On one side of the fire was a turnspit, which my wife had contrived by driving two forked pieces of wood into the ground, and placing a long even stick, sharpened at one end, across them. By this invention she was enabled to roast fish, or other food, with the help of little

Francis, who was entrusted with the care of turning it round from time to time. On the occasion of our return, she had prepared us the treat of a goose, the fat of which ran down into some oyster-shells placed there to serve the purpose of a dripping-pan. There was, besides, a dish of fish, which the little ones had caught; and the iron pot was upon the fire, provided with a good soup, the odour of which increased our appetite. By the side of these most exhilarating preparations stood one of the casks which we had recovered from the sea, the head of which my wife had knocked out, so that it exposed to our view a cargo of the finest sort of Dutch cheeses, contained in round tins. All this display was made to excite the appetite of the two travellers, who had fared but scantily during the day; and I must needs observe, that the whole was very little like such a dinner as one should expect to see on a desert island.

'What you call a goose,' said my wife, 'is a kind of wild bird, and is the booty of Ernest, who calls him by a singular name, and assures me that it is good to eat.'

'Yes, father, I believe that the bird which I have caught is a kind of penguin, or we might distinguish him by the surname of *Stupid*. He showed himself to be a bird so destitute of even the least degree of intelligence, that I killed him with a single blow with my stick.'

'What is the form of his feet, and of his beak?' asked I.

'His feet are formed for swimming; in other words, he is what is called web-footed; the beak is long, small, and a little curved downwards: I have preserved his head and neck, that you might examine it yourself; it reminds me exactly of the penguin, described as so stupid a bird in my book of natural history.'

'You now then perceive, my son, of what use it is to read, and to extend our knowledge, particularly of the productions of nature: by this study and knowledge, we are enabled to recognize at the moment, the objects which chance throws in our way, whether we have seen them before or not. Tell me now what birds there are with feet like those which you have just described, and which are so formed to enable the creature to strike the water and prevent himself from sinking?'

'There are the man-of-war bird, cormorants, and pelicans, father.'

'By what mark do you distinguish the kind to which you just now said the *penguin* or *Stupid* belonged?'

'Upon my word,' interrupted his mother, 'I must give the answer myself; and it shall be a petition, that you will take some other time for your catechism on birds: when once you begin a subject, one never sees the end of it. Now to my mind there is a time for every thing: Ernest killed the bird, and was able to tell his kind; we on our parts shall eat him; what more therefore is necessary? Do you not see, husband, that the poor child is thinking all the while of his coconuts? Let me intercede on his behalf, and prevail upon you to let him have the pleasure of examining and tasting them.'

'Ah! thank you, my good mother; I shall be very glad if papa will consent.'

Father.—Well, well, you have my full permission. But first you will be obliged to learn from Fritz the best manner of opening them, so as to preserve the milk: and one word more; I recommend to you not to forget the young monkey, who has no longer his mother's milk for food.

Jack.—I cannot prevail upon him to taste a bit: I have offered him every thing we have.

Father.—This is not surprising, for he has not yet learned how to eat; you must feed him with the milk of coconuts till we can procure something more suitable.

Jack.—I will give the poor little creature my share with all my heart.

Ernest.—I have, however, the greatest desire to taste this milk myself, just to know what it is like.

'And so have I,' said the little Francis.

'However, gentlemen, the monkey must live,' cried Jack a little maliciously.

'And we and our children must live too,' answered their mother: 'Come then, the supper is ready, and the coconuts shall be for the dessert.'

We seated ourselves on the ground; my wife had placed each article of the repast in one of our new dishes, the neat appearance of which exceeded all our expectations. My sons had not patience to wait, but had broken the coconuts, and already convinced themselves of their delicious flavour; and then they fell to making spoons with the fragments of the shells. The little monkey, thanks to the kind temper of Jack, had been served the first, and each amused himself with making him suck the corner of his pocket handkerchief, dipped in the milk of the coconut. He appeared delighted with the treatment he received, and we remarked with satisfaction, that we should most likely be able to preserve him.

The boys were preparing to break some more of the nuts with the hatchet, after having drawn out the milk through the three little holes, when I pronounced the word *halt*, and bade them bring me a saw;—the thought had struck me, that by dividing the nuts carefully with this instrument, the two halves, when scooped, would remain with the form of teacups or basins already made to our hands. Jack, who was on every occasion the most active, brought me the saw. I performed my undertaking in the best manner I could, and in a short time each of us was provided with a convenient receptacle for food. My wife put the share of soup which belonged to each into the new basins. The excellent creature appeared delighted that we should no longer be under the necessity, as before, of scalding our fingers by dipping into the pot; and I firmly believe, that never did the most magnificent service of china occasion half the pleasure to its possessor, as our utensils, manufactured by our own hands from gourds and coconuts, excited in the kind heart of my wife. Fritz asked me if he might not invite our company to taste his fine champagne, which he said would not fail to make us all the merrier.—'I have not the least objection,' answered I, 'but remember to taste it yourself before you serve it to your guests.'—He ran to draw out the stopple and to taste it—'How unfortunate!' said he, 'it is already turned to vinegar.'

'What, is it vinegar!' exclaimed my wife: 'How lucky! it will make the most delicious sauce for our bird, mixed with the fat which has fallen from it in roasting, and will be as good a relish as a salad.' No sooner said than done. This vinegar produced from coconut proved a corrective of the wild and fishy flavour of the penguin. The same sauce improved our dish of fish also. Each boasted most of what he himself had been the means of procuring: it was Jack and Francis who had caught the fish in one of the shallows, while Ernest was employed with very little trouble to himself in securing his penguin *the Stupid*. My poor wife had herself performed the most difficult task of all, that of rolling the cask of Dutch cheeses into the kitchen, and then knocking out its head.

By the time we had finished our meal, the sun was retiring from our view; and recollecting how quickly the night would fall upon us, we were in great haste to regain our place of rest. My wife had considerately collected a tenfold quantity of dry grass, which she had spread in the tent, so that we anticipated with joy the prospect of stretching our limbs on a substance some-what approaching to the quality of mattresses, while, the night before, our bodies seemed to touch the ground. Our flock of fowls placed themselves as they had done the preceding even-ing; we said our prayers, and, with an improved serenity of mind, lay down in the tent, taking the young monkey with us, who was become the little favourite of all. Fritz and Jack con-tended for a short time which should enjoy the honour of his company for the night; and it was at last decided that he should be laid between them; after which, each would have a hand in covering him carefully, that he might not catch cold. We now all lay down upon the grass, in the order of the night before, myself remaining last to fasten the sailcloth in front of the tent; when, heartily fatigued by the exertions of the day, I, as well as the rest, soon fell into a profound and refreshing sleep.

But I had not long enjoyed this pleasing state, when I was awaked by the motion of the fowls on the ridge of the tent, and by a violent barking of our vigilant safeguards, the dogs. I was instantly on my legs; my wife and Fritz, who had also been alarmed, got up also: we each took a gun, and sallied forth.

The dogs continued barking with the same violence, and at intervals even howled. We had not proceeded many steps from the tent, when to our surprise we perceived by the light of the moon a terrible combat. At least a dozen of jackals* had sur-rounded our brave dogs, who defended themselves with the stoutest courage. Already the fierce champions had laid three or four of their adversaries on the ground, while those which remained began a timid kind of moan, as if imploring pity and forbearance.—Meanwhile they did not the less endeavour to entangle and surprise the dogs, thus thrown off their guard, and so secure to themselves the advantage. But our watchful combatants were not so easily deceived; they took good care not to let the enemy approach them too nearly.

I, for my part, had apprehended something worse than jackals.—'We shall soon manage to set these gentlemen at rest,' said I. 'Let us fire both together, my boy; but let us take care how we aim, for fear of killing the dogs; mind how you fire, that you may not miss, and I shall do the same.' We fired, and two of the intruders fell instantly dead upon the sands. The others made their escape; but we perceived it was with great difficulty, in consequence, no doubt, of being wounded. Turk and Flora afterwards pursued them, and put the finishing stroke to what we had begun; and thus the battle ended: but the dogs, true Caribbees* by nature, made a hearty meal on the flesh of their fallen enemies. My wife, seeing all quiet, entreated us to lie down again and finish our night's sleep: but Fritz asked me to let him first drag the jackal towards the tent, that he might exhibit him the next morning to his brothers. I however observed to Fritz, that if Turk and Flora were still hungry, we ought to give them this last jackal in addition, as a recompense for their courageous behaviour.

We had now done with this affair. The body of the jackal was left on the rock, by the side of the tent, in which were the little sleepers, who had not once awaked during the whole of the scene which had been passing. Having, therefore, nothing further to prevent us, we lay down by their side till day began to break, and till the cocks, with their shrill morning salutation, awoke us both.—The children being still asleep, afforded us an excellent opportunity to consult together respecting the plan we should pursue for the ensuing day.

5

Return to the Wreck

I BROKE a silence of some moments, with observing to my wife, that I could not but view with alarm the many cares, and exertions to be made!—'In the first place, a journey to the vessel. This is of absolute necessity; at least, if we would not be deprived of the cattle and other useful things, all of which from moment to moment we risk losing by the first heavy sea. What ought we to resolve upon? For example, should not our very first endeavour be the contriving a better sort of habitation, and a more secure retreat from wild beasts, also a separate place for our provisions? I own I am at a loss what to begin first.'

'All will fall into the right order by degrees,' observed my wife; 'patience and regularity in our plans will go as far as actual labour. I cannot, I confess, help shuddering at the thought of this voyage to the vessel; but if you judge it to be of absolute necessity, it cannot be undertaken too soon. In the meanwhile, nothing that is immediately under my own care shall stand still, I promise you. Let us not be over anxious about tomorrow: "sufficient unto the day is the evil thereof." These were the words of the true friend of mankind, and let us use so wise a counsel for our own benefit.'

'I will follow your advice,' said I, 'and without further loss of time. You shall stay here with the three youngest boys; and Fritz, being so much stronger and more intelligent than the others, shall accompany me in the undertaking.'

At this moment I started from my bed, crying out loudly and briskly, 'Get up, children, get up; it is almost light, and we have some important projects for today; it would be a shame to suffer the sun to find us still sleeping, we who are to be the founders of a new colony!'

At these words Fritz sprang nimbly out of the tent, while the young ones began to gape and rub their eyes, to get rid of their sleepiness. Fritz ran to visit his jackal, which during the night had become cold and perfectly stiff. He fixed him upon his legs, and placed him like a sentinel at the entrance of the tent, joyously anticipating the wonder and exclamations of his brothers at so unexpected an appearance. But no sooner had the dogs caught a sight of him, than they began a howl, and set themselves in motion to fall upon him instantly, thinking he was alive. Fritz had enough to do to restrain them, and succeeded only by dint of coaxing and perseverance.

In the meantime, their barking had awaked the younger boys, and they ran out of the tent, curious to know what could be the occasion. Jack was the first who appeared, with the young monkey on his shoulders; but when the little creature perceived the jackal, he sprang away in terror, and hid himself at the furthest extremity of the grass which composed our bed, and covered himself with it so completely, that scarcely could the tip of his nose be seen.

The children were much surprised at the sight of a yellow-coloured animal standing without motion at the entrance of the tent.—'Oh heavens!' exclaimed Francis, and stepping back a few paces for fear; 'it is a wolf!'—'No, no,' said Jack, going near the jackal, and taking one of his paws; 'it is a yellow dog, and he is dead; he does not move at all.'—'It is neither a dog nor a wolf,' interrupted Ernest in a consequential tone; 'do you not see that it is the golden fox?'—'Best of all, most learned professor!' now exclaimed Fritz. 'So you can tell an agouti, when you see him, but you cannot tell a jackal; for jackal is the creature you see before you, and I killed him myself in the night!'

Ernest.—In the night, you say, Fritz. In your sleep, I suppose——

Fritz.—No, Mr Ernest; not in my sleep, as you so good-naturedly suppose, but broad awake, and on the watch to protect you from wild beasts! But I cannot wonder at this mistake in one who does not know the difference between a jackal and a golden fox!

Ernest.—You would not have known it either, if papa had not told you——

'Come, come, my lads, I will have no disputes,' interrupted I. 'Fritz, you are to blame in ridiculing your brother for the mistake he made. Ernest, you are also to blame for indulging that little peevishness of yours. But as to the animal, you all are right and all are wrong; for he partakes at once of the nature of the dog, the wolf, and the fox.' The boys in an instant became friends; and then followed questions, answers, and wonder in abundance.—'And now, my boys, let me remind you, that he who begins the day without first addressing the Almighty, ought to expect neither success nor safety in his undertakings. Let us therefore acquit ourselves of this duty before we engage in other occupations.'

Having finished our prayers, the next thing thought of was breakfast; for the appetites of young boys open with their eyes. Today their mother had nothing to give them for their morning meal but some biscuit, which was so hard and dry, that it was with difficulty we could swallow it. Fritz asked for a piece of cheese to eat with it, and Ernest cast some searching looks on the second cask we had drawn out of the sea, to discover whether it also contained Dutch cheeses. In a minute he came up to us, joy sparkling in his eyes: 'Father,' said he, 'if we had but a little butter spread upon our biscuit, do you not think it would improve it?'

'That indeed it would; but—*if*,—*if*, these never-ending *ifs* are but a poor dependence. For my part, I had rather eat a bit of cheese with my biscuit at once, than think of *ifs*, which bring us so meagre a harvest.'

Ernest.—Perhaps, though, the *ifs* may be found to be worth something, if we were to knock out the head of this cask.

Father.—What cask, my boy? and what are you talking of?

Ernest.—I am talking of this cask, which is filled with excellent salt butter. I made a little opening in it with a knife; and see, I got out enough to spread nicely upon this piece of biscuit.

'That glutton instinct of yours for once is of some general use,' answered I. 'But now let us profit by the event. Who will have some butter on his biscuit?' The boys surrounded the cask in a moment, while I was in some perplexity as to the best method of getting at the contents. Fritz was for taking off the topmost hoop, and thus loosening one of the ends. But this I objected to, observing that the great heat of the sun would not fail to melt the butter, which would then run out, and be wasted. The idea occurred to me, that I would make a hole in the bottom of the cask, sufficiently large to take out a small quantity of butter at a time; and I set about manufacturing a little wooden shovel, to use for the purpose. All this succeeded vastly well, and we sat down to breakfast, some biscuits and a cocoa-nut shell full of salt butter being placed upon the ground, round which we all assembled. We toasted our biscuit, and, while it was hot, applied the butter, and contrived to make a hearty breakfast.

'One of the things we must not forget to look for in the vessel', said Fritz, 'is a spiked collar or two for our dogs, as a protection to them should they again be called upon to defend themselves from wild beasts, which I fear is too probable will be the case.'

'Oh!' says Jack, 'I can make spiked collars, if my mother will give me a little help.'

'That I will, most readily, my boy; for I should like to see what new fancy has come into your head', cried she.

'Yes, yes,' pursued I, 'as many new inventions as you please; you cannot better employ your time; and if you produce something useful, you will be rewarded with the commendations of all. But now for work. You, Mr Fritz, who, from your superior age and discretion, enjoy the high honour of being my privy counsellor, must make haste and get yourself ready, and we will undertake today our voyage to the vessel, to bring away whatever may be possible. You younger boys will remain here, under the wing of your kind mother: I hope I need not mention, that I rely on your perfect obedience to her will, and general good behaviour.'

While Fritz was getting the boat ready, I looked about for a pole, and tied a piece of white linen to the end of it: this I drove into the ground, in a place where it would be visible from the vessel; and I concerted with my wife, that in case of any accident that should require my prompt assistance, they should take down the polé and fire a gun three times as a signal of distress, in consequence of which I would immediately turn back. But I gave her notice, that there being so many things to accomplish on board the vessel, it was probable that we should not otherwise return at night; in which case I, on my part, also promised to make signals. My wife had the good sense and the courage to consent to my plan. She, however, extorted from me a promise that we should pass the night in our tubs, and not on board the ship. We took nothing with us but our guns and a recruit of powder and shot, relying that we should find provisions on board; yet I did not refuse to indulge Fritz in the wish he expressed, to take the young monkey, as he wished to see how the little creature would like some milk from the cow, or from a goat.

We embarked in silence, casting our anxious looks on the beloved objects we were quitting. Fritz rowed steadily, and I did my best to second his endeavours, by rowing from time to time, on my part, with the oar which served me for a rudder. When we had gone some distance, I remarked a current which was visible a long way. To take advantage of this current, and to husband our strength by means of it, was my first care. Little as I knew of the management of sea affairs, I succeeded in keeping our boat in the direction in which it ran, by which means we were drawn gently on, till at length the gradual diminution of its force obliged us again to have recourse to our oars; but our arms having now rested for some time, we were ready for new exertions. A little afterwards we found ourselves safely arrived at the cleft of the vessel, and fastened our boat securely to one of its timbers.

Fritz the first thing went with his young monkey on his arm to the main deck, where he found all the animals we had left on board assembled. I followed him, well pleased to observe the generous impatience he showed to relieve the wants of the poor abandoned creatures, who, one and all, now saluted us by the sounds natural to its species! It was not so much the want of food, as the desire of seeing their accustomed human companions, which made them manifest their joy in this manner, for they had a portion of the food and water we had left them still remaining. The first thing we did was to put the young monkey to one of the goats, that he might suck; and this he did with such evident pleasure, and such odd grimaces, that he afforded us much amusement.—We next examined the food and water of the other animals, taking away what was half spoiled, and adding a fresh supply, that no anxiety on their account might interrupt our enterprise. Nor did we neglect the care of renewing our own strength by a plentiful repast.

While we were seated, and appeasing the calls of hunger, Fritz and I consulted what should be our first occupation; when, to my surprise, the advice he gave was, that we should contrive a sail for our boat.—'In the name of Heaven,' cried I, 'what makes you think of this at so critical a moment, when we have so many things of indispensable necessity to arrange?' —'True, father,' said Fritz; 'but let me confess that I found it very difficult to row for so long a time, though I assure you I did my best, and did not spare my strength.—I observed that, though the wind blew strong in my face, the current still carried us on. Now, as the current will be of no use in our way back, I was thinking that we might make the wind supply its place. Our boat will be very heavy when we have loaded it with all the things we mean to take away, and I am afraid I shall not be strong enough to row to land: so do you not think that a sail would be a good thing just now?'

'Ah ha, Mr Fritz! You wish to spare yourself a little trouble, do you? But seriously, I perceive much good sense in your argument, and feel obliged to my privy counsellor for his good advice. The best thing we can do is, to take care and not overload the boat, and thus avoid the danger of sinking, or of being obliged to throw some of our stores overboard. We will, however, set to work upon your sail; it will give us a little trouble. But come, let us begin.'

I assisted Fritz to carry a pole strong enough for a mast, and another not so thick, for a sailyard.* I directed him to make a hole in a plank with a chisel, large enough for the mast to stand upright in it. I then went to the sail-room, and cut a large sail down to a triangular shape: I made holes along the edges, and passed cords through them. We then got a pulley, and with this and some cords, and some contrivance in the management of our materials, we produced a sail.

Fritz, after taking observations through a telescope of what was passing on land, and which we had already done several times, imparted the agreeable tidings that all was still well with our dear family. He had distinguished his mother walking tranquilly along the shore. He soon after brought me a small streamer, which he had cut from a piece of linen, and which he entreated me to tie to the extremity of the mast, as much delighted with the streamer as with the sail itself. He gave to our machine the name of *The Deliverance;* and in speaking of it, instead of calling it a *boat,* it had now always the title of *the little vessel.*

'But now, father,' said Fritz, looking kindly on me as he spoke, 'as you have eased me of the labour of rowing, it is *my* turn to take care of *you.* I am thinking to make you a better-contrived rudder; one that would enable you to steer the boat both with greater ease and greater safety.'—'Your thought would be a very good one,' said I, 'but that I am unwilling to lose the advantage of being able to proceed this way and that, without being obliged to veer. I shall therefore fix our oars in such a manner as to enable me to steer the raft from either end.' Accordingly, I fixed bits of wood to the stem and stern of the machine, in the nature of grooves, which were calculated to spare us a great deal of trouble.

During these exertions the day advanced, and I saw that we should be obliged to pass the night in our tubs, without much progress in our task of emptying the vessel. We had promised our family to hoist a flag as a signal, if we passed the night from home, and we found the streamer precisely the thing we wanted for this purpose.

We employed the remnant of the day in emptying the tubs of the useless ballast of stones, and putting in their place what would be of service, such as nails, pieces of cloth, and different kinds of utensils, etc. etc. The Vandals* themselves could not have made a more complete pillage than we had done. The prospect before us of an entire solitude, made us devote our attention to the securing as much powder and shot as we could, as a means of catching animals for food, and of defending ourselves against wild beasts to the latest moment possible. Utensils for every kind of workmanship, of which there was a large provision in the ship, were also objects of incalculable value to us. The vessel, which was now a wreck, had been sent out as a preparation for the establishment of a colony in the South Seas, and had been provided with a variety of stores not commonly included in the loading of a ship. Among the rest, care had been taken to have on board considerable numbers of European cattle; but so long a voyage had proved unfavourable to the oxen and the horses, the greatest part of which had died, and the others were in so bad a condition, that it had been found necessary to destroy them. The quantity of useful things which presented themselves in the store-chambers made it difficult for me to select among them, and I much regretted that circumstances compelled me to leave some of them behind. Fritz, however, already meditated a second visit; but we took good care not to lose the present occasion for securing knives and forks and spoons, and a complete assortment of kitchen utensils. In the captain's cabin we found some services of silver, dishes and plates of high-wrought metal, and a little chest filled with bottles of many sorts of excellent wine. Each of these we put into our boat. We next descended to the kitchen, which we stripped of gridirons, kettles, pots of all kinds, a small roasting-jack,* etc. Our last prize was a chest of choice eatables, intended for the table of the officers, containing Westphalia hams,* Bologna sausages, and other savoury food. I took good care not to forget some little sacks of maize,* of wheat, and other grain, and some potatoes. We next added such implements for husbandry as we could find;—shovels, hoes, spades, rakes, harrows, etc. etc. Fritz reminded me that we had found sleeping on the ground both cold and hard, and prevailed upon me to increase our cargo by some hammocks, and a certain number of blankets: and as guns had hitherto been the source of his pleasures, he added such as he could find of a particular costliness or structure, together with some sabres and clasp-knives. The last articles we took were a barrel of sulphur, a quantity of ropes, some small string, and a large roll of sailcloth. The vessel appeared to us to be in so wretched a condition, that the least tempest must make her go to pieces. It was then quite uncertain whether we should be able to approach her any more.

Our cargo was so large, that the tubs were filled to the very brim, and no inch of the boat's room was lost. The first and last of the tubs were reserved for Fritz and me to seat ourselves in and row the boat, which sunk so low in the water that, if the sea had not been quite calm, we should have been obliged to ease her of some of the loading: we, however, used the precaution of putting on our swimming-jackets, for fear of any misfortune.

It will easily be imagined that the day had been laboriously employed. Night suddenly surprised us, and we lost all hope of returning to our family the same evening. A large blazing fire on the shore soon after greeted our sight,—the signal agreed upon for assuring us that all was well, and to bid us close our eyes in peace. We returned the compliment, by tying four lanterns with lights in them to our mast-head. This was answered, on their part, by the firing of two guns; so that both parties had reason to be satisfied and easy.

After offering up our earnest prayers for the safety of all, and not without some apprehension for our own, we resigned ourselves to sleep in our tubs, which appeared to us safer than the vessel. Our night passed tranquilly enough: my boy Fritz slept as soundly as if he had been in a bed; while I, haunted by the recollection of the nocturnal visit of the jackals, could neither close my eyes, nor keep them from the direction of the tent. I had, however, great reliance that my valiant dogs would do their duty, and was thankful to Heaven for having enabled us to preserve so good a protection.

6

A Troop of Animals in Cork Jackets

EARLY the next morning, though scarcely light, I mounted the vessel, hoping to gain a sight of our beloved companions through a telescope. Fritz prepared a substantial breakfast of biscuit and ham; but before we sat down, we recollected that in the captain's cabin we had seen a telescope of a much superior size and power, and we speedily conveyed it to the deck. While this was doing, the brightness of the day had come on. I fixed my eye to the glass, and discovered my wife coming out of the tent and looking attentively towards the vessel, and at the same moment perceived the motion of the flag upon the shore. A load of anxiety was thus taken from my heart; for I had the certainty that all were in good health, and had escaped the dangers of the night.—'Now that I have had a sight of your mother,' said I to Fritz, 'my next concern is for the animals on board; let us endeavour to save the lives of some of them, at least, and to take them with us.'

'Would it be possible to make a raft, to get them all upon it, and in this way get them to shore?' asked Fritz.

'But, what a difficulty in making it, and how could we induce a cow, an ass, and a sow, either to get upon a raft, or, when there, to remain motionless and quiet? The sheep and goats one might perhaps find means to remove, they being of a more docile temper; but for the larger animals, I am at a loss how to proceed.'

'My advice, father, is to tie a long rope round the sow's neck, and throw her without ceremony into the sea: her immense weight will be sure to sustain her above water; and we can draw her after the boat.'

'Your idea is excellent; but unfortunately it is of no use but for the pig; and she is the one I care the least about preserving.'

'Then here is another idea, father: let us tie a swimming-jacket round the body of each animal, and contrive to throw one and all into the water; you will see that they will swim like fish, and we can draw them after us in the same manner.'

'Right, very right, my boy; your invention is admirable: let us therefore not lose a moment in making the experiment.'

We hastened to the execution of our design: we fixed a jacket on one of the lambs, and threw it into the sea; and full of anxious curiosity, I followed the poor beast with my eyes. He sunk at first, and I thought him drowned; but he soon re-appeared, shaking the water from his head, and in a few seconds he had learned completely the art of swimming. After another interval, we observed that he appeared fatigued, gave up his efforts, and suffered himself to be borne along by the course of the water, which sustained and conducted him to our complete satisfaction.—'Victory!' exclaimed I, hugging my boy with delight: 'these useful animals are all our own; let us not lose a moment in adopting the same means with those that remain; but take care not to lose our little lamb.' Fritz now would have jumped into the water to follow the poor creature, who was still floating safely on the surface: but I stopped him till I had seen him tie on a swimming-jacket. He took with him a rope, first making a slip knot in it, and, soon overtaking the lamb, threw it round his neck, and drew him back to our boat; and then took him out of the water.

We next got four small water-butts. I emptied them, and then carefully closed them again; I united them with a large piece of sailcloth, nailing one end to each cask. I strengthened this with a second piece of sailcloth, and this contrivance I destined to support the cow and the ass, two casks to each, the animal being placed in the middle with a cask on either side. I added a thong of leather, stretching from the casks across the breast and haunches of the animal, to make the whole secure; and thus, in less than an hour, both my cow and my ass were equipped for swimming.

It was next the turn of the smaller animals: of these, the sow gave us the most trouble; we were first obliged to put her on a muzzle to prevent her biting; and then we tied a large piece of cork under her body. The sheep and goats were more accom-modating, and we had soon accoutred them for our adventure. And now we had succeeded in assembling our whole company on the deck, in readiness for the voyage: we tied a cord to either the horns or the neck of each animal, and to the other end of the cord a piece of wood similar to the mode used for marking nets, that it might be easy for us to take hold of the ropes, and so draw the animal to us if it should be necessary. We struck away some more of the shattered pieces of wood from the fissure of the vessel, by which we were again to pass. We began our experiment with the ass, by conducting him as near as possible to the brink of the vessel, and then suddenly shoving him off. He fell into the water, and for a moment disappeared; but we soon saw him rise, and in the action of swimming between his two barrels, with a grace which really merited our commendation.

Next came the cow's turn; and as she was infinitely more valuable than the ass, my fears increased in due proportion. The ass had swum so courageously, that he was already at a considerable distance from the vessel, so that there was sufficient room for our experiment on the cow. We had more difficulty in pushing her overboard, but she reached the water in as much safety as the ass had done before; she did not sink so low in it, and was no less perfectly sustained by the empty barrels; and she made her way with gravity, and, if I may so express it, a sort of dignified composure. According to this method we proceeded with our whole troop, throwing them one by one into the water, where by and by they appeared in a group floating at their ease, and seemingly well content. The sow was the only exception; she became quite furious, set up a loud squalling, and struggled with so much violence in the water, that she was carried to a considerable distance, but fortunately in a direction towards the landing-place we had in view. We had now not a moment to lose. Our last act was to put on our cork-jackets; and then we descended without accident through the cleft, took our station in the boat, and were soon in the midst of our troop of quadrupeds. We carefully gathered all the floating bits of wood, and fastened them to the stern of the machine, and thus drew them after us. When every thing was adjusted, and our company in order, we hoisted our sail, which soon filling with a favourable wind, conducted us all safe to the land.

We now perceived how impossible it would have been for us to have succeeded in our enterprise without the aid of a sail; for the weight of so many animals sunk the boat so low in the water, that all our exertions to row to such a distance would have been ineffectual; while, by means of the sail, she proceeded completely to our satisfaction, bearing in her train our company of animals; nor could we help laughing heartily at the singular appearance we made. Proud of the success of so extraordinary a feat, we were in high spirits, and seated ourselves in the tubs, where we made an excellent dinner. Fritz amused himself with the monkey while I was occupied in thinking of those I had left on land, and of whom I now tried to take a view through my telescope. My last act on board the vessel had been to take one look more at those beloved beings, and I perceived my wife and the three boys all in motion, and seeming to be setting out on some excursion; but it was in vain that I endeavoured, by any thing I saw, to conjecture what their plan might be. I therefore seized the first moment of quiet to make another trial with my glass, when a sudden exclamation from Fritz filled me with alarm.—'O Heavens!' cried he, 'we are lost! a fish of an enormous size is coming up to the boat.'—'And why lost?' said I, half angry, and yet half partaking of his fright. 'Be ready with your gun, and the moment he is close upon us, we will fire upon him.' He had nearly reached the boat, and with the rapidity of lightning had seized the foremost sheep: at this instant Fritz aimed his fire so skilfully, that the balls of the gun were lodged in the head of the monster, which was an enormous shark. The fish half turned himself round in the water and hurried off to sea, leaving us to observe the lustrous smoothness of his belly, and that as he proceeded he stained the water red, which convinced us he had been severely wounded. I determined to have the best of our guns at hand the rest of the way, lest we should be again attacked by the same fish, or another of his species.

The animal being now out of sight and our fears appeased, I resumed the rudder; and as the wind drove us straight towards the bay, I took down the sail, and continued rowing till we reached a convenient spot for our cattle to land. I had then only to untie the end of the cords from the boat, and they stepped contentedly on shore. Our voyage thus happily concluded, we followed their example.

I had already been surprised and uneasy at finding none of my family looking out for us on the shore; we could not, however, set out in search of them, till we had disencumbered our animals of their swimming apparatus. Scarcely had we entered upon this employment, when I was relieved by the joyful sounds which reached our ears, and filled our hearts with rapture. It was my wife and the youngest boys who uttered them, the latter of whom were soon close up to us, and their mother followed not many steps behind, each and all of them in excellent health, and eager for our salutations. When the first burst of happiness at meeting had subsided, we all sat down on the grass, and I began to give them an account of our occupations in the vessel, of our voyage, and of all our different plans and their success, in the order in which they occurred. My wife could find no words to express her surprise and joy at seeing so many useful animals round us; and the hearty affection she expressed for them, in language the most simple and touching, increased my satisfaction at the completion of our enterprise.

'Yes,' said Fritz, a little consequentially, 'for this once the privy counsellor has tried his talents at invention.'

'This indeed is very true,' replied I; 'in all humility have I to confess, that to Fritz alone all praise belongs, and that to his sagacity it is that we are indebted for our success.' His mother could not refrain from giving him a hearty kiss. 'Our gratitude is due to both,' said she; 'for both have laboured to give us the possession of this troop of animals, an acquisition beyond any other, agreeable and serviceable to us in the situation in which it has pleased Providence to place us.'

Ernest and Jack now ran to the boat, and began to shout their admiration of the mast, the sail, and the flag, desiring their brother to explain to them how all the things they saw had been effected and what he himself did of them. In the meantime we began to unpack our cargo, while Jack stole aside and amused himself with the animals, took off the jackets from the sheep and goats, bursting from time to time into shouts of laughter at the ridiculous figure of the ass, who stood before them adorned with his two casks and his swimming apparatus, and braying loud enough to make us deaf.

By and by I perceived, with surprise, that Jack had round his waist a belt of metal covered with yellow skin, in which were fixed two pistols. 'In the name of Heaven,' exclaimed I, 'where did you procure this curious costume, which gives you the look of a smuggler?'

'From my own manufactory,' replied he; 'and if you cast your eyes upon the dogs, you will see more of my specimens.'

Accordingly I looked at them, and perceived that each had on a collar similar to the belt round Jack's waist, with, however, the exception of the collars being armed with nails, the points of which were outwards, and exhibited a formidable appearance. 'And is it you, Mr Jack,' cried I, 'who have invented and executed these collars and your belt?'

'Yes, father, they are indeed my invention, with a little of my mother's assistance when it was necessary to use the needle.'

'But where did you get the leather and the thread and the needle?'

'Fritz's jackal furnished the first,' answered my wife; 'and as to the last, a good mother of a family is always provided with them. Then have I not an enchanted bag, from which I draw out such articles as I stand in need of? So, if you have a particular fancy for any thing, you have only to acquaint me with it.' I tenderly embraced her, to express my thanks for this effort to amuse by so agreeable a raillery, and Jack too came in for his share both of the caresses and our hearty commendations. But Fritz was both discontented and angry on finding that Jack had taken upon him to dispose of his jackal, and to cut his beautiful skin into strips. He, however, concealed his ill-humour as well as he could; but presently he called out suddenly, holding his nose as he spoke, 'What a filthy smell! Does it perchance proceed from you, Mr Currier?* Is this the perfume we may expect from your manufactory?'—'It is rather yours than mine,' replied Jack in a resentful tone; 'for it was your jackal which you hung up in the sun to dry.'—'And which would have been dried in a whole skin, if it had not pleased your sublime fancy to cut it to pieces, instead of leaving me the power to do what I please with my own booty,' answered his brother.

'Son Fritz,' said I, in a somewhat angry tone, 'this is not generous on your part. Of what importance is it who cut up the skin of the jackal, if by so doing it has contributed to our use? My dear children, we are here in this desert island, in just such a situation as that of our first parents when they were driven out of the garden of Eden; it was still in their power to enjoy happiness in the fertile land in which God permitted them to live; and this happiness was to proceed from their obedience, from the work of their hands, and the sweat of their brow: a thousand and a thousand blessings were granted for their use, but they suffered the passions of jealousy, envy, and hatred to take root in their bosoms: Cain killed his brother Abel,* and thus plunged his unhappy parents into the deepest affliction, so that he and his race were cursed by God. This is the horrid crime to which the habit of disputing may conduct. Let us then avoid such an evil, let us share one with the other in every benefit bestowed upon us, and from this moment may the words *yours* and *mine* be banished from our happy circle! What is discovered or procured by one of you, should be equally for the service of all, and belong to all, without distinction. It is quite certain, Jack, that the belt round your waist, not being dry, has an offensive smell; the pleasure of wearing what you had ingeniously contrived makes you willing to bear with the inconvenience: but we should never make our own pleasure the pain of another. I therefore desire that you will take it off and place it in the sun to dry, and take care that it does not shrink during the operation; and then you can join your brothers, and assist them to throw the jackal into the sea.'

Fritz's ill-humour was already over; but Jack, whose temper was less docile, still retained the belt, and walked about in it with somewhat of an air of resistance. His brothers continued their warfare, pretending to avoid him, and crying out—'What a smell! What a smell!' till at length Jack, tired with the part he had been acting, suddenly stripped off the belt, and joined the others in dragging the dead jackal to the sea, where he no longer offended anyone.

Perceiving that no preparations were making for supper, I told Fritz to bring us the Westphalia ham. The eyes of all were now fixed upon me with astonishment, believing that I could only be in jest; when Fritz returned, displaying with exultation a large ham, which we had begun to cut in the morning. 'A ham!' cried one and all; 'a ham! and ready dressed! What a nice supper we shall have!' said they, clapping their hands to give a hearty welcome to the bearer of so fine a treat.—'It comes quite in the nick of time too,' interrupted I; 'for, to judge by appearances, a certain careful steward I could name

seems to have intended to send us supperless to bed, little
thinking, I suppose, that a long voyage by water is apt to
increase the appetite.'

'I will tell you presently', replied my wife, 'what it was that
prevented me from providing a supper for you all at an early
hour: your ham, however, makes you ample amends; and I have
something in my hand with which I shall make a pretty side-
dish; in the twinkling of an eye you shall see it make its entrance.' She now showed us about a dozen of turtle's eggs, and
then hurried away to make an omelette of some of them.

'Look, father,' said Ernest, 'if they are not the very same
which Robinson Crusoe found in his island!* See, they are like
white balls, covered with a skin like wetted parchment! We
found them upon the sands along the shore.'

'Your account is perfectly just, my dear boy,' said I: 'by what
means did you make so useful a discovery?'—'Oh, that is part
of our history,' interrupted my wife; 'for I also have a history
to relate, when you will be so good as to listen to it.'

'Hasten then, my love, and get your pretty side-dish ready,
and we will have the history for the dessert. In the meantime
I will relieve the cow and the ass from their jackets. Come
along, boys, and give me your help.'—I got up, and they all
followed me gaily to the shore. We were not long in effecting
our purpose with the cow and the ass, who were animals of a
quiet and kind temper; but when it was the sow's turn, our
success was neither so easy nor so certain; for no sooner had
we united the rope than she escaped from us, and ran so fast
that none of us could catch her. The idea occurred to Ernest
of sending the two dogs after her, who caught at her ears, and
sent her back, while we were half deafened with the hideous
noise she made; at last she suffered us to take off her cork
jacket. We now laid the accoutrements across the ass's back,
and returned to the kitchen; our slothful Ernest highly delighted that he was likely in future to have our loads carried by
a servant.

In the meanwhile the kind mother had prepared the omelette, and spread a table-cloth on the end of the cask of butter,
upon which she had placed some of the plates and silver
spoons we had brought from the ship. The ham was in the
middle, and the omelette and the cheese opposite to each

other; and altogether made a figure not to be despised by the
inhabitants of a desert island. By and by the two dogs, the fowls,
the pigeons, the sheep, and the goats, had all assembled round
us, which gave us something like the air of sovereigns of the
country. It did not please the geese and ducks to add themselves to the number of these our loyal subjects: they deserted
us for a marshy swamp, where they found a kind of little
crabs in great abundance, and which furnished a delicious
food for them, and relieved us of the care of providing for their
support.

When we had finished our repast, I bade Fritz present our
company with a bottle of Canary wine,* which we had brought
from the captain's cabin, and I desired my wife to indulge us
with the promised history.

7
Second Journey of Discovery, Performed by the Mother of the Family

'YOU pretend', said my wife, with a little malicious smile, 'to
be curious about my history, yet you have not let me speak a
single word in all this time; but the longer a torrent is pent up,
the longer it flows when once let loose. Now then that you are
in the humour to listen, I shall give vent to a certain little
movement of vanity which is fluttering at my heart.—Not,
however, to intrude too long upon your patience, we will skip
the first day of your absence, in the course of which nothing
new took place, except my anxiety on your account, which
confined me for the most part to the spot from whence you
embarked, and from which I could see the vessel. But this
morning, when I was made happy by the sight of your signal,
and had set up mine in return, I looked about, before the boys
were up, in hopes to find a shady place where we might now
and then retire from the heat of the sun; but I found not a
single tree. This made me reflect a little seriously on our
situation.—It will be impossible, said I to myself, to remain in
this place with no shelter but a miserable tent, under which
the heat is even more excessive than without. Courage then!
pursued I; my husband and my eldest son are at this moment
employed for the general good; why should not I be active and

enterprising also? Why not undertake, with my youngest sons, to do something that shall add some one comfort to our existence? I will pass over with them to the other side of the river, and with my own eyes examine the country respecting which my husband and Fritz have related such wonders. I will try to find out some well-shaded agreeable spot, in which we may all be settled. I now cast another look towards the vessel; but perceiving no sign of your return, I determined to share a slight dinner with the boys, and then we set out resolutely, on a journey of discovery for a habitation better sheltered from the sun.

'In the morning, Jack had slipped to the side of the tent where Fritz had hung the jackal, and with his knife, which he sharpened from time to time upon the rock, he cut some long strips of skin from the back of the animal, and afterwards set about cleaning them. Ernest discovered him in this uncleanly occupation; and as he is, as we all know, a little delicate, and afraid to soil his fingers, he not only refused to give Jack any assistance, but thought fit to sneer a little at the currier-like trade which he had engaged in. Jack, who, as we also know, has not the most patient temper in the world, raised his hand to give him a little cuff. Ernest made his escape, more alarmed, I believe, by Jack's dirty hands, than by the expected blow; while I, for my part, ran to set them right, and to give a mother's reproof to both. Jack persisted that he had a justification full and undeniable in the great usefulness of the said dirty work; "for", observed he, "it is intended to make some collars, which I shall arm with spikes, and the dogs will wear them for our defence". I saw in an instant that Ernest had been the aggressor, and on him fell the reproof: I represented how little a squeamishness like his suited with the difficulties of our situation, in which one and all were called upon to assist in any employment that should promise to contribute to the general good.

'Jack returned to his strips of skin, the cleaning of which he completed very cleverly. When he had finished this part of his undertaking, he looked out from the chest of nails those that were longest, and which had the largest and flattest heads; these he stuck through the bits of skin intended for the collars, at small distances. He next cut a strip of sailcloth the same breadth as the leather, and, laying it along on the heads of the nails, politely proposed to me the agreeable occupation of sewing them together, to prevent the heads of the nails from injuring the dogs. I begged to be excused; but seeing the good humour with which he tried to sew them for himself, and that, with all his goodwill, it was too hard a task, I rewarded him by doing it myself;—few mothers refuse the sacrifice of a little personal convenience, to afford delight to a virtuous child.

'But now having yielded the first time, I found I had made myself liable to further claims. The next thing was a belt for himself, which he had manufactured of the same materials, and was impatient to see completed, it being intended to contain his pistols. "We shall see", said he, strutting about as he spoke, "if the jackals will dare to attack us now."—"But, dear Jack, you do not foresee what will happen;—a piece of skin not entirely dry is always liable to shrink when exposed to the heat; so, after all, you will not be able to make use of it." My little workman, as I said this, struck his forehead, and betrayed other marks of impatience.—"What you say is true," said he, "and I had not well considered; but I know of an effectual remedy." He then took a hammer and some nails, and stretched his strips of leather on a plank, which he laid in the sun to dry quickly, thus preventing the possibility of their shrinking. I applauded his invention, and promised him I would not fail to give you a full account of his proceedings.

'I next assembled them round me, and informed them of my plans for an excursion, and you may believe I heard nothing like a dissenting voice. They lost not a moment in preparing themselves; they examined their arms, their game-bags, looked out the best clasp-knives, and cheerfully undertook to carry the provision-bags; while I, for my share, was loaded with a large flask of water and a hatchet, for which I thought it likely we might find a use. I also took the light gun which belongs to Ernest, and gave him in return a carbine, which might be loaded with several balls at once. We took some refreshment, and then sallied forth, attended by the two dogs for our escort. Turk, who had already accompanied you in the same direction, seemed well aware that he knew the way, and proceeded at the head of the party in quality of a conductor. We arrived at the place at which you had crossed the river, and succeeded in passing over, though not without difficulty.

'As we advanced, I reflected that our safety depended in some measure on the two boys, because it was they only who knew how to use the guns. I now for the first time began to feel how fortunate it was that you had accustomed them from infancy to face danger of every kind: but I am now convinced that the parent who adopts a hardy scheme of education acts the wisest part. But now for the passing of the river.

'Ernest was first in reaching the other side. The little Francis entreated me to carry him on my back, which was difficult enough. At length we found means to manage pretty well, thanks to Jack, who relieved me of my gun and the hatchet. But for himself, finding he was scarcely able to stand under his added weight, he resolved to go straight into the water at once, rather than run the risk of slipping, by stepping on the loose wet pieces of stone so heavily loaded. I myself had great difficulty to keep myself steady with the dear little burden at my back, who joined his hands round my neck, and leaned with all his weight upon my shoulders. After having filled my flask with river water, we proceeded on our way till we had reached to the top of the hill which you described to us as so enchanting, and where I partook of the pleasure you had experienced. I continued for some time to look around and admire in silence; and for the first time since the event of our dreadful accident at sea, I felt my heart begin to open to a sense of enjoyment and of hope.

'In casting my eyes over the vast extent before me, I had observed a small wood of the most inviting aspect. I had so long sighed for a little shade, that I resolved to bend our course towards it: for this, however, it was necessary to go a long way through a strong kind of grass which reached above the heads of the little boys; an obstacle which, on trial, we found too difficult to overcome. We therefore resolved to walk along the river, and turn at last upon the wood. We found traces of your footsteps, and took care to follow them till we had come to a place which seemed to lead directly to it; but here again we were interrupted by the height and thickness of the grass, which nothing but the most exhausting endeavours could have enabled us to get through. Jack was now loitering a little behind, and I frequently turned round to observe what he could be doing: at last I saw him tearing off some handfuls of grass, and wiping his clothes with it, and then shake his pocket-handkerchief, which was wet, and lay it on his shoulders to dry. I hastened back to enquire what had happened.

' "Oh, mother," said he, "I believe all the water of the river we have crossed has got into my pockets: only see, every thing I had in them is wet, pistols, turfs, every thing."

' "Good Heavens!" interrupted I in great alarm, "had you put your pistols in your pocket? They were not loaded, I hope?"

' "I am sure I do not know, mother; I only put them there while my belt was drying, that I might always have them about me."

' "Thoughtless, yet fortunate boy!" exclaimed I. "Do you know what an escape you have had? If with the suddenness of your motions the pistols had gone off, they would infallibly have killed you. Take care, I entreat you, not to commit such an imprudence in future."—"There is nothing, I believe, to fear, mother, for this time," replied he, holding the pistols so as to let the water run out of them. And in reality I perceived, by the condition they were in, that there was little danger of their going off. While we were talking of what had happened, our attention was interrupted by a sudden noise, and looking about, we perceived a large bird rising from the thickest part of the grass, and mounting in the air. Each of the boys prepared to fire, but before they could be ready, the bird was out of the reach of shot. Ernest was bitterly disappointed, and instantly exchanged the gun for the carbine I had given him, crying, "What a pity! If I had but had the lightest gun! if the bird had not got away so fast, I would lay any wager I should have killed him."

' "The mischief was, no doubt, that you did not let him know beforehand, that it was your pleasure he should wait till you could be quite ready," observed I, laughing.

' "But, mother, how could I possibly suppose that the bird could fly away in less than the twinkling of an eye? Ah, if one would but come at this very moment!"

' "A good sportsman, Ernest, always holds himself in readiness, this being, as I understand, one of his great arts; for you

must know, that birds do not send messages to give notice of
their coming."

' "I wish I could but know", said Jack, "what bird it was;
I never saw any the least like it."

' "I am sure it was an eagle," said the little Francis, "for I
have read in my book of fables, that an eagle can carry off a
sheep; and this bird was terribly large."

' "O yes," said Ernest scoffingly, "as if all large birds must
be eagles! Why do you not know that there are some birds
much larger even than eagles? The ostrich, for example, which
travellers sometimes name the Condor* or the Candor.—I
must confess it would have afforded me the highest pleasure
to have examined this bird minutely."

' "If you had had time to examine him, you would have had
time to kill him," said I; "but as the opportunity is gone, let us
look for the place in the grass from which he mounted; we may
judge at least of his size by the mark he will have left there."
The boys now all scampered away to the place, when suddenly
a second bird, exactly like the first, except that he was a little
larger, rushed out with a great noise and mounted above their
heads.

'The boys remained stupid with astonishment, following
him with their eyes and open mouths without speaking a word,
while for my own part I could not help laughing heartily. "Oh!
such fine sportsmen as we have here!" cried I: "they will never
let us be in want of game, I plainly perceive. *Ah! if one would
but come at this very moment!*" Ernest, always a little disposed to
vent uneasiness by crying, now began to whimper; while Jack,
with a curious mixture of tragi-comic bravery upon his features,
his eyes darting upon the mountain traveller, takes off his hat,
makes a profound bow, and roars out, as if for the bird to hear:
"Have the goodness, Mr Traveller, to indulge me once more
with a little visit, only for a single minute: you cannot imagine
what good sort of people we are: I entreat that we may have
the pleasure of seeing you once again———." We now minutely
examined the place from which the birds had mounted, and
found a kind of large nest formed of dry plants, of clumsy
workmanship; the nest was empty, with the exception of some
broken shells of eggs. I inferred from this, that their young had
lately been hatched; and observing at this moment a rustling

motion among some plants of shorter growth, at some distance from the spot on which we stood, I concluded that the young covey were scampering away in that direction; but as the motion soon ceased, we had no longer a guide to conduct us to their retreat. We next reached a little wood; and here our son Ernest had an opportunity of recognizing many of the originals of the engravings in his books of natural history, and of displaying his knowledge, or his ignorance, to his heart's content. A prodigious quantity of unknown birds were skipping and warbling on the branches of the trees, without betraying the least alarm at our vicinity. The boys wanted to fire on them; but this I absolutely forbade, and with the less scruple, as the trees were of so enormous a height as to be out of gun-shot reach.—No, my dear husband, you cannot possibly form an idea of the trees we now beheld! You must somehow have missed this wood; or so extraordinary a sight could not have escaped your observation. What appeared to us at a distance to be a wood, was only a group of about fourteen of them, the trunks of which seemed to be supported in their upright position by arches on each side, these arches being formed by the roots of the tree.

'Jack climbed with considerable trouble upon one of these arch-formed roots, and with a packthread in his hand measured the actual circumference of the tree itself. He found that it measured more than 15 braches (the brache* is equal to 22.5 inches). I made 32 steps in going round one of those giant productions at the roots; and its height, from the ground to the place where the branches begin to shoot, may be about 36 braches. The twigs of the tree are strong and thick; its leaves moderately large in size, and bearing some resemblance to the hazel tree of Europe; but I was unable to discover that it bore any fruit. The soil immediately round and under its branches produced in great abundance a short thick kind of plant, unmixed with any of the thistle kind, and of a perfectly smooth surface. The large breadth of shade which presented itself, seemed to invite us to make this spot the place of our repose; and my predilection for it grew so strong, that I resolved to go no further, but to enjoy its delicious coolness till it should be time to return. I sat down in this verdant elysium with my three sons around me. We took out our provision-bags: a charming stream formed to increase the coolness and beauty of the scene, flowed at our feet, and supplied us with a fresh and salutary beverage. Our dogs were not long in reaching us; they had remained behind, sauntering about the skirts of the wood. To my great surprise, they did not ask for anything to eat, but lay down quietly, and were soon asleep at our feet. For my own part, I felt that I could never tire of beholding and admiring this enchanting spot; it occurred to me, that if we could but contrive a kind of tent that could be fixed in one of the trees, we might safely come and make our abode here. I had found nothing in any other direction that suited us so well in every respect; and I resolved to look no further. When we had shared our dinner among us, and well rested from our fatigue, we set out on our return, again keeping close to the river, half expecting to see along the shore some of the pieces or other vestiges of the vessel, which the waves might have washed there.

'But before we left our enchanting retreat, Jack entreated me to stay, and finish sewing the linen strips to his leather belt. The little coxcomb had so great an ambition to strut about and exhibit himself in this new ornament, that he had taken the trouble to carry the piece of wood, on which he had nailed his skin to dry, along with him through the whole of our expedition. Finding that the skin was really dry, I granted his request, preferring, since work I must, to do it now when I had the advantage of being in the shade. When I had finished, he eagerly fastened on the belt, and placed his pistols in it; he set himself before us in a marching step, with the knuckles of his hand turned back upon his hip, leaving to Ernest the care of putting on the dogs' collars; which he insisted should be done, for it would give them, he said, a martial air. The self-imagined hero was all impatience for you and Fritz to see him in his new accoutrement; so that I had enough to do to walk quick enough to keep sight of him; for in a country where no track of the foot of man is to be found, we might easily lose each other. I became more tranquil respecting him when we had got once more together on the sea-shore; for, as I expected, we found there pieces of timber, poles, large and small chests; and other articles which I knew had come from the vessel. None of us, however, were strong enough to bring them away; we therefore contented ourselves with dragging all we could reach to the dry sands, beyond the reach of the waves at high water. Our dogs, for their part, were fully employed in catching crabs, which they drew with their paws to the shore as the waves washed them up, and on which they made an excellent repast. I now understood that it was this sort of prey which had appeased their hunger before they joined us at dinner. Heaven be praised, cried I, that our animals have found means to procure sustenance at so cheap a rate! for I really began to think that, with their enormous appetites, they might some day have taken it into their heads to eat their masters.

'We now suddenly cast our eyes on Flora, whom we perceived employed in turning over a round substance she had found in the sands, some pieces of which she swallowed from time to time. Ernest also perceived her motions, and did us the favour, with his usual composure, to pronounce just these words:—"They are turtle's eggs!"

' "Run, my children," cried I, "and get as many of them as you can; they are excellent, and I shall have the greatest pleasure in being able to regale our dear travellers on their return with so new and delicious a dish." We found it difficult to make Flora leave the eggs, to which she had taken a great fancy. At length, however, we succeeded in collecting near two dozen of them, which we secured in our provision-bags. When we had concluded this affair, we by accident cast our eyes upon the sea, and to our astonishment perceived a sail, which seemed to be joyfully approaching towards the land. I knew not what to imagine; but Ernest exclaimed that it was you and Fritz; and we soon had the happiness of being convinced that it was indeed our well-beloved! We ran eagerly towards the river, which Jack and Ernest recrossed as before, by leaping from one great stone to another; while I also resumed my burden of little Francis at my back, and in this manner soon arrived at the place of your landing, when we had nothing further to do but to throw ourselves into your arms!'

'And you think we could set up a tent in one of those giant trees at a distance of 66 feet from the ground! And by what means are we to ascend this tree? for at present I have no clear view of this important part of the subject.'

I perceived a tear stealing into my wife's eye, that she could not prevail upon me to think as she wished of her discovery, and that I treated the subject of her giant trees with so little respect: I therefore endeavoured to soothe and relieve her somewhat wounded sensibility.

'Do you recollect', said she, 'the large lime-tree in the public walk of the town we lived in; and the pretty little room which had been built among its branches, and the flight of stairs which led to it? What should hinder us from effecting such a contrivance in one of my giant trees, which afford even superior facilities in the enormous size and strength of their branches, and the peculiar manner of their growth?'

'Well, well, we shall see about it. In the meanwhile, my boys, let us extract a little lesson in arithmetic, from the subject of these marvellous trees; for this, at least, will be deriving a real benefit from them. Tell me, learned Mr Ernest, how many feet there are in 36 braches? for that, your mother assures us, is the height of the trees.'

Ernest.—To answer this question, I must know first how many feet or inches the brache contains.

Father.—The brache, or half-ell, contains 1 foot 10 inches, or 22 inches. Now then make your calculation.

Ernest.—I do not find it so easy as I thought. You must help me, Fritz: you are older than I am.

Fritz.—With all my heart. First we take 36 braches; then multiply 36 by 22, the number of inches each brache contains, and you have 792; divide this by 12, the number of inches in a foot, and it will give us 66 for the number of feet. Is that right, father?

Father.—Yes, quite right. So, my dear wife, you will have every evening to climb 66 feet to get to bed, which, as we have no ladder, is not the easiest thing imaginable. Now then let us see how many feet the tree is in circumference, taking it round the roots. Your mother found that she walked round it in 32 steps. Tell us then, Ernest, how many feet do you think these 32 steps would make?

Ernest.—You always ask me the things that I know nothing at all about: you should tell me, at least, how many feet there are in a step.

Father.—Well, say 2.5 feet to each step.

Ernest.—Twice 32 makes 64; the half of 32 is 16; which added to 64 makes 80 feet.

Father.—Very well. Tell me now, if you recollect the proper term in geometry for the circumference of a circle, or say of a tree, since we are talking of trees.

Ernest.—Oh, you may be sure that I could not forget that it is called the periphery.

Father.—Right. And what is the term for any line which may be drawn from one point of the periphery to another, passing through the centre? Now, Jack, you may show us what a great geometrician you intend to be.

Jack.—I believe it is called the diameter.

Father.—So far right. Next, can you tell me what is the diameter of a periphery of 80 feet, and what distance there is between the extremities of the roots of the giant tree and its trunk?

The boys all began to reckon, and soon one said one number, one another, at random; but Fritz called out louder than the rest, that the distance was 26 feet.

Father.—You are pretty near. Tell me, did you make a calculation, or was it a mere guess?

Fritz.—No, father, not a guess; but I will tell you: in the town in which we lived, I have often taken notice that the hatter, when he was about to bind the edge of a hat, always measured three times the length of the diameter, and a trifle over, for the quantity of ribbon he should use.

Father.—So; height from the ground to the branches, 66 feet; thickness, 8 feet in diameter, and 28 feet distance from the extremities of the roots to the trunk; they really, with propriety, may be called giant trees.

We now performed our devotions, and retired to rest, grateful to find ourselves once more together, and in health. We soon closed our eyes, and enjoyed tranquil slumbers till break of day.

8
Construction of a Bridge

WHEN my wife and I awaked the next morning, we resumed the question of our change of abode. I observed to her, that it was a matter of difficulty, and that we might have reason to repent such a step. 'My own opinion is', said I, 'that we had better remain here, where Providence seems to have conducted us; the place is favourable to our personal safety, and is near the vessel, from which we may continue to enrich ourselves: we are on all sides protected by the rocks; it is an asylum inaccessible but by sea, or by the passage of the river, which is not easily accomplished. Let us then have patience yet a little longer at least, till we have got all that can be removed, or that would be useful to us, from the ship.'

My wife replied, that the intense heat of the sands was insupportable; that by remaining, we lost all hope of procuring fruits of any kind, and must live on oysters, or on such wild birds as that we found so unpalatable. 'As for the safety you boast of,' pursued she, 'the rocks did not prevent our receiving a visit from the jackals; nor is it improbable that tigers or other animals might follow their example. Lastly, as to the treasures we might continue to draw from the vessel, I renounce them with all my heart. We are already in possession of provisions and other useful things; and, to say the truth, my heart is always filled with distressing apprehensions, when you and Fritz are exposed to the danger of that perfidious element the sea.'

'We will then think seriously of the matter; but let us have a well-digested scheme of operation before we leave this spot for your favourite wood. First, we must contrive a store-house among the rocks for our provisions and other things, and to which, in case of invasion in the wood, we can retreat and defend ourselves.—This agreed, the next thing is to throw a bridge across the river, if we are to pass it with all our family and baggage.'

'A bridge!' exclaimed my wife: 'Can you possibly think of such a thing? If we stay while you build a bridge, we may consider ourselves as fixed for life. Why should we not cross the river as we did before? The ass and the cow will carry all we possess upon their backs.'

'But do you recollect, that to keep what they carry dry, they must not perform their journey as they did from the vessel? For this reason, then, if for no other, we must contrive a bridge. We shall want also some sacks and baskets to contain our different matters; you may therefore set about making these, and I will undertake the bridge, which, the more I consider, the more I find to be of indispensable necessity; for the stream will, no doubt, at times increase, and the passage become impracticable in any other way. At this moment it would be found so for our shortest-legged animals, and I am sure you would not wish to see them drowned.'

'Well, then, a bridge let there be,' said my wife, 'and you will leave our stock of gunpowder here, I hope; for I am never easy with it so near us: a thunder-storm, or some thoughtless action of one of the boys, might expose us to serious dangers.'

'You are right, my love; and I will carefully attend to your suggestion. We will keep on hand only a sufficient quantity for daily use; I will contrive a place in the rock for the rest, where it will be safe from the chance of fire or dampness. It is an article which, according to the use which is made of it, may become, on the one hand, a most dangerous enemy, and, on the other, a most useful friend.'

Thus, then, we decided the important question of removing to a new abode; after which we fixed upon a plan of labour for the day, and then awaked the boys. Their delight on hearing of our project may easily be conceived, but they expressed their fear that it would be a long while before a bridge could be built; a single hour appearing an age to them, with such a novelty in view as the prospect of removing to the wood, to live under the giant trees. They, in the fullness of their joy, entreated that the place might be called *The Promised Land.*

We now began to look about for breakfast; Fritz taking care not to neglect his monkey, who sucked one of the goats as contentedly as if she had been its mother. My wife undertook to milk another, and then the cow, and afterwards gave some of the milk to each of the children; with a part of what remained she made a sort of soup with biscuits, and the rest she put into one of the flasks, to accompany us in our expedition. During this time, I was preparing the boat for another journey to the vessel, to bring away a sufficient quantity of planks and timbers for the bridge. After breakfast we set out; and now I took with me Ernest as well as Fritz, that we might accomplish our object in a shorter time.

We rowed stoutly till we reached the current, which soon drew us on beyond the bay; but scarcely had we passed a little islet, lying to one side of us, than we perceived a prodigious quantity of seagulls and other birds. I had a curiosity to discover what could be the reason of such an assemblage of these creatures. I steered for the spot; but, finding that the boat made but little way, I hoisted my sail.

To Ernest our expedition afforded the highest delight. He was in ecstasies at seeing the sail begin to swell, and the motion of the streamer in the air. Fritz, on his part, did not for a moment take his eyes from the islet where the birds were. Presently he suddenly exclaimed, 'I see what it is; the birds are all pecking, tooth and beak, at a monstrous fish, which lies dead upon the soil.'

I approached near enough to step upon the land, and after bringing the boat to an anchor with a heavy stone, we stole softly up to the birds. We soon perceived that the object which attracted them was in reality an enormous fish, which had been thrown there by the sea. So eagerly were they occupied with the feast, that not one of them attempted to fly off. We observed with astonishment the extreme voracity of this plumed group; each bird was so intent upon its prey, that we might have killed great numbers of them with our sticks alone. Fritz did not cease to express his wonder at the monstrous size of the animal, and asked me by what means he could have got there?

'I believe', answered I, 'you were yourself the means: there is every appearance that it is the very shark you wounded yesterday. See, here are the two balls which you discharged at its head.'

'Yes, yes, it is the very same,' said my young hero, skipping about for joy: 'I well remember I had two balls in my gun, and here they are, lodged in his hideous head.'

'I grant it is hideous enough,' continued I; 'its aspect even when dead makes one shudder, particularly when I recollect how easy it would have been for him to have devoured us. See what a huge mouth he has, and what a rough and prickly skin! one might almost use it for a file; and his length must be above 20 feet. We ought to be thankful to Providence, and a little to our Fritz also, for having delivered us from such a monster! But let us take away with us some pieces of his skin, for I have an idea that it may in some way or other be useful to us. But how to get at him is the difficulty.'

Ernest drew out the iron ramrod from his gun, and by striking with it to right and left among the birds, soon dispersed them. Fritz and I then advanced and cut several long strips of the skin from the head of the shark, with which we were proceeding to our boat, when I observed, lying on the ground, some planks and timbers which had recently been cast by the sea on this little island. On measuring the longest, we perceived they would answer our purpose; and, with the assistance of the crow and a lever which we had brought with us, found means to get them into the boat, and thus spare ourselves the trouble of proceeding to the vessel. With great exertion of our strength, we contrived to bind the timbers together, with the planks upon them, in the manner of a raft, and tied them to the end of the boat; so that, through this adventure, we were ready to return in four hours from the time of departure, and might boast of having done a good day's work. I accordingly pushed again for the current, which soon drove us out to sea; then I tacked about, and resumed the direct rout for the bay. All this succeeded to my utmost wishes; I unfurled my sail, and a brisk wind soon conveyed us to our landing-place.

While we were sailing, Fritz, at my request, had nailed the strips of skin we cut from the shark to the mast to dry; and he now observed to me that this was wrong, as they had taken its round shape in drying, and could not be made flat again.

'That was precisely my intention,' replied I; 'they will be more useful to us round than flat; besides, you have still some left, which you may dry flat; and then we shall have a fine provision of shagreen,* if we can find a good method to rub off the sharp points, and afterwards to polish it.'

'I thought,' said Ernest, 'that shagreen was made of ass's skin.' 'And you were not mistaken,' rejoined I; 'the best shagreen is made in Turkey, Persia, and Tartary, from skin taken from the back of the ass and the horse. While the skin is yet moist, it is stretched upon a kind of hard fat; they then beat the skin, by which means the fat is incorporated, and gives the surface the appearance of a kind of file: but very good shagreen is also made from the skin of sea-fish, particularly in France.'

Ernest asked his brother if he knew why the mouth of the shark is not, as in other animals, placed in the middle of the snout, but directly under. Fritz confessed his inability to answer this question.

'I suppose,' rejoined Ernest, 'that the mouth of the shark is thus placed, with the intention of preventing him from depopulating the sea and the land. With so excessive a voraciousness of appetite as he possesses, nothing would escape him if he had the power to seize his prey without turning his body; but as it is, there is time enough for a smaller animal to make his escape.'

'Well reasoned, my young philosopher,' cried I; 'and though we should not always be able to comprehend the intention of the Creator in the objects which surround us, at least the conjectures we are induced to form respecting them cannot fail of being a useful exercise to the mind.'

We were once more landed safely on our shore, but no one of our family appeared. We called to them as loud as we could, which was answered by the same sounds in return, and in a few minutes my wife appeared between her two little boys returning from the river, a rising piece of ground having concealed her from our sight: each carried a handkerchief in hand, which appeared filled with some new prize; and little Francis had a small fishing-net formed like a bag and strung upon a stick, which he carried on his shoulder. No sooner did they hear our voices, than they flew to meet us, surprised at our quick return. Jack reached us before the rest; and his first act was to open the handkerchief he held, and pour out a large number of lobsters at our feet: their mother and little Francis produced each as many more, forming all together a prodigious heap, and all alive; so that we were sure of excellent dinners for some days at least. Some of the animals tried to escape in different directions; and the boys, in following them, were kept in full chase, sometimes pleased and sometimes angry; sometimes laughing, sometimes scolding at the bootless trouble they were engaged in; for no sooner had they seized on the deserter, than ten more had followed his example.

'Now, have I not been very lucky, papa?' said little Francis; 'for you must know it was I who found them out. Look, there are more than 200 of them, and see how large they are, and what fine claws they have! I am sure they will be quite delicious!'

Father.—Excellent indeed, my little fellow, and particularly if it was your industry that first discovered them.

Jack.—Yes, father, it was Francis who saw them first; but it was I who ran to tell mamma, and it was I who fetched the net and put it to rights, and it was I who went up to my knees in water to catch them.

Father.—You make a charming story of it together, my boys; but as it is an interesting subject, you may tell me as many particulars as you please; it is indeed an event of some importance for our kitchen, and I have great pleasure in looking forward to partaking of a dish of your providing.

Jack.—Well then, papa, as soon as you were gone, mamma sat down outside the tent and began to work, while Francis and I took a little walk towards the river, to find out a proper place for you to begin the bridge.

Father.—Bravo, Mr Architect; but joking apart, I am much gratified to find that careless head of yours for once employed upon a useful subject. Did you find a proper place for me to begin the bridge?

Jack.—Yes, father, yes. But listen, and you will know all. When we reached the river we saw a large stone just at the edge, and little Francis kneeling down, and touching it, suddenly cried out, 'Jack, Jack, Fritz's jackal is covered all over with lobsters! Run as fast as you can.' I sprang to him in an instant, and saw not only the jackal covered with them, but legions more coming in with the stream. I ran to tell mamma, who quickly got the net you brought from the vessel. Partly with this net, and partly with our hands, we caught those you see in a very few minutes; and we should have caught a much larger number if we had not heard you call, for the river is quite full of them.—'You took quite enough for once, my boy,' said I: 'A little at a time is the maxim that suits us best, and I should even advise your taking the smallest of them back to the river, where they will grow larger; we shall still have sufficient for several magnificent repasts.'—This, then, said I to myself, is a new source for our support: even here, in these arid regions, we find means to procure not only the necessaries of life, but even luxuries. May we never cease to evince our gratitude to Providence, by the exercise of a more than ordinary care and industry!

After giving in our turn an account of our voyage, my wife set about dressing some of the lobsters, and in the meantime

Fritz and I employed ourselves in untying the raft of timbers and planks, and in moving them from the boat. I then imitated the example of the Laplanders,* in harnessing their reindeer for drawing their sledges. Instead of traces, halters, etc. I put a piece of rope, with a running knot at the end, round the neck of the ass, and passed the other end between its legs, to which I tied the piece of wood which I wished to be removed. The cow was harnessed in the same manner, and we were thus enabled to carry our materials, piece by piece, to the spot which architect Jack had chosen at the river, as the most eligible for our bridge: to say the truth, I thought his judgement excellent; it was a place where the shore on each side was steep, and of equal height; there was even on our side an old trunk of a tree lying on the ground, which I foresaw would have its use.

'Now then, boys,' said I, 'the first thing is to see if our timbers are long enough to reach to the other side: by my eye, I should think they are; but if I had a surveyor's plane, we might be quite sure, instead of working at a venture.'

'But my mother has some balls of packthread,* with which she measured the height of the giant tree,' interrupted Ernest, 'and nothing would be more easy than to tie a stone to the end of one of them, and throw it to the other side of the river; then we could draw it to the very brink, and thus obtain the exact length that would be required for our timbers.'

'Your idea is excellent,' cried I; 'nothing gives me more pleasure than to see you exercise your invention: run quickly and fetch the packthread.' He returned without loss of time; the stone was tied to its end, and thrown across as we had planned; we drew it gently back to the river edge, marking the place where the bridge was to rest: we next measured the string, and found that the distance from one side to the other was 18 feet. It appeared to me, that to give a sufficient solidity to the timbers, I must allow 3 feet at each end of extra length for fixing them, making therefore in all 24; and I was fortunate enough to find that many of those we had brought did not fall short of this length. There now remained the difficulty of carrying one end across the stream; but we determined to discuss this part of the subject while we ate our dinner, which had been waiting for us more than an hour.

We all now proceeded homewards, and entering the kitchen, we found our good steward had prepared for us a large dish of lobsters; but before tasting them, she insisted we should look at something she had been employed about: she produced two sacks intended for the ass, which she had seamed with packthread; the work, she assured us, had with difficulty been accomplished, since, for want of a needle large enough to carry packthread, she had been obliged to make a hole with a nail for every stitch; we might therefore judge by her perseverance in such a task, of the ardour with which she longed to see her plan of a removal executed. She received on this occasion, as was well her due, abundance of compliments and thanks from her companions, and also a little good-humoured raillery. For this time we hurried through our meal, each being deeply interested in the work we were about to undertake, and thinking only of the part which might be assigned him towards the execution of the *Nonsuch*; for this, for mutual encouragement, was the name we gave our bridge, even before it was in existence.

Having consulted as to the means of laying our timbers across the river, the first thing I did was to attach one of them to the trunk of the tree, of which I have already spoken, by a strong cord, long enough to turn freely round the trunk; I then fastened a second cord to the other end of the timber, and tying a stone to its extremity flung it to the opposite bank. I next passed the river as I had done before, furnished with a pulley, which I secured to a tree; I passed my second cord through the pulley, and recrossing the river with this cord in my hand, I contrived to harness the ass and cow to the end of the cord. I next drove the animals from the bank of the river; they resisted at first, but I made them go by force of drawing. I first fixed one end of the beam firm to the trunk of the tree, and then they drew along the other end, so as gradually to advance over the river: presently, to my great joy, I saw it touch the other side, and at length become fixed and firm by its own weight. In a moment Fritz and Jack leaped upon the timber, and, in spite of my paternal fears, crossed the stream with a joyful step upon this narrow but effective bridge.

The first timber being thus laid, the difficulty was considerably diminished; a second and a third were fixed in succession, and with the greatest ease. Fritz and I, standing on opposite sides of the river, placed them at such distances from each other as was necessary to form a broad and handsome bridge: what now remained to be done was to lay some short planks across them quite close to each other, which we executed so expeditiously, that our construction was completed in a much shorter time than I should have imagined possible. The reader should have seen our young workmen, to form the least conception of the delight they felt: they jumped, danced, played a thousand antics, and uttered a thousand joyful sounds upon their bridge. For my own part, I could hardly restrain myself from joining in these demonstrations of their perfect happiness; and my wife, who had been the mover of all our operations, was as little disposed to a silent calm enjoyment of our success as any of the rest: she ran to one, and then to another, embracing each in turn, and was never tired of passing and repassing on our piece of workmanship, which was everywhere safe and even, and at least 10 feet in breadth. I had not fastened the cross planks to each other, for they appeared to be close and firm without it; and besides, I recollected that in case of danger from any kind of invasion, we could with the greater ease remove them, and thus render the passage of the river more difficult. Our labour however had occasioned us so much fatigue, that we found ourselves unable for that day to enter upon new exertions; and the evening beginning to set in, we returned to our home, where we partook heartily of an excellent supper, and went to bed.

9
Change of Abode

As soon as we were up and had breakfasted, the next morning, I assembled all the members of my family together, to take with them a solemn farewell of this our first place of reception from the awful disaster of the shipwreck. I confess that for my own part I could not leave it without regret; it was a place of greater safety than we were likely again to meet with; it was also nearer to the vessel. I thought it right to represent strongly to my sons the danger of exposing themselves, as they had done the evening before, along the river.—'We are now going', continued I, 'to inhabit an unknown spot, which is not so well protected by nature as that we are leaving: we are unacquainted both with the soil and its inhabitants, whether human creatures or beasts; much caution is therefore necessary, and take care not to remain separate from each other.' Having unburdened my mind of this necessary charge, we prepared for setting out. I directed my sons to assemble our whole flock of animals, and to leave the ass and the cow to me, that I might load them with the sacks as before concerted; I had filled these, and made a slit longways in the middle of each, and to each side of the slits I tied several long pieces of cord, which crossing each other, and being again brought round and fastened, served to hold the sacks firmly on the back of the animal. We next began to put together all the things we should stand most in need of for the two or three first days in our new abode: working imple-ments, kitchen utensils, the captain's service of plate, and a small provision of butter, etc. etc. I put these articles into the two ends of each sack, taking care that the sides should be equally heavy, and then fastened them on. I afterwards added our hammocks to complete the load, and we were about to begin to march, when my wife stopped me.—'We must not', said she, 'leave our fowls behind, for fear they should become the prey of the jackals. We must contrive a place for them among the luggage, and also one for our little Francis, who cannot walk so far, and would interrupt our speed. There is also my enchanted bag, which I recommend to your particular care,' said she, smiling, 'for who can tell what may yet pop out of it for your good pleasure.'

I now placed the child on the ass's back, fixing the en-chanted bag in such a way as to support him, and I tied them together with so many cords, that the animal might even have galloped without danger of his falling off.

In the meanwhile, the other boys had been running after the cocks and hens and the pigeons, but had not succeeded in catching one of them; so they returned empty-handed and in ill humour.—'Little blockheads!' said their mother, 'see how you have heated yourselves in running after these untractable creatures! I could have put you in a way to catch them in a moment; come with me and see.'—She now stepped into the tent, and brought out two handfuls of peas and oats, and by pronouncing a few words of invitation in the accustomed tone, the birds flocked round her. She then walked slowly before them, dropping the grain all the way, till they had followed her into the tent. When she saw them all inside, and busily employed in picking up the grain, she shut the entrance, and caught one after the other without difficulty. The boys looked at each other half ashamed though much amused with the adventure. The fowls were then tied by the feet and wings, put into a basket covered with a net, and placed in triumph on the top of our luggage.

We packed and placed in the tent everything we were to leave, and, for greater security, fastened down the ends of the sailcloth at the entrance, by driving stakes through them into the ground. We ranged a number of vessels, both full and empty, round the tent, to serve as a rampart, and thus we confided to the protection of heaven our remaining treasures. At length, we set ourselves in motion each of us, great and small, carried a gun upon his shoulder, and a game bag at his back. My wife led the way with her eldest son, the cow and the ass immediately behind them; the goat conducted by Jack came next; the little monkey was seated on the back of his nurse, and made a thousand grimaces. After the goats, came Ernest, conducting the sheep, while I, in my capacity of general superintendent, followed behind and brought up the rear; the dogs for the most part pranced backwards and forwards, like adjutants to a troop of soldiers. Our march was slow, and there was something solemn and patriarchal in the spectacle we exhibited; I fancied we must resemble our forefathers journeying in the deserts, accompanied by their families and their possessions.—'Now then, Fritz,' cried I, 'you have the specimen you wished for of the patriarchal mode of life; what do you think of it?'—'I like it much, father,' replied he: 'I never read the Bible, without wishing I had lived in those good times.'

'And I too,' said Ernest, 'I am quite delighted with it; I cannot help fancying myself not merely a patriarch, but a Tartar,* or an Arab, and that we are about to discover I know not how many new and extraordinary things. Is it not true, father, that the Tartars and the Arabs pass their lives in journeying from one place to another, and carrying all they have about them?'

'It is certainly for the most part true,' replied I, 'and they are denominated wandering tribes; but they generally perform their journeys attended by horses and camels, by means of which they can proceed a little faster, than if, like us, they had only an ass and a cow. For my part, I should not be sorry if I were quite sure that the pilgrimage we are now making would be our last.'—'And I too am of your way of thinking,' cried my wife, 'and I hope that in our new abode we shall be so well satisfied with the shade of such luxuriant trees, that we shall not be inclined to further rambles.'

We had now advanced half-way across our bridge, when the sow for the first time took the fancy of joining us. At the moment of our departure she had shown herself so restive and indocile, that we were compelled to leave her behind us; but seeing that we had all left the place, she had set out voluntarily to overtake us; taking care, however, to apprise us, by her continual grunting, that she disapproved of our migration.

On the other side of the river we experienced an inconvenience wholly unexpected. The tempting aspect of the grass, which grew here in profusion, drew off our animals, who strayed from us to feed upon it; so that, without the dogs, we should not have been able to bring them back to the line of our procession. The active creatures were of great use to us on this occasion; and when everything was restored to proper order, we were able to continue our journey. For fear, however, of a similar occurrence, I directed our march to the left, along the seaside, where the produce of the soil was not of a quality to attract them.

But scarcely had we advanced a few steps on the sands, when our two dogs, which had strayed behind among the grass, set up a sort of howl, as if engaged in an encounter with some formidable animal. Fritz in an instant raised his gun to his cheek, and was ready to fire; Ernest, always somewhat timid, drew back to his mother's side; Jack ran bravely after Fritz with his gun upon his shoulder; while I, fearing the dogs might be attacked by some dangerous wild beast, prepared myself to advance to their assistance. But youth is always full of ardour; and in spite of my exhortations to proceed with caution, the boys, eager for the event, made but three jumps to the place from which the noise proceeded. In an instant Jack had turned to meet me, clapping his hands and calling out 'Come quickly, father, come quickly, here is a monstrous porcupine!'

I soon reached the spot, and perceived that it was really as they said, bating a little exaggeration. The dogs were running to and fro with bloody noses about the animal; and when they approached too near him, he made a frightful noise, and darted his quills so suddenly at them, that a great number had penetrated the skins of the valiant creatures, and remained sticking in them; and it was no doubt the pain they occasioned which made them howl so violently.

While we were looking on, Jack determined on an attack, which succeeded well. He took one of the pistols which he carried in his belt, and aimed it so exactly at the head of the porcupine, that he fell dead the instant he fired, and before we had a notion of what he was about. This success raised Jack to the height of joy and vanity; while Fritz, on the other hand, felt a sensation of jealousy almost to shedding tears.—'Is it right, Jack,' said he, 'that such a little boy as you should venture to fire off a pistol in this manner? How easily might you have wounded my father or me, or one of the dogs, by so rash an action!'—'Oh yes, to be sure, and what do you suppose hindered me from seeing that you were all behind me? Do you think I fired without taking care of that? Do you take me for an idiot? The porcupine could tell you about that, brother Fritz, if he could but speak. My first fire—pop—dead as a herring! This is something like, brother Fritz, and you would be glad enough to have had such a chance yourself!'

Fritz only replied by a motion of his head. He was out of humour because his younger brother had deprived him of the honour of the day; and he sought a subject of complaint against him, as the wolf did with the poor little lamb.* 'Come, come, boys,' said I, 'let me hear no envious speeches and no reproaches; luck for one today, for another tomorrow; but all for the common good. Jack was, perhaps, a little imprudent, but you must allow that he showed both skill and courage; let us not therefore tarnish the glory of his exploit.' We now all got round the extraordinary animal, on whom nature has bestowed a strong defence, by arming his body all over with long spears. The boys were at a loss what means to use for carrying away his carcass. They thought of dragging it along the ground; but as often as they attempted to take hold, there was nothing but squalling, and running to show the marks made by his quills on their hands.—'We must leave him behind,' said they; 'but it is a great pity.'

While the boys were talking, my wife and I had hastened to relieve the dogs, by drawing out the quills and examining their wounds. Fritz had run on before with his gun, hoping he should meet with some animal of prey. What he most desired was to find one or two of those large bustards* which his mother had described to him. We followed him at our leisure, taking care not to expose our health by unnecessary fatigue; till at last, without further accident or adventure, we arrived at the place of the giant trees. Such, indeed, we found them, and our astonishment exceeded all description—'Good heavens! what trees! what a height! what trunks! I never heard of any so prodigious!' exclaimed one and all.—'Nothing can be more rational than your admiration', answered I, measuring them with my eyes as I spoke. 'I must confess I had not myself formed an idea of the reality. To you be all the honour, my dear wife, for the discovery of this agreeable abode, in which we shall enjoy so many comforts and advantages. The great point we have to gain, is the fixing a tent large enough to receive us all, in one of these trees, by which means we shall be perfectly secure from the invasion of wild beasts. I defy even one of the bears, who are so famous for mounting trees, to climb up by a trunk so immense, and so destitute of branches.'

We began now to release our animals from their burdens, having first thrown our own on the grass. We next used the precaution of tying their two forelegs together with a cord, that they might not go far away, or lose themselves. We restored the fowls to liberty; and then seating ourselves upon the grass, we held a family council on the subject of our future establishment. I was myself somewhat uneasy on the question of our safety during the ensuing night; for I was ignorant of the nature of the extensive country I beheld around me, and what chance there might be of our being attacked by different kinds of wild beasts. I accordingly observed to my wife, that I would make an endeavour for us all to sleep in the tree that very night. While I was deliberating with her on the subject, Fritz, who longed to take his revenge for the porcupine adventure, had stolen away to a short distance, and we heard the report of a gun. This would have alarmed me, if, at the same moment, we had not recognized Fritz's voice crying out, 'I touched him! I touched him!' and in a moment we saw him running towards us, holding a dead animal of uncommon beauty by the paws. —'Father, father, look, here is a superb tiger cat,' said he, proudly raising it in the air, to show it to the best advantage.—'Bravo! bravo!' cried I; 'bravo, Nimrod* the undaunted! Your exploit will call forth the gratitude of our cocks, hens, and pigeons, for you have rendered them what they cannot fail to think an important service. If you had not killed this animal, he would no doubt have demolished in one night our whole stock of poultry. I charge you look about in every direction, and try to destroy as many of the species as fall in your way, for we cannot have more dangerous intruders.'

Ernest.—I wish, father, you would be so good as to tell me why God created wild beasts, since man seems to be appointed to destroy them.

Father.—This indeed is a question I cannot answer, and we must be contented with taking care to arm ourselves against them: neither can I explain to you why many other things, which to us appear to have only injurious qualities, have been created. With respect to beasts of prey, I am inclined to believe, that one of the ends of Providence, in giving them existence, is their embellishing and varying the works of the creation; of maintaining a necessary equilibrium among creatures endowed with life; and lastly, to furnish man, who comes naked into the world, with materials for protecting himself from the cold, by the use of their skins, which become the means of exchange and commerce between different nations. We may also add, that the care of protecting himself from the attacks of ferocious animals invigorates the physical and moral powers of man, supports his activity, and renders him inventive and courageous. The ancient Germans, for example, were rendered robust and valiant warriors, through their habitual exercises in the field, which enabled them, at a time of need, to defend their country and their liberty with as little difficulty as they would have experienced in killing a wolf or a bear.—But let us return to the animal Fritz has killed. Tell me all the particulars of your adventure. How did you kill him?

Fritz.—With my pistol, father, as Jack killed the porcupine.

Father.—Was he on this tree just by us?

Fritz.—Yes, father, I had been observing that something moved among the branches. I went softly as near as I could; and on seeing him I knew him for a tiger cat. I fired, when he fell at my feet, wounded and furious; and then I fired a second time and killed him.

Father.—You were very fortunate, for he might easily have devoured you. You should always take care, in aiming at animals of this kind, to be at a greater distance.

Fritz.—Why so, father? I might have missed him if I had been further off. I, on the contrary, tried to be as near him as possible, and fired close to his ears.

Father.—This was acting in the same way as your brother Jack, whom you so much derided for his want of care, and may serve you as a lesson not to blame in your brothers, what you would yourself be perhaps obliged to do in the same situation; also not to interrupt their joy with unkind reflections, but rather to partake with them the pleasure of their success.

Fritz.—Well, father; all I now ask of Jack is, that he will be so good as not to spoil the beautiful skin of this animal as he did that of the jackal. Only observe what beautiful figures it is marked with, and the fine effect of the black and yellow spots; the most richly manufactured stuff could not exceed it in magnificence. What is the exact name of the animal?

Father.—You may for the present give it the name of the tiger cat. I do not, however, think that it is the animal which is so denominated at the Cape of Good Hope; I rather think it is the margay,* a native of America, an animal of extremely vicious dispositions and singular voraciousness; he attacks all the birds of the forest, and neither a man, a sheep, or goat, that should fall in his way, could escape his rapacity. In the name of humanity, therefore, we ought to be thankful to you for having destroyed him.

Fritz.—All the recompense I ask, father, is, that you will let me keep the skin; and I wish you would tell me what use I can make of it.

Father.—One idea occurs to me; skin the animal, carefully, so as not to injure it, particularly the parts which cover the forelegs and the tail. You may then make yourself a belt with it, like your brother Jack's. The odd pieces will serve to make some cases to contain our utensils for the table, such as knives, forks, spoons. Go then, boy, and put away its bloody head, and we will see how to set about preparing the skin.

The boys left me no moment of repose till I had shown them how to take off the skins of animals without tearing them. In the meanwhile Ernest looked about for a flat stone as a sort of foundation for a fireplace, and little Francis collected some pieces of dry wood for his mother to light a fire. Ernest was not long in finding what he wanted, and then he ran to join us and give us his assistance, or rather to reason, right or wrong, on the subject of skinning animals; and then on that of trees, making various comments and enquiries respecting the real name of those we intended to inhabit.—'It is my opinion,' said he, 'that they are, really and simply, enormously large hazel trees; see if the leaf is not of exactly the same form'—'But that is no proof,' interrupted I:'for many trees bear leaves of the same shape, but nevertheless are of different kinds.'

Ernest.—I thought, father, that the mango tree* only grew on the sea-shore, and in marshy soils?

Father.—You were not mistaken: it is the black mango tree which loves the water. But there is, besides, the red mango, which bears its fruit in bunches, something like our currant bushes. This kind of the mango tree is found at a considerable distance from the sea, and its wood is used for dyeing red. There is a third sort, which is called the mountain mango, or yellow wood, and this is the kind whose roots produce the beautiful arches you now see around us.

Presently little Francis came running, with his mouth crammed full of something, and calling out, 'Mamma, mamma, I have found a nice fruit to eat, and I have brought you home some of it!'

'Little glutton!' replied his mother, quite alarmed, 'what have you got there? For Heaven's sake, do not swallow, in this imprudent manner, the first thing that falls in your way; for by this means you may be poisoned, and then you would die.' She made him open his mouth, and took out with her finger what he was eating with so keen a relish. With some difficulty she drew out the remains of a fig.—'A fig!' exclaimed I: 'where did you get this fig?'

Francis.—I got it among the grass, papa; and there are a great many more. I thought it must be good to eat, for the fowls and the pigeons, and even the pig, came to the place and ate them in large quantities.

Father.—You see then, my dear, said I to my wife, that our beautiful trees are fig-trees,* at least the kind which are thus named at the Antilles.* I took this occasion to give the boys another lesson on the necessity of being cautious, and never to venture on tasting anything they met with, till they had seen it eaten by birds and monkeys. At the word monkeys, they all ran to visit the little orphan, whom they found seated on the root of a tree, and examining with the oddest grimaces the half-skinned tiger cat, which lay near him. Francis offered him a fig, which he first turned round and round, then smelled at it, and concluded by eating it voraciously.—'Bravo, bravo! Mr Monkey,' exclaimed the boys, clapping their hands; 'so then these figs are good to eat! Thank you, Mr Monkey, for, after your wise decision, we shall make a charming feast on them.'

In the meanwhile my wife had been busy in making a fire, putting on the pot, and preparing for our dinner. The tiger cat was bestowed upon the dogs, who waited impatiently to receive it. While our dinner was dressing, I employed my time in making some packing-needles* with some of the quills of the porcupine, which the boys had contrived to draw from his skin, and bring home. I put the point of a large nail into the fire till it was red-hot; then taking hold of it with some wet linen in my hand, by way of guard, I with great ease perforated the thick end of the quills with it. I had soon the pleasure of presenting my wife with a large packet of long, stout needles, which were the more valuable in her estimation, as she had formed the intention of contriving some better harness for our animals, and had been perplexed how to set about them without some larger needles. I, however, recommended to her to be frugal in the use of her packthread, for which I should soon have so urgent a need, in constructing a ladder for ascending the tree we intended to inhabit.

I had singled out the highest fig-tree; and while we were waiting for dinner, I made the boys try how high they could throw a stick or stone into it. I also tried myself; but the lowest branches were so far from the ground, that none of us could

touch them. I perceived, therefore, that we should want some
new inventions for fastening the ends of my ladder to them.
I allowed a short pause to my imagination, during which I
assisted Jack and Fritz in carrying the skin of the tiger cat to a
near rivulet, where we confined it under water with some large
stones. After this we returned and dined heartily on some slices
of ham and bread and cheese, under the shade of our favourite
trees.

10
Construction of a Ladder

OUR repast ended, I observed to my wife, that we should be
obliged to pass the night on the ground. I desired her to begin
preparing the harness for the animals, that they might go to
the sea-shore, and fetch pieces of wood, or other articles which
might be useful to us. I, in the meantime, set about suspending
our hammocks to some of the arched roots of the trees. I next
spread a piece of sailcloth large enough to cover them, to
preserve us from the dew and from the insects. I then hastened
with the two eldest boys to the sea-shore, to choose out such
pieces of wood as were most proper for the steps of my ladder.
Ernest was so lucky as to discover some bamboo canes in a sort
of bog. I took them out, and, with his assistance, completely
cleared them from the dirt; and stripping off their leaves, I
found, to my great joy, that they were precisely what I wanted.
I then instantly began to cut them with my hatchet, in pieces
of four or five feet long; the boys bound them together in
faggots, and we prepared to return with them to our place of
abode. I next secured some of the straight and most slender of
the stalks, to make some arrows with, of which I knew I should
stand in need. At some distance from the place where we stood,
I perceived a sort of thicket, in which I hoped to find some
young pliant twigs, which I thought might also be useful to me;
we proceeded to the spot; but apprehending it might be the
retreat of some dangerous reptile, or animal, we held our guns
in readiness. Flora, who had accompanied us, went before. We
had hardly reached the thicket before she made several jumps,
and threw herself furiously into the middle of the bushes; when
a troop of large-sized flamingos sprang out, and with a loud

rustling noise mounted into the air. Fritz fired, when two of the birds fell among the bushes: one of them was quite dead; the other was only slightly wounded in the wing, and finding that he could not fly, he ran so fast towards the water, that we were afraid he would escape us. Fritz, in the joy of his heart, plunged up to his knees in the water, to pick up the flamingo he had killed, and with great difficulty was able to get out again; while I, warned by his example, proceeded more cautiously in my pursuit of the wounded bird. Flora came to my assistance, and running on before, caught hold of the flamingo, and held him fast till I reached the spot and took him into my protection. All this was effected with considerable trouble; for the bird made a stout resistance, flapping its wings with violence for some time. But at last I succeeded in securing him.

Fritz was not long in extricating himself from the swamp; he now appeared holding the dead flamingo by the feet: but I had more trouble in the care of mine, as I had a great desire to preserve him alive. I had tied his feet and his wings with my handkerchief; notwithstanding which, he still continued to flutter about to a distressing degree, and tried to make his escape. I held the flamingo under my left arm, and my gun in my right hand. I made the best jumps I was able to get to the boys, but at the risk of sinking every moment in the mud, which was extremely deep, and from which it would have been difficult to release me.

The joy of the boys was excessive, when they saw that my flamingo was alive.—'If we can but cure his wound and contrive to feed him, what a happiness it will be!' said they. 'Do you think he will like to be with the other fowls?' 'I know,' answered I, 'that he is a bird that may be easily tamed; but he will not thank you for such food as we give our fowls; he will make his humble petition to you for some small fish, a few worms, or insects.'

Ernest.—Our river will furnish him with all these: Jack and Francis can catch as many as he will want; and very soon, with such long legs as he has, he may learn the way to the river and find them for himself. But, father, are all flamingos like this, of such a beautiful red colour, and the wings so exquisitely tinted with purple? I think I have seen the flamingo in my

Natural History, and the colours were not like these: so perhaps this is not a flamingo at last.

Father.—I believe it is a flamingo, Ernest, and that this difference in the plumage denotes the age of the bird: when very young they are gray; at a more advanced age they are white; and it is only when they are full grown, that they are adorned with this beautiful tinted plumage. But one of you must hold our live flamingo, while I repeat my visit to the canes, for I have not done with them yet. I accordingly selected some of the oldest of the stalks, and cut from them their hard pointed ends, to serve for the tips of my arrows, for which they are also used by the savages of the Antilles. Lastly, I looked for two of the longest canes, which I cut, for the purpose of measuring the height of our giant tree, about which I felt so deep an interest. When I told my sons the use I intended to make of the two longest canes, they indulged themselves in a hearty laugh at me, and maintained, that though I should lay ten such canes up the trunk of the tree, the last would not reach even the lowest branch. I requested they would oblige me by having a little patience; and I reminded them, that it was not long ago that they defied their mother to catch the fowls, because they themselves had not known how to set about it. We now thought of returning. Ernest took the charge of the canes; Fritz carried the dead flamingo, and I resumed the care of the living one.

We had now reached the spot where we had left the three bundles of bamboo-canes; and as my sons were sufficiently loaded, I took charge of them myself.

We at length arrived once more at our giant trees, and were received with a thousand expressions of interest and kindness. All were delighted at the sight of our new captures. My wife, with her usual anxiety about the means for subsisting, asked where we should get food enough for all the animals we brought home?—'You should consider,' said I, 'that some of them feed us, instead of being fed; and the one we have now brought you need not give much uneasiness, if, as I hope, he proves able to find food for himself.' I now began to examine his wound, and found that only one wing was injured by the ball, but that the other had also been slightly wounded by the dog laying hold of him. I applied some ointment to both, which seemed immediately to ease the pain. I next tied him by one of his legs, with a long string, to a stake I had driven into the ground, quite near to the river, that he might go in and wash himself when he pleased.

In the meantime, my little railers had tied the two longest canes together, and were endeavouring to measure the tree with them; but when they found that they reached no further than the top of the arch formed by the roots, they all burst into immoderate fits of laughter, assuring me, that if I wished to measure the tree, I must think of some other means. I however sobered them a little, by recalling to Fritz's memory some lessons in land-surveying he had received in Europe, and that the measure of the highest mountains, and their distance from each other, may be ascertained by the application of triangles and supposed lines.* I instantly proceeded to this kind of operation, fixing my canes in the ground, and making use of some string, which Fritz guided according to my directions. I found that the height of the lower branches of our tree was 40 feet: a particular I was obliged scrupulously to ascertain, before I could determine the length of my ladder. I now set Fritz and Ernest to work, to measure our stock of thick ropes, of which I wanted no less than 80 feet for the two sides of the ladder: the two youngest I employed in collecting all the small string we had used for measuring, and carrying it to their mother. For my own part I sat down on the grass, and began to make some arrows with a piece of the bamboo, and the short sharp points of the canes I had taken such pains to secure. As the arrows were hollow, I filled them with the moist sand, to give them a little weight; and lastly, I tipped them with a bit of feather from the flamingo, to make them fly straight. Scarcely had I finished my work, than the boys came jumping round me, uttering a thousand demonstrations of joy:—'A bow, a bow, and some real arrows!' cried they, addressing each other, and then running to me.—'Tell us father,' continued they, 'what you are going to do with them; do let me shoot one;—and me; and me too', cried one and all as fast as they could speak.

Father.—'Have patience, boys; I say, have patience. Have you, my dear, any strong thread?' said I to my wife; 'I want some immediately.'—'We shall see', said she, 'what my enchanted bag, which has never yet refused its aid, can do for you.' She then threw open its mouth.—'Come,' said she, 'pretty bag, give me what I ask for; my husband wants some thread, and it must be very strong——See now, did I not promise you should have your wish?—Here is a large ball of the very thread you want.'

Ernest.—But I do not see much magic, however, mother, in taking out of a bag exactly what we had before put into it.

Father.—If we are to discuss the matter seriously, Ernest, I cannot but allow that your observation is a just one; but in a moment of dreadful apprehension, such as we experienced on leaving the vessel, to think of a variety of little things that might be useful to one or all of us, was an act that we may truly call magical; only from the best of wives and mothers, could it have proceeded: it is, then, something like a truth, that your mother is a good fairy, who constantly provides for all our wants: but you young giddy things think little of the benefit you thus enjoy.

Just at this moment Fritz joined us, having finished measuring the string: he brought me the welcome tidings that our stock, in all, was about 500 fathoms, which I knew to be more than sufficient for my ladder. I now tied the end of the ball of strong thread to an arrow, and fixing it to the bow, I shot it off in such a direction, as to make the arrow pass over one of the largest branches of the tree, and fall again to the ground. By this method I lodged my thread securely, while I had the command of the end and the ball below. It was now easy to tie a piece of rope to the end of the thread, and draw it upwards, till the knot should reach the same branch. Having thus made quite sure of being able to raise my ladder, we all set to work with increased zeal and confidence. The first thing I did was to cut a length of about 100 feet from my parcel of ropes, an inch thick; this I divided into two equal parts, which I stretched along on the ground in two parallel lines, at the distance of a foot from each other. I then directed Fritz to cut portions of sugar-cane, each 2 feet in length. Ernest handed them to me, one after another; and as I received them, I inserted them into my cords at the distance of twelve inches respectively; fixing them with knots in the cord, while Jack, by my order, drove into each a long nail at the two extremities, to hinder them from slipping out again. Thus, in a very short time, I had formed a ladder of 40 rounds in length, and, in point of execution, firm and compact, and which we all beheld with a sort of joyful astonishment. I now tied it with strong knots to the end of the rope which hung from the tree, and pulled it by the other, till our ladder reached the branch, and seemed to rest so well upon it, that the joyous exclamations of the boys and my wife resounded from all sides. All the boys wished to be the first to ascend upon it; but I decided that it should be Jack, he being the nimblest and of the lightest figure among them.—Accordingly, I and his brothers held the ends of the rope and of the ladder with all our strength, while our young adventurer tripped up the rounds with perfect ease, and presently took his post upon the branch; but I observed that he had not strength enough to tie the rope firmly to the tree. Fritz now interfered, assuring me that he could ascend as safely as his brother: but as he was much heavier, I was not altogether without apprehension. I fastened the end of the ladder with forked stakes to the ground, and then gave him instructions how to step in such a way as to divide his weight, by occupying four rounds of the ladder at the same time, with his feet and hands. It was not long before we saw him side by side with Jack, 40 feet above our heads, and both saluting us with cries of exultation. Fritz set to work to fasten the ladder, by passing the rope round and round the branch; and this he performed with so much skill and intelligence, that I felt sufficient reliance to determine me to ascend myself, and well conclude the business he had begun. But first I tied a large pulley to the end of the rope, and carried it with me. When I was at the top, I fastened the pulley to a branch which was within my reach, that by this means I might be able the next day to draw up the planks and timbers I might want for building my aerial castle. I executed all this by the light of the moon, and felt the satisfaction of having done a good day's work. I now gently descended my rope ladder, and joined my wife and children.

Finding an inconvenience in being three together on the branch, I had directed the boys to descend first. My astonishment, therefore, on reaching the ground, where neither Fritz nor Jack had made their appearance, it is easier to conceive than to describe. While I was endeavouring to conjecture where they could be, we suddenly heard the sound of voices which seemed to come from the clouds, and which chanted an

evening hymn. I soon perceived the trick our young rogues had
played, who, seeing me busily employed in the tree, instead of
descending as I had desired, had climbed upwards from
branch to branch, till they had reached the very top. My heart
was now lightened of my apprehensions for their safety. I called
out to them as loudly as I could to take great care in coming
down. It was almost night, and the light of the moon scarcely
penetrated the extreme thickness of the foliage. They presently
descended without any accident, when they told us, that
scarcely had my voice reached to the great height at which they
were. I now directed them to assemble all our animals, and to
get what dry wood we should want for making fires, which I
looked to as our defence against the attacks of wild beasts.
I explained to them my reasons for this; informing them that
in Africa, a country remarkable for its prodigious numbers of
ferocious animals, the natives secure themselves from their
nocturnal visits by lighting large fires, which all these creatures
are known to dread and avoid.

My wife now presented me with the day's work she had
performed: it was some traces, and a breast-leather each for
the cow and the ass. I promised her, as a reward for her zeal
and exertion, that we should all be completely settled in the
tree the following day, and we then assembled to supper.

All our animals came round us, one after the other. My wife
threw some grain to the fowls, to accustom them to draw
together in a particular spot; and when they had eaten it, we
had the pleasure of seeing our pigeons take their flight to the
top of the giant tree, and the cocks and hens perching and
settling themselves, and cackling all the time, upon the rounds
of the ladder. The quadrupeds we tied to the arched roots of
the tree, quite near to our hammocks, where they quietly lay
on the grass to ruminate in tranquillity. Our beautiful flamingo
was not forgotten, Fritz having fed him with some crumbs of
biscuit soaked in milk, which he ate very heartily; and
afterwards putting his head under his right wing, and raising
his left foot, he abandoned himself with confidence to sleep.

And now the gaping of one and the outstretched arms of
another, gave us notice that it was time for our young labourers
to retire to rest. We performed our evening devotions. I set fire
to several of the heaps, and then threw myself contentedly

upon my hammock. My young ones were already cased in
theirs, and we were soon greeted with their murmurs at being
obliged to lie so close to each other that they could not move
their limbs.—'Ah, gentlemen,' cried I, 'you must try to be
contented. No sailor is ever better accommodated than you are
now, and you must not expect beds to drop from the clouds
on your behalf!' I directed them how to put themselves in a
more convenient posture, and to swing their hammock gently
to and fro. 'And see', added I, 'if sleep will not visit you as soon
in a hammock as on a bed of down.' They profited by my
advice, and all, except myself, were soon asleep.

11
The Settling in the Giant Tree

I HAD thought it necessary to keep watch during this first night.
Every leaf that stirred gave me the apprehension that it was the
approach of a jackal or a tiger, who might attack us. As soon
as one of the heaps was consumed, I lighted another; and at
length, finding that no animal appeared, I by degrees became
assured, and fell into a sound sleep. The next morning we took
our breakfast, and fell to work. My wife, having finished her
daily occupation of milking the cow and preparing the breakfast, set off with Ernest, Jack, and Francis, attended by the ass,
to the sea-shore; they had no doubt of finding some more
pieces of wood, and they thought it would be prudent to
replenish our exhausted store. In her absence, I ascended the
tree with Fritz, and made the necessary preparations for my
undertaking, for which I found it in every respect convenient;
for the branches grew close to each other, and in an exactly
horizontal direction. Such as grew in a manner to obstruct my
design, I cut off either with the saw or hatchet, leaving none
but what presented me with a sort of foundation for my work.
I left those which spread themselves evenly upon the trunk,
and had the largest circuit, as a support for my floor. Above
these, at the height of 46 feet, I found others, upon which to
suspend our hammocks; and higher still, there was a further
series of branches, destined to receive the roof of my tent,
which for the present was to be formed of nothing more than
a large surface of sailcloth.

The progress of these preparations was considerably slow.
It was necessary to raise certain beams to this height of 40 feet,
that were too heavy for my wife and her little assistants to lift
from the ground. I had, however, the resource of my pulley,
which served to excellent purpose, and Fritz and I contrived
to draw them up to the elevation of the tent, one by one. When
I had already placed two beams upon the branches, I hastened
to fix my planks upon them; and I made my floor double, that
it might have sufficient solidity if the beams should be warped
from their places. I then formed a wall of staves of wood like
a park-paling,* all round, for safety. This operation, and a third
journey to the sea-shore to collect the timber necessary, filled
our morning so completely, that not one of us had thought
about dinner. For this once we contented ourselves with a bit
of ham and some milk, which we ate, and returned to finish
our aerial palace, which began to make an imposing appearance. We unhooked our hammocks from the projecting roots,
and by means of my pulley, contrived to hoist them up the tree.
The sailcloth roof was supported by the thick branches above;
and as it was of great compass, and hung down on every side,
the idea occurred to me of nailing it to the paling on two sides,
thus getting not only a roof, but two walls also; the immense
trunk of the tree forming a third side, while in the fourth was
the entrance to our apartment; and in this I left a large aperture, both as a means of seeing what passed without, and admit-
ting a current of air to cool us in this burning temperature.
We also on this side enjoyed an extensive view of the vast ocean,
and its lengthening shore. The hammocks were soon hung on
the branches, and everything was ready for our reception that
very evening. Well satisfied with the execution of my plan, I
descended with Fritz, who had assisted me throughout the
whole; and as the day was not far advanced, and I observed we
had still some planks remaining, we set about contriving a large
table, to be placed between the roots of the tree, and surrounded with benches; and this place, we said, should be called
our dining-parlour. For this time, we performed our task im-
perfectly, for I confess I was much fatigued. The table, however,
was such as might be well endured, and my wife expressed her
approbation as she looked on, busied with preparations for our
supper. In the meantime, the three youngest boys collected all

the pieces of wood we had thrown down from the tree, and a quantity of small wood, to dry in a heap, at a small distance from our fireplace.

Exhausted by the fatigues of the day, I threw myself on a bank, and my wife having seated herself near me, I thanked her for the tender care she was ever imposing on herself; and then I observed to her, that the many blessings we enjoyed led the thoughts naturally to the beneficent giver of them all; and tomorrow being a Sabbath-day, we would rest from work, in obedience to his command, and otherwise keep it holy. We then summoned our young family, and prepared them for the intended solemnity. I called their recollection to the nature of the Sabbath-day; to the gratitude due from us to an Almighty being, who had saved and comforted us in the hour of peril, and the duty of our prayers and acknowledgements. I informed them that, after performing with them the service of the church, I should read to them a paper I had composed for the occasion, and to which I had given the name of a parable of the Great King. The children expressed their approbation of what I had said, each in his own way, and we now assembled round our table to supper, my wife holding in her hand an earthen pot, which we had before observed upon the fire, and the contents of which we were all curious to be informed of. She took off the cover, and with a fork drew out of it the flamingo which Fritz had killed. She informed us that she had preferred dressing it this way, to roasting, because Ernest had assured her that it was an old bird, which would prove hard and tough, and had advised her to improve it by stewing. We rallied our glutton boy on this foible of his character, and his brothers gave him the name of the *cook*. We, however, had soon reason to know that he had conferred upon us an important obligation; for the bird which, roasted, we perhaps should not have been able to touch, now appeared excellent, and was eaten up to the very bones.

While we were enjoying our repast, the live flamingo stalked up to the place where we were sitting, in the midst of our flock of fowls, to receive his part of the repast. He had now become so tame that we had released him from the stake. He took his walks gravely from place to place, and looked perfectly contented with his company. His fine plumage was a most pleasing sight; while, on the other hand, the sportive tricks and the grimaces of our little monkey afforded the most agreeable spectacle imaginable. The little animal had become quite familiar with us; jumped from the shoulder of one to that of another; always caught adroitly the meat we threw him, and ate it in so pleasant a way as to make us laugh heartily.

The boys now, by my direction, lighted one of the heaps of wood. We tied long ropes loosely round the necks of our dogs, purposing to mount to our tent with the ends in my hand, that I might be able to let them loose upon the enemy at the first barking I should hear. Everyone was eager to retire to rest, and the signal for ascending the ladder was given. The three eldest boys were up in an instant; then came their mother's turn, who proceeded slowly and cautiously, and arrived in perfect safety. My own ascension was last, and the most difficult; for I carried little Francis on my back, and the end of the ladder had been loosened at the bottom, that I might be able to draw it up in the tent during the night: every step, therefore, was made with the greatest difficulty, in consequence of its swinging motion. At last, however, I got to the top, and, to the admiration of the boys, drew the ladder after me. It appeared to them that we were in one of the strong castles of the ancient cavaliers,* in which, when the drawbridge is raised, the inhabitants are secured from every attack of the enemy. Notwithstanding this apparent safety, I kept our guns in readiness for whatever event might require their use. We now abandoned ourselves to repose; our hearts experienced a full tranquillity; and the fatigue we had all undergone induced so sound a sleep, that daylight shone full in the front of our habitation before our eyes had opened.

12
The Sabbath and the Parable

ON awaking in the morning, we were all sensible of an unusual refreshment, and a new activity of mind. 'Well, young ones,' cried I, jocosely, 'you have learned, I see, how to sleep in a hammock: I heard not a single complaint all the night; no disputing about room from any one of you; all was still and tranquil.'—'Ah,' answered they, stretching and yawning as they spoke, 'we were so heartily fatigued yesterday, that it is no wonder we slept soundly.'

Father.—Here, then, my children, is another advantage derived from labour; that of procuring a sweet and peaceful sleep.

My wife could not help wishing we had some place like a church for the worship of the Sabbath, till I said to her, 'There is no place in the world that may not serve for a church, because we may entertain pious sentiments everywhere; and this majestic arch of Heaven, the immediate work of the Almighty, ought more effectually to raise the soul and touch the heart, than an edifice of stone made by the hand of man!'

We descended the ladder, and breakfasted on warm milk; we served the animals also with their meal, and then we all sat down on the tender grass; the boys full of impatient curiosity; their mother absorbed in silent reflection, her hands joined, and her eyes sometimes turned towards the sky; while I was penetrated with the most lively desire to impress upon the young minds of my children, a subject I considered of the highest importance for their well-being, both in this world and in that which is to come.

All now standing up, I repeated aloud the church service, which I knew by heart, and we sang some verses from the hundred-and-nineteenth psalm, which the boys had before learned; after which we sat down, and I began as follows:

'My dear children, there was once a Great King, whose kingdom was called The Country of Light and Reality, because the purest and softest light of the sun reigned there continually, which caused the inhabitants to be in a perpetual state of activity. On the furthest borders of this kingdom, northward, there was another country, which also belonged to the Great King, and the immense extent of which was unknown to all but himself. From time immemorial, a plan the most exact of this country had been preserved in the royal archives. This second kingdom was called The Kingdom of Obscurity or of Night, because everything in it was gloomy and inactive.

'In the most fertile and agreeable part of his empire of Reality, this Great King had a residence called the Heavenly City, in which he lived and kept his court; which was the most brilliant that the imagination can form an idea of. Millions of guards, and servants high in dignity, remained for ever round him, and a still larger number held themselves in readiness to receive his commands. The first of these were clothed in robes of cloth that was lighter than silk, and white as snow; for white, the image of purity, was the favourite colour of the Great King. Others of his attendants carried flaming swords in their hands, and their garments displayed the most brilliant colours of the rainbow; each of these stood in waiting to execute the will of the King, with the rapidity of lightning, on receiving from him the slightest sign. All were happy to be admitted into his presence; their faces shone with the mildest joy: there was but one heart and one soul among them; the sentiment of paternal concord so united these beings, that no envy or jealousy ever arose among them. The common centre of all their thoughts, and all their sentiments, was devotion to their sovereign: it would have been impossible either to see or converse with them, without desiring to obtain their friendship, and to partake their lot. Among the rest of the inhabitants of the Heavenly City, there were some less close in their attendance upon the Great King; but they were all virtuous, all happy, all had been enriched by the beneficence of the monarch, and, what is of still higher price, had received constant marks of his paternal care; for his subjects were all equal in his eyes, and he loved them and treated them as if they had been his children.

'The Great King had, besides the two kingdoms I have been describing, an uninhabited island of considerable extent: it was his wish to people and cultivate this island, for all within it was a kind of chaos: he destined it to be for some years the abode of such future citizens as he intended to receive finally into his residence, to which only such of his subjects were admitted, as had rendered themselves worthy by their conduct. This island was called Earthly Abode, he who should have passed some time in it, and by his virtues, his application to labour, and the cultivation of the land, should have rendered himself worthy of reward, was afterwards to be received into the Heavenly City, and made one of its happy inhabitants.

'To effect this end, the Great King caused a fleet to be equipped, which was to transport the new colonists to this island. These he chose from the kingdom of Night, and for his first gift bestowed upon them the enjoyment of light, and the view of the lovely face of nature, of which they had been deprived in their gloomy and unknown abode. It will easily be imagined that they arrived joyful and happy, at least they became so when they had been for a short time accustomed to the multitude of new objects which struck their feeble sight. The island was rich and fertile when cultivated. The beneficent King provided each individual who was disembarked upon it, with all the things he could want in the time he had fixed for their stay in it, and all the means for obtaining the certainty of being admitted as citizens of his magnificent abode, when they should leave the Earthly Island. All that was required to entitle them to this benefit was, that they should occupy themselves unceasingly in useful labour, and strictly obey the commands of the Great King, which he made known to them. He sent to them his only son, who addressed them from his father in the following terms:

' "My dear children, I have called you from the kingdom of Night and Insensibility to render you happy by the gifts of life, of sentiment, and of activity. But your happiness for the most part will depend upon yourselves. You will be happy if you wish to be so. If such is your sincere desire, you must never forget that I am your good King, your tender father; and you must faithfully fulfil my will in the cultivation of the country I have confided to your care. Each of you shall receive, on his arriving at the island, the portion of land which is intended for him; and my further commands respecting your conduct, will be soon communicated to you. I shall send you wise and learned men, who will explain to you my commands; and that you may of yourselves seek after the light necessary for your welfare, and remember my laws at every instant of your lives, it is my will that each father of a family shall keep an exact copy of them in his house, and read them daily to all the persons who belong to him. Further, each first day of the week I require to be devoted to my service. In each colony, all the people shall assemble together as brothers in one place, where shall be read and explained to them the laws contained in my archives. The rest of this day shall be employed in making serious reflections on the duties and destination of the colonists, and on the best means to fulfil the same: thus it shall be possible to all to receive instruction concerning the best manner and most effectual means of improving the land which has been confided to your care: thus you will each day learn to manure, to sow, to plant, to water, and cleanse the land from tares,* and from all evil weeds that may choke the good seed. On this same day, each of you may present his supplications, may tell me what he stands in need of, and what he desires to have, to forward the perfection of his labour: all these requests will appear before me, and I shall answer, by granting such as I shall think reasonable, and tending to a salutary end. If your heart tells you that the various benefits you enjoy, deserve your gratitude, and if you will testify it by doubling your activity, and by consecrating to me the day I have chosen for myself, I will take care that this day of rest, instead of being an injury to you, shall become a benefit, through the salutary repose of your body, and that of the animals given you to assist your labours, and who, as well as yourself, should enjoy repose on that day, to recruit their strength. Even the wild animals of the field, and of the forests, ought on that day to be protected from the pursuit of the hunter.

' "He who, in his Earthly Abode, shall most strictly have observed my will, who shall have best fulfilled the duties of a brother towards his fellow inhabitants, who shall have preserved his land in the best order, and shall show the largest produce from it, shall be recompensed for his deeds, and shall become an inhabitant of my magnificent residence in the Heavenly City. But the neglectful and the idle man, and the wicked man, who shall have spent their time in interrupting the useful labours of others, shall be condemned to pass their lives in slavery, or, according to the degree of their wickedness, shall be condemned to live in subterraneous mines, in the bowels of the earth.

' "From time to time, I shall send ships to fetch certain individuals from the Earthly Island, to reward or punish them, according as they have done well or ill; and as none will be warned beforehand, of the time of the coming of my messenger, it will be well for you to keep watch, that you may be ready to perform the voyage, and worthy to be received into the Heavenly City. It will not be permitted for anyone to pass by stealth on board the ship, and leave his abode without my orders; for such a one shall be severely punished. I shall have the most certain knowledge of all that passes in the Earthly

Island, and no one will be able to deceive me. A magical mirror will at all times show me the actions of each individual in the island, and you shall be judged according to your most secret thoughts and actions."

'All the colonists were well satisfied with the discourse of the Great King, and made him the most sacred promises. After a short time allowed for repose from the fatigue of the voyage, a portion of land, and the proper instruments for labour, were distributed to each of the strangers. They received also seeds, and useful plants, and young trees, for producing them refreshing fruits. Each was then left at liberty to act as he pleased, and increase the value of what was confided to his care. But what happened? After some time, each followed the suggestions of his fancy: one planted his land with arbours, flowery banks, and sweet-smelling shrubs; all pleasing to the sight, but which brought forth nothing. Another planted wild apple-trees, instead of the good fruit, as the Great King had commanded; contenting himself with giving high-sounding names to the worthless fruit he had caused to be brought forth. A third had indeed sown good grain; but not knowing how to distinguish the tares that grew up along with it, he pulled up the good plants before they were mature, and left only the tares in his ground. But the greater part let their land lie fallow, and bestowed no labour upon it, having spoiled their implements, or lost their seed, either from negligence or idleness, or liking better to amuse themselves than to labour; many of them had wilfully misunderstood the instructions of the Great King, and sought by subtle turns to change their meaning.

'Few, very few, worked with diligence and courage, and seeking to improve their land, according to the orders they had received. The great fault of these was, that they would not believe what the Great King had sent to tell them. All the fathers of families had indeed a copy of the laws of the Sovereign, but most of them omitted to read in the book: some saying that it was useless to read it, for they knew it by heart, while they never employed their thoughts upon it. Others pretended that these laws were good for times past, but were no longer beneficial for the present state of the country. Some had even the audacity to assert, that it contained many inexplicable contradictions; that the laws it prescribed were merely

supposed or falsified, and that they had therefore a right to
deviate from them. Others among them maintained, that the
magical mirror was a mere fable; that the King was of too
merciful a nature to keep galleys; that there was no such place
as the subterranean mines; and that all would at last enter the
Heavenly City. From habit they continued to celebrate the first
day of the week, but by far the smallest part of it was consecrated to the honour of the Great King. Great numbers of
them dispensed with going to the general assembly, either from
idleness, or to employ themselves in occupations which had
been expressly forbidden. By far the greater part of the people
considered this day of repose as intended for pleasure, and
thought of nothing but adorning and amusing themselves as
soon as daylight appeared. There were only then a small
number of persons who kept the day according to the decree;
and even of those who frequented the assembly, many had their
thoughts absent, or were sleepy, or engaged in forming empty
projects, instead of listening to the words which fell from the
lips of the minister of the Sovereign. The Great King, however,
observed unalterably the laws he had laid down and announced, respecting them. From time to time, some frigates
appeared on their coasts, each bearing the name of some
disastrous malady; and these were followed by a large ship of
the line, named the Grave, on board of which, the admiral,
whose name was Death, caused his flag of two colours, green
and black, to be constantly floating in the air. He showed the
colonists, according to the situation in which he found them,
either the smiling colour of Hope, or the gloomy colour of
Despair.

'This fleet always arrived without being announced, and
seldom gave any pleasure to the inhabitants. The admiral sent
the captains of his frigates, to seize the persons he was ordered
to bring back with him. Many who had not the smallest inclination, were suddenly embarked, while others, who had prepared
everything for the harvest, and whose land was in the best
condition, were also seized. But these last took their departure
cheerfully, and without alarm; well knowing that nothing but
happiness awaited them. It was those who were conscious they
had neglected to cultivate their land, who felt the most regret.
It was even necessary to employ force, to bring them under

subjection. When the fleet was ready for departure, the admiral sailed for the port of the Royal Residence; and the Great King, who was present on their arrival, executed with strict justice both the rewards and punishments which had been promised to them. All the excuses alleged by those who had been idle, were of no avail. They were sent to the mines and to the galleys, while those who had obeyed the Great King, and well cultivated their land, were admitted into the Heavenly City, clothed in robes of brilliant colours, one exceeding the other according to the degree of merit.'—Here, my dear children, ends my parable. May you have thoroughly understood its meaning, and may you reap the advantage it is capable of affording you! Make it the subject of your reflections the whole of this day. You, Fritz, I see, are thoughtful: tell me what struck you most in my narration.

Fritz.—The goodness of the Great King, and the ingratitude of the colonists, father.

Father.—And you, Ernest, what is your thought?

Ernest.—For my part, I think they were great fools to have made so bad a calculation. What did they get by conducting themselves as they did? With a little pains they might have passed a very agreeable sort of life in the island, and would have been sure of going afterwards to the Heavenly City.

Jack.—To the mines, gentlemen, away with you! You have well deserved it.

Francis.—For my part, I should have liked best to have lived with the men who were dressed in the colours of the rainbow. How beautiful they must have looked!

Father.—This is well, my boys. I perceive that each of you, according to his age and character, has seized the meaning of my parable. I have by this image endeavoured to represent to you the conduct of God towards man, and that of man towards God: let us see now if you have completely seized the sense.—I then put different questions to them, and explained what they had not perfectly comprehended; and after a short review of the principal parts of my discourse, I concluded by a moral application.

'Human creatures', said I, 'are the colonists of God; we are required to perform the business of probation for a certain period, and, sooner or later, are destined to be taken hence.

Our final destination is Heaven, and a perfect happiness with the spirits of just men made perfect, and in the presence of the bountiful Father of us all. The piece of land entrusted to each is the soul; and according as he cultivates and ennobles it, or neglects or depraves it, will be his future reward or punishment. At present, dear children, that you know the true sense of my parable, each of you should make the application of it according to his own consciousness. You, Fritz, should think of the subjects who planted the wild apples, and wished to make them pass for sweet savoury fruit of a superior kind. These represent persons who make a parade of the natural virtues belonging to their character, and which are consequently exercised without any trouble to themselves; such as courage, strength, etc.; who prefer them to more essential qualities acquired by others, with sacrifices and labour to themselves; and who, full of presumption and arrogance, consider themselves as irreproachable, because nature has given them personal courage, and bodily strength, and a certain skill in the use of these qualities.

'You, Ernest, should think of the subjects of the Great King, who cultivated their land so as to produce arbours, flowery banks, and sweet-smelling shrubs, and such productions in general as would please the eye, but which produced no fruit. These are they, who give their whole attention to the acquiring unfruitful knowledge, sciences, etc. and consider with a sort of contempt the things more immediately required for the conduct of life; who exert themselves solely for the understanding, and neglect the heart; whose principal aim it is, to obtain self-indulgences, and who neglect what is useful in society.

'You, Jack and you, Francis, should apply to yourselves the case of those men who let their land lie fallow, or, in their thoughtlessness, mistook the grain, and sowed tares instead of wheat. These are the neglectful subjects, who neither think nor learn, but give to the winds what is taught them, or entirely forget instruction; who reject virtuous sentiments, and let the bad ones grow in their hearts. But for ourselves, one and all, we will adopt the model of the good and zealous labourers; and should our exertions be a little painful, we shall think of the reward which awaits us, when we shall have adorned our souls with all that is good, just, and praiseworthy. Thus, when death, which cannot fail to come at last, shall summon us, we may follow him with joy to the throne of the Good and Great King, to hear him pronounce these sweet and consoling words: "O good and faithful servant! thou hast been tried, and found faithful in many things; enter thou into the joy of thy Lord." —With these words, and a short prayer of benediction, I concluded the solemnity of our Sunday; and I had the satisfaction of seeing, that my four sons had not only listened attentively, but that they were struck with the application I had made to each of them.—They remained for a short time reflecting in silence. Jack was the first to break it:—'You have explained to us every part of the parable, father, except the copy of the laws of the Great King, which was to be kept and read in every family: have you one of these copies? for you never read it to us.'

Father.—My children, this copy is the Holy Bible, which contains all God's laws, and which we ought constantly to study. I cannot forgive myself for not having thought of bringing it from the vessel. Should we not be able to go another voyage, we shall forever be deprived of this divine doctrine.

My Wife.—Have you then forgot my enchanted bag, which I have promised shall furnish everything you can desire? You wish for a Bible. In a minute I will put one into your hands; and heartily do I rejoice in having the power to procure you so great a satisfaction.

Father.—Most excellent of women! Give me then the inestimable book, these laws of the Great King, which from this moment we will take for the rule of our lives. She opened her bag, and with joy I received from her the book of life. I opened it, and read some passages from it to my family. In this solitude, in which for so long a time we had heard only our own thoughts expressed in an appropriate language, we were singularly affected with the voice from Heaven, which now seemed to address us: we felt forcibly that, notwithstanding our exile, we were still connected with the community of mankind by the invisible tie of the same religion, and the same Father: we were forever numbered among the children of God, to whom he enjoins laws, and on whom he bestows his care, no less in a desert, than in an immense capital. I explained with the utmost care what I read to them, and I gave the book in turn to each of the boys, that they might have the pleasure of reading for themselves. I chose in preference, such passages as were applicable to our circumstances. We then raised our hearts to God, to thank him for so signal a benefit as the preservation of our Bible. My young folks still remained thoughtful and serious; but by and by the gaiety natural to their age prevailed, and each slipped away to seek the recreation he liked best.

The next morning, the boys assembled round me with a petition that I would show them how to use arrows. We accordingly sat down on the grass; I took out my knife, and, with the remains of a bamboo cane, began to make a bow. I was well satisfied to observe them one and all take a fancy to shooting with an arrow, having been desirous to accustom them to this exercise, which constituted the principal defence of the warriors of old, and might possibly become our only means of protection and subsistence: our provision of powder must at last be exhausted; we might even, from moment to moment, be deprived of it by accident; it therefore was of the utmost importance to us, to acquire some other means of killing animals, or attacking our enemies. The Caribbees, I recollected, were taught at a very tender age, to strike an object at the distance of 30 or 40 steps; they hit the smallest birds perched on the top of the tallest trees. Why then should it not be possible for my boys to learn to do the same?

While I was silently reflecting on the subject, employed in finishing a bow, Ernest, who had been observing me for some time, slipped suddenly away; and Fritz coming up at the same moment, with the wetted skin of the tiger cat in his hand, I paid no attention to the circumstance. I began my instructions to my eldest boy respecting the trade of a tanner. I told him the method of getting rid of the fat of the skin, by rubbing it over with sand, and placing it in running water till it had no longer any appearance of flesh, or any smell; next to rub it with soft butter, to make it supple, and then to stretch the skin in different directions; and also to make use of some eggs in the operation, if his mother could spare them. You will not at first produce such excellent workmanship as I have seen of this kind from England; but with a little patience, regretting neither your time nor your labour, you will have completed some decent-looking cases, which will give you the more pleasure, from being the work of your own hands. When your skin shall have thus been prepared, cut certain small cylinders of wood of the size and length required; scoop these cylinders hollow, so as to form a convenient case for a knife, a fork, or a spoon; then stretch your softened skin upon the surface of the cylinders, in such a manner, that the skin may reach a little beyond the extremity of the wood, and close at the top; you have nothing more to do, than to let the skin cling to, and dry upon these moulds.

At this moment we heard the firing of a gun, which proceeded from our tent in the tree, and two birds at the same time fell dead at our feet. We were at once surprised and alarmed, and all eyes were turned upwards to the place. There we saw Ernest standing outside the tent, a gun in his hand, and heard him triumphantly exclaiming, 'Catch them! catch them there! I have hit them; and you see I did not run away for nothing.' He descended the ladder joyfully, and ran with Francis to take up the two birds; while Fritz and Jack mounted to our castle, hoping to meet with the same luck.

One of the dead birds proved to be a sort of thrush, and the other was a very small kind of pigeon, which in the Antilles is called an ortolan: they are very fat, and of a delicious taste. We now observed, for the first time, that the wild figs began to ripen, and that they attracted these birds. I foresaw, in consequence, that we were about to have our table furnished with a dish which even a nobleman might envy us. I gave the boys leave to kill as many of them as they liked. I knew that, half roasted and put into barrels with melted butter thrown over them, they would keep a long time, and might prove an excellent resource. My wife set about stripping off the feathers of the birds, to dress them for our dinner. I seated myself by her side, and proceeded in my work of arrow-making.

Thus finished another day. Supper ended, and prayers said, we ascended the ladder in procession; and each got into his hammock to taste the sweets of a tranquil sleep.

13
Conversation, a Walk, and Important Discoveries

JACK had finished the trial of his arrows: they flew to admiration; and he practised his new art incessantly. Little Francis waited with impatience for the moment when he should try also, and followed with his eyes every stroke I made. But when I had finished my bow, and prepared some little arrows for him, I must next under-take to make him a quiver; I took some bark from the branch of a tree, which came off in a round form; and folding the edges over each other, I stuck them together with some glue produced from our soup-cakes. I next stuck on a round piece to serve for the bottom; and then tied to it a loop of string which I hung round his neck. He put his arrows into it; and, quite happy, took his bow in his hand, and ran to try his skill by the side of his brother. Fritz had also cleaned and prepared his materials for the cases, when his mother summoned us to dinner. We cheerfully placed ourselves under the shade of our tree, round the table I had manufactured. At the end of the repast, I made the following proposition to the boys, which I was sure would give them pleasure.

'What think you, my good friends,' said I, 'of giving a name to the place of our abode, and to the different parts of the country which are known to us? I do not mean a general name to the whole island, but to the objects we are most concerned with: this will make us better understand each other, when conversing about them; and also present to us the soothing illusion, of inhabiting a country already known and peopled.'

They all exclaimed, joyfully, that the idea was excellent.

Jack.—Oh! pray, father, let us invent some very long names, and that are very difficult to be pronounced. I should be glad that those who shall read about us, should be a little puzzled to remember the names of the places and things that belonged to us. What pains has it not cost me to remember their *Monomotapa*, their *Zanguebar*, their *Coromandel*, and many other still more difficult. Ah! now we shall take our revenge of them.

Father.—This would be well, if it were probable that our history in this country, and the names we shall have bestowed, were likely to be objects of public curiosity; but in the meanwhile you forget that our own organs will be fatigued, by frequently pronouncing such barbarous words as you propose.

Jack.—How shall we manage, then? What pretty names can we find?

Father.—We will do as all sorts of nations have done before us. We will call the places by different words from our own language, that shall express some particular circumstance with which we have been concerned.

Jack.—Well, so we will: I shall like this still better. Where shall we begin?

Father.—We shall naturally begin with the bay by which we entered this country. What shall we call it? What say you, Fritz? You must speak first, for you are the eldest.

Fritz.—Let us call it *Oyster Bay*: you remember what quantities of oysters we found in it.

Jack.—Oh, no! let it rather be called *Lobster Bay*: for you cannot have forgot what a large one it was that caught hold of my leg, and which I carried home to you.

Ernest.—Why then we may as well call it the *Bay of Tears*, for you must remember that you blubbered loud enough for all of us to hear you.

My Wife.—My advice would be that, out of gratitude to God, who conducted us hither in safety, we ought to call it *Providence Bay*, or the *Bay of Safety*.

Father.—These words are both appropriate and sonorous, and please me extremely. But what name shall we give to the spot where we first set up our tent?

Fritz.—Let us call it simply *Tent House*.

Father.—That will do very well. And the little islet at the entrance of *Providence Bay*, in which we found so many planks and beams that enabled us to make our bridge, how shall it be named?

Ernest.—It may be called *Sea-Gull Island*, or *Shark Island*; for it was here we saw those animals.

Father.—I am for the last of these names, *Shark Island*: for it was the shark that was the cause of the sea-gulls being there; and thus we shall also have a means of commemorating the courage and the triumph of Fritz, who killed the monster.

Jack.—For the same reason, we will call the marsh, in which you cut the canes for our arrows, *Flamingo Marsh*.

Father.—Quite right, I think; and the plain, through which we passed on our way to this place, Porcupine Field, in memory of your skilful encounter with the animal. But now comes the great question,—What name shall we give to our present abode?

Ernest.—It ought to be called, simply, *Tree Castle.*

Fritz.—No, no, that will not do at all; that is the same as if, when we wanted to name a town, we called it *The Town.* Let us invent a more noble name.

Jack.—Yes, so we will. I say *Fig Town.*

Fritz.—Ha, ha, ha! a noble name, it must be confessed! Let us call it *The Eagle's Nest,* which I am sure has a much better sound. Besides, our habitation in the tree, is really much more like a nest, than a town, and the eagle cannot but ennoble it, since he is the king of birds.

Father.—Will you let me decide the question for you? I think our abode should be called *The Falcon's Nest;* for, you are not arrived at the dignity of eagles, but are, too truly, poor simple birds of prey; and like the falcon, you also are, I trust, obedient, docile, active, and courageous. Ernest can have no objection to this; for, as he knows, falcons make their nests in large trees.

All exclaimed, clapping their hands, 'Yes, yes, we will have it *The Falcon's Nest!* the sound is quite chivalrous; so health to *Falcon's Nest Castle!*' cried they, all looking up to the tree, and making low bows. I poured out a small quantity of sweet wine, and presented it to each, to solemnize our baptism.—'Now then,' said I, 'for the promontory, where Fritz and I in vain wearied our eyes, in search of our companions of the vessel? I think it may properly be called *Cape Disappointment.*'

All.—Yes, this is excellent. And the river with the bridge——

Father.—If you wish to commemorate one of the greatest events of our history, it ought to be called *The Jackall's River;* for these animals crossed it when they came and attacked us, and it was there that one of them was killed. The bridge I should name *Family Bridge,* because we were all employed in its construction, and all crossed it together in our way to this place. Let me ask you all, if it will not be a great pleasure to converse about the country we inhabit, now that we have instituted names as if everything belonged to us?

Ernest.—It will be just as if we had farms and country houses, all dependent upon our castle.

Francis.—It is the same as if we were kings.

My Wife.—And the queen-mother is not without hope, that her little slips of majesty will conduct themselves mercifully towards their subjects, the birds, the agoutis, the geese, and the flamingos; the——What more shall I say? for I do not know the family name of all your vassals. Let me therefore end, by hoping that you will not depopulate your kingdom.

Fritz.—No, mother, we will take care of that. We will endeavour to extirpate only those among our subjects who are wicked.

In this pleasing kind of chat, the time of dinner passed agreeably away. We settled the basis of a geography of this our new country; and amused ourselves with saying, that it must go by the first post to Europe.

As the evening advanced, and the intense heat of the day began to diminish, I invited all my family to take a walk. 'Leave your work for this time, my boys,' said I, 'and let us make a short excursion; let us seek, in the beautiful face of nature, the traces of the wisdom and goodness of the Creator. Which way shall we direct our steps?'

Fritz.—Let us go to Tent House, father; we are in want of powder and shot for the little consumers of our figs; nor must we miss our dinner for tomorrow, or forget that we are to secure a supply for winter.

My Wife.—I too vote for Tent House; my butter is nearly gone, for Fritz took an unreasonable share for his new trade of tanning; also, I have never failed to observe, that those who most zealously preach a life of frugality and economy, are at least as well satisfied as the rest, when I present them with a savoury dinner.

Ernest.—If we go to Tent House, let us try to bring away some of the geese and ducks with us: they will look very well swimming about in our stream here, by Falcon's Nest.

Jack.—I will undertake to catch them, if any one will help to bring them home.

Francis.—And I will catch my handkerchief full of lobsters in the Jackal's River, and we will put them into Falcon's Stream, where, no doubt, they will thrive to admiration.

Father.—You really all of you assign such good reasons, that I see I must yield to them. To Tent House, then, we will go; but we will not take our accustomed road along the sea-shore, but rather vary our pleasure, by trying to explore some other way. We will keep along our own little stream as far as the wall of rocks: it will be easy for us to cross it, by jumping from stone to stone, and so to get to Tent House: we will return with our provisions by the road of Family Bridge, and along the sea-shore. This new route may possibly furnish some additional discoveries.

My idea was highly applauded, and all was soon arranged for our setting out. Fritz was adorned with his fine tiger-cat-made belt. Jack had his belt also armed with two pistols, round his waist. Each carried a gun and a game bag; even little Francis had his bow in his hand, and his quiver on his shoulder. Their mother was the only person not burdened with a gun; but she carried her large butter-pot, to fill it at our large store-house. Turk marched before us, with his coat of mail studded with spikes, but it was apparent that he felt intimidated and ill at ease; his step was therefore slow and quiet.

Our route along the stream was at first extremely agreeable, being sheltered by the shade of large trees, while the ground under our feet was a short and soft kind of grass. To prolong the pleasure of our walk, we proceeded slowly, amusing ourselves with looking about us to the right and left; the eldest boys made frequent escapes on before, so that we sometimes lost sight of them. In this manner we reached the end of the wood; but the country now appearing to be less open, we thought it would be prudent to bring our whole company together. On looking forward, we saw the boys approaching us full gallop, and this time, for a wonder, the grave Ernest was first. He reached me panting for breath, and so full of joy and eagerness, that he could not pronounce a single word distinctly; but he held out his hand, which contained three little balls of a light green colour.

'We have found a prize, indeed, father,' cried he at last, when he had recovered his voice; 'we have found some potato seed!'

'What say you? Potato seed?' inquired I joyfully; 'Have you really been so fortunate? Come near, every one of you, and let me look at your little balls'; for I scarcely dared believe in so happy an event, as the discovery of a plant which would place us forever beyond the reach of hunger, and even of apprehension.

We all hastened to the place where these tubercles had been gathered, and, with extreme joy we found there a large plantation of potato plants; a number of them were covered with their lilac and yellow blossoms, the sight of which conveyed more pleasure to our hearts than if they had been the most fragrant roses. Jack bawled out, jumping for joy, 'They are really potatoes! and though it was not I who discovered them, at least it shall be I who will dig them up.' Saying this, he knelt down and began to scratch them up from the earth with his hands; the rest of us, unwilling to be idle spectators, set to work also: with our knives and sticks we soon procured a sufficient number to fill our bags and our pockets.

'There are', observed I, 'different kinds of vegetables, more succulent and more delicate than the potato; but it is this plain sustenance, that can be eaten for the longest time together, without satiety: accordingly, food of this nature, such as bread, rice, potatoes, obtains on the whole, a preference over provisions possessing a higher flavour. Can you tell me boys, the reason of this?'

Ernest.—I know; it is because they are more wholesome.

Jack.—And because they occasion no disgust: I could eat potatoes every day of my life, without being tired of them.

Father.—All you say is true; in future they will serve us for bread, and often indeed for our whole dinner. But let us for the present dismiss the subject of our unexpected good fortune, and resume our expedition.

14
Continuation of the Preceding Chapter; and More Discoveries

CONVERSING on different subjects, we reached the long chain of rocks, over which our pretty Falcon Stream made its escape in a cascade, delighting at once the eye and the ear in its progress. We thus reached Jackal's River, and from thence to Tent House, having with difficulty pushed through the high grass which presented itself. Our fatigue, however, was relieved by the uncommon beauty of the scenery around: on the right hand was a boundless sea; on the left, the island, with the bay by which it was accessible, and the chain of rocks, forming altogether an assemblage of the picturesque, equal to what the liveliest fancy could desire. We distinguished different families of grasses, many of them of the thorn-leaved species, and stronger than those cultivated in the greenhouses of Europe. There was also in abundance the Indian fig, with its large broad leaf; aloes of different forms and colours; the superb prickly candle, or cactus, bearing straight stalks, taller than a man, and crowned with long straight branches, forming a sort of star. The broad plantain spread along the rocks its innumerable boughs twisted with each other, hanging down perpendicularly, and ornamented with flowers, which grew in large tufts, and were of the brightest rose-colour, while that which pleased us best, and which was found there in great abundance, was the king of fruits, both for figure and relish, the crowned pine-apple, of which we all partook with avidity.

Soon after, I was fortunate enough to discover among the multitude of plants which grew either at the foot or in the clefts of the rock, the karata* (the Bromelia Karata of Linnæus*), many of which were now in blossom. Travellers have given so perfect a description of this plant, that it was impossible I should mistake it. I pointed out to the boys the immense size of its leaves, hollowed in the middle like a saucer, in which rain is for a long time preserved; also, its beautiful red flowers. As I was acquainted with the properties of this useful plant, the pith of which is used as tinder by the Negroes, who also make a strong kind of thread from the fibres of its leaves, I was not less satisfied with the discovery than I had been with that of the potatoes. Wishing to exhibit one of its uses to my children, I desired Ernest to take out my flint and steel.

I took a dried stalk of the tree, stripped off the bark, and there appeared a kind of dry spongy substance, which I laid upon the flint; and then striking it with a steel, it instantly caught fire. The boys looked on with astonishment, and soon began to caper about, exclaiming: 'Long live the tinder-tree!'

'Here, then,' said I, 'we have an article of greater usefulness than if it served merely to gratify the appetite. Your mother will next inform us what materials she will use for sewing your clothes, when her provision of thread from the enchanted bag is exhausted.'

My Wife.—I have long been uneasy upon this very subject, and would willingly exchange our greatest luxury for some hemp or flax.

Father.—And your wish shall be accomplished. If you examine, you will find some excellent thread under the leaves of this extraordinary plant, where all-provident nature has placed a store-house of this valuable article, though the lengths of thread will be found not longer than the leaf. I accordingly drew out of one of the leaves a strong piece of thread of a red colour, which I gave to my wife. 'How fortunate it is for us,' said she, 'that you have had the habit of reading and of study! None of us would have had a thought about this plant, or have conceived that it could be of any use:—but will it not be difficult to draw out the lengths of thread through the prickles that surround them?'

Father.—Not in the least; we shall put the leaves to dry, either in the sun, or by a gentle fire. The useless part of the leaf will then separate by being beaten, and the mass of thread will remain.

Fritz.—I see clearly, father, that we ought not to trust to appearances; but one may, I suppose, assert that there are no good qualities in the prickly plants, which are growing here in all directions, and wounding the persons who go near them: of what use can they possibly be?

Father.—The greatest part of these possess medicinal virtues; great use is made in pharmacy of the aloe,* which produces such abundance of beautiful flowers; in greenhouses in Europe, some have been seen to bear more than 3,000 blossoms. At Carlsbad, upon the estates of Count de Limbourg, there was an aloe-tree 26 feet in height; it had 28 branches, which branches bore more than 3,000 blossoms in the space of a month. At Paris, at Leyden, in Denmark, there have been also seen some exceedingly curious specimens of this tree; many of them are full of a resinous sort of sap, of which valuable gums are made. But look, here, too, is the Indian fig, or prickly pear, a vegetable of no common interest; it grows in the poorest soils, and, as you see, upon the rocks; the poorer the soil, the more luxuriant and succulent its leaves; I should be tempted to believe that it was nourished by the air rather than by the earth. It is also called the racket-tree, from the resemblance of its long, thick, flat leaves to that well-known instrument. The plant bears a kind of fig, which is said to be sweet and palatable when ripened in its native sun, and it is a salutary and refreshing food. This, then, is another plant of great utility. I next instructed them how to gather this prickly fruit without injury to their fingers. I threw up a stone, and brought down a fig, which I caught upon my hat; I cut off one end, and was thus enabled to hold it on a knife while I peeled off the skin. I then resigned it to the curiosity of my young companions.

The novelty, rather than the taste, of the fruit, made them think it excellent: they all found means to gather some of the figs, and each was busied in inventing the best method of taking off the skins. In the meantime, I perceived Ernest holding a fig upon the end of his knife, turning it about in all directions, and bringing it close to his eye with a look of curious enquiry.—'I wish I could know', said at length our young observer, 'what little animals these are in the fig, which feed so eagerly upon it, and are of quite a scarlet colour.'

Father.—Ha, ha! this too will perhaps turn out a new discovery, and an additional source of usefulness. Let me look at your fig; I will wager that it is the insect called the cochineal.*

Jack.—The cochineal! what a droll name! What is the cochineal, father?

Father.—It is an insect of the kind called *suckers*, or *kermes.* He feeds upon the Indian fig, which, no doubt, is the cause of his beautiful colour, so much esteemed in dyeing; for nothing else produces so fine a scarlet. In America, they stretch pieces of linen under the branches, and then shake the tree; and when the insects have fallen in great numbers, the ends of the linen are folded together to enclose them; the insects are sprinkled with vinegar or cold water, and then dried, and sent to Europe, where a high price is paid for them. But I have not yet mentioned a still superior usefulness, peculiar to the Indian fig-tree:—what if I should assert that it can be used as a protection to man?

Fritz.—As a protection to man! Why, how can that be, father?

Father.—It is well adapted for enclosing the dwellings of man; for you see, that besides the prickles, there is a large thorn at each of the knots in the stalk, well calculated for repelling the attacks of animals or men. This, then, you see, is a third usefulness the Indian fig-tree can boast, and of which I was not at first aware. You must perceive of what importance these enclosures are; and the rather, as they are made with so little trouble; for if you plant only one of its leaves in the ground, it immediately takes root, and grows with astonishing rapidity.

Jack, the thoughtless, here cried out, that with the assistance of a knife, or even a stick, it would be easy to get over such a hedge; and he began to cut down with his clasp-knife a pretty large plant, striking to right and left with all his might, till one of the divided leaves fell with such violence on his leg, that the thorns struck into the flesh, and Jack roared out piteously, and quickly sat down to draw them out. I could not, as I assisted him, refrain from laughing a little at his adventure. I observed to him, how difficult it must be for savages, who wear no clothes, to force such a barrier as they formed; and for this once, I had the pleasure of convincing him.

Ernest.—Ah, father, do let us make a hedge of these plants round our tree; we shall then have no further occasion for fires to preserve us from wild beasts, or even from the savages, who may arrive in their canoes, as they did on Robinson Crusoe's Island.

Fritz.—And we could, then, easily gather the cochineal, and try to make the same beautiful scarlet colour.

Father.—We shall have time enough for many things, my dear children; but for the present, it is sufficient to prove to you, that God has not made anything to be wholly useless; and that it is the duty of man, on whom he has bestowed the gifts of wisdom and intelligence, to employ those faculties in discovering the utility of the different productions he has allowed to exist.

Jack.—For my part, I have done with the Indian fig-tree, its fruit, its cochineal, and its ugly thorns, and I will never go near it again.

Father.—If the plant could speak, it would most likely say, That little boy shall not come near me any more. Without any reason, or any necessity, but purely out of contradiction to his father, he attacks and destroys me; me, who would have done him service, if he would but have treated me with kindness, and have been careful in coming near me.—And now, Jack, if your leg is still painful, apply a leaf of the karata to it, for I recollect that the plant possesses the property of curing wounds. He accordingly took my advice, and in a few minutes was able to join us on our road to Tent House.

'Now then,' said Ernest, 'I have had an opportunity of learning the valuable properties of the karata tree, and of the Indian fig-tree; but I wish I could also be informed what those tall plants are which look like sticks covered with thorns, that I perceive everywhere about us; I see neither fruit nor insects on them: of what use, then, father, do you think they can be?'

Father.—It is not in my power to explain to you the uses of all the plants in the world; I presume that many exist which have no other than that of contributing to the sustenance of different kinds of animals; and, as I have already told you, it is for man, by his superior intelligence, to discover those that can be applied to his own use. Many possess medicinal qualities of which I am ignorant, and which will become better known as the world advances in age. The plant you speak of is perhaps the prickly candle, described by Bruce,* in his *Travels to Abyssinia*, and of which he gives a drawing; the only difference that I perceive being the size. 'They serve', says he, 'for food to the elephant and the rhinoceros; the first with his strong teeth, or his trunk, and the latter with his horn, lays hold of this seeming stick, and rips it up from one end to the other; they then devour the pith, and sometimes the rind.'

Ernest.—The palate of these animals must surely be made of iron, to be able to chew such a thorny substance without injury.

Fritz.—Why so? Camels and asses are very fond of thistles, and appear to digest them extremely well. It is probable, therefore, that the stomach of these animals is so formed, that these prickly substances occasion in it only an agreeable excitation, favourable to their appetite and their digestion.

Father.—Your idea is not a bad one; and if it be not true, it is at least probable.

Fritz.—Will you tell me, father, the precise difference between *true* and *probable*?

Father.—Your question is one of those which have occupied the attention of philosophers for countless years, and would be too tedious for discussion at this moment; I will, however, endeavour to make my answer such, as to be of use to you, in the science of logic, or the art of reasoning. Let us see if you will understand it—What we call *true*, is that which cannot in any way be contradicted, and which exactly agrees, in every point, with the idea we conceive of a certain object, or as it really exists before our eyes: for example, when I make an impression with my seal on some warm wax, it is absolutely *true* that the figure impressed on the wax, is the same as that on the seal. A thing is *probable*, when we have a variety of motives for believing it true, without, however, being able to bring any proof. Again, we call *false*, that which is in positive contradiction to all our notions, our reason, and our experience. Is it *true*, *probable*, or *false*, that a man can fly up into the air?

All.—It is false, absolutely false.

Father.—How so?

Jack.—Because the thing is impossible.

Father.—Very well, my young philosopher, and why is the thing impossible?

Jack.—Because it is not possible.

Father.—Ha, ha, ha! here is a pretty round of *possible* and *impossible. It is false because no such thing can be done, and no such thing can be done, because it is not possible.* Presently you will tell me that it is impossible because it is false. Try again, my lads, we must have some better reasons. What say you, Ernest?

Ernest.—I say, that the thing cannot be done, because it is not in the nature of man to fly; that having no wings, he is not formed for flying.

Father.—Well, but if someone should assert, that a man is able to make a machine, by the assistance of which he can raise and support himself in the air without wings, and without the machine resting upon anything; would this be *probable* or *improbable*? What think you Fritz?

Fritz.—I think I should have said *improbable*, if I had not known that people have accomplished what you describe, by the invention of balloons.

Father.—And why should you have thought it *improbable*?

Fritz.—Because man is, in his nature, heavier than the air; and I should have supposed, that a machine of whatever kind, instead of diminishing, would only add to his weight.

Father.—Very well reasoned. But you would be told that this machine is of large dimensions, and composed of a close, light kind of silk, and that it is filled with air chemically prepared, which being much lighter than atmospheric air, tends perpetually to ascend, and supports the man in the air, as bladders support you upon the water. Do you understand all this, my boy? And what have you to say in answer?

Fritz.—Yes, father, I understand it; and I perceive how it might be probable, that since man has discovered a means to be sustained upon water, he might also find the means to raise and sustain himself in the air.

Father.—And when a multitude of persons of veracity, and of different ages, shall declare that with their own eyes they saw a balloon, to which a parachute* was fastened filled with men, and that all mounted in the air together, and disappeared above the clouds; should you still maintain that it is false that a man can fly?

All.—No, to be sure, we should say that it is quite true that he can fly.

Father.—And yet you all said, but a minute ago, that it was absolutely false.

Fritz.—Ah! but we said that, father, of a man by himself, independently of any machine he might construct; for though nature has refused him wings, she has not failed to bestow on him an inventive mind, which more than compensates for that deficiency.

Father.—Your observation is perfectly just, and I hope you will not fail to profit by it. With the aid of his intelligence, and his reason, there is scarcely anything which man cannot attain to. But to return to our example: you will find in it the definition of the words which you ask me about: it is *false*, that a man of himself can fly; it is *probable*, that by the aid of a machine of his own invention he may be enabled to mount and sustain himself in the air; and it is also *absolutely true*, that this has been effected by man, though without his having yet found a certain means of guiding these factitious wings; a defect which, in a great measure, renders his discovery useless.

At this point of our discourse, we reached Jackal's River, which we crossed, stepping with great care from stone to stone, and shortly arrived at our old habitation, where we found everything as we had left it; and each went in pursuit of what he intended to take away. Fritz loaded himself with powder and shot: I and my wife and Francis employed ourselves in filling our pot with butter, the carrying of which on our return it was agreed was to fall on me. Ernest and Jack looked about for the geese and ducks; but as they were become somewhat savage, the boys could not succeed in catching one of them. The idea then occurred to Ernest, of taking a small bit of cheese, and tying it to the end of a piece of string, and holding it to float in the water. The voracious animals hastened eagerly to seize it. In this way, Ernest drew them towards him, one by one, with the cheese in its mouth, till he had caught the whole: each bird was then tied in a pocket handkerchief, leaving the head at liberty, and fastened one to each game-bag, so that all had a share in carrying them.

We had a thought of taking back a provision of salt; but the sacks being occupied with potatoes, we could only throw a small quantity loose into one of them, to lie between the potatoes: in this way we secured a tolerable supply.

We now set out loaded on our return. The ducks and geese, with their heads and necks stretching out at our shoulders, cackling with all their might, gave us a truly singular and ludicrous appearance, and we could not help laughing immoderately as we passed the bridge, one after another, accoutred in so strange a fashion. Our mutual jokes, and the general good humour which prevailed, served to shorten the length of the walk, and none complained of fatigue, till seated under our tree at Falcon's Stream. My wife now prepared to console us, by putting some of the potatoes which we so eagerly desired to taste, immediately on the fire. She next milked the cow and the goat, and gave us a draught of their warm milk. The kind creature, fatigued at least as much as any of us, made no attempt to rest herself, till she had provided us with all she had to give for our refreshment. Having dined heartily on our potatoes, we concluded the day with evening prayers, and then joyfully climbed our ladder to seek the blessing of repose in our aerial castle.

15
Hopes of a Sledge; Some Short Lessons in Useful Things

I HAD observed along the shore many pieces of wood, of which I thought I could make a kind of conveyance for our cask of butter and other provisions from Tent House to Falcon's Stream, and had secretly determined to go early the next morning, before my family should be awake, to the spot. I had fixed upon Ernest for my assistant, thinking that his indolent temper required to be stimulated to exertion. I made him feel as a great favour the preference I gave him, and he promised to be ready at a very early hour. I was also desirous to leave Fritz with the family, as, being the tallest and strongest, he was more able to afford protection.

At the first dawn of morning I quietly awoke Ernest. He got up, and we descended the ladder without being perceived by the rest, who continued to sleep soundly. We roused the ass, and I made him draw some large branches of a tree, which I wanted for my undertaking.

We were not long in finding the pieces of wood, and set to work to cut them the proper length, and we then laid them cross-ways on the branches, which we thus converted into a kind of vehicle. We added to the load a little chest, which we found half buried in the sands, quite close to the waves, and then we set out on our return to Falcon's Stream. When we reached our abode, the chest we had brought was soon opened by a strong hatchet, for all were eager to see what was within. It contained only some sailors' dresses and some linen: and both were wet with the sea.

We then sat down tranquilly to breakfast; and I next inspected the booty of the young sportsmen, who had shot, in all, no less than fifty ortolans* and thrushes, and had used so large a quantity of powder and shot, that when they were about to resume their sport, my wife and I stopped them, recommending a more frugal use of those valuable materials. I taught them how to make some snares, to be suspended from the branches of the fig-tree, and advised them to use the thread of the karata, which is as strong as horse-hair, for the purpose. What is new always amuses young persons, and the boys accordingly took a great fancy to this mode of sporting. Jack succeeded in his very first attempt; I left Francis to assist him, and took Fritz and Ernest to help me in making the new carriage.

As we were hard at work, a prodigious clatter was heard among the fowls; the cock crowed louder than the rest, and the hens ran to and fro, as if pursued. 'I wonder what is the matter with the creatures,' cried my wife, rising; 'every day I hear the hens clucking as if they had been laying eggs.' At this moment Ernest happened to look at the monkey, and remarked that he fixed his piercing eyes on the hens; and when he saw my wife approaching, driving the hens before her, he jumped quickly into a hollow place under one of the roots of the tree, and hid himself. Ernest was at the place as soon as he, and caught him with a new-laid egg in his paw, which he was going to conceal. The monkey sprang away to another hole, and Ernest followed; here also he found some eggs, and brought them in his hat to his mother, who received them with great pleasure. The monkey, greedy of such food, had seized the eggs as soon as the hens had laid them. We inflicted no other punishment upon him for this little piece of knavery, than that of tying him up when the hens were about to lay. My wife expressed her joy at this new acquisition, and soon collected a great number of eggs, and we waited with impatience for the time when the hens would sit, in the hope of seeing their species multiplied.

In the meanwhile, Jack had got up into the tree, and had suspended some of the snares to the branches, to catch the little devourers of our figs; he came down again to bring us the acceptable intelligence, that our pigeons had made a sort of nest there of some dry grass, and that it already contained several eggs. I therefore forbade the boys from firing any more in the tree, for fear of alarming or wounding these gentle creatures. I also directed that the snares should be frequently examined, to see that the pigeons were not caught in them, as they might be strangled in their efforts to get loose. My sons had all murmured a little at my prohibition of the gunpowder; and little Francis with his innocent face came running to tell me, that he was going to ask his brother to help him to sow some gunpowder, that they might have plenty. We all laughed heartily at the idea, and Professor Ernest did not overlook the occasion to display his science.

During these arrangements, the boys and I had been busily employed: our work was completed. Two bent pieces of wood, the segments of a circle, which I fixed in their places by a straight piece of wood placed across, and firmly fixed to the bent pieces in the middle, and at the rear, formed the outline of my machine. I then fastened two ropes in front, and here was a sledge as perfect as could be desired. As I had not raised my eyes from my work, I did not know what my wife and the two youngest boys had been about. On looking up, I perceived that they had been stripping off the feathers from a quantity of birds which the boys had killed, and that they afterwards spitted them on an officer's sword, which Fritz had fancied and brought from the ship, and which my wife had turned into this useful kitchen utensil. I approved of the idea; but I blamed her profusion, in dressing more birds at once than we could eat. She reminded me, that I had myself advised her to half roast the birds before putting them into the butter, to be preserved for future use. She was in hopes, she said, that as I had now a sledge, I should not fail of going to Tent House after dinner, to fetch the cask of butter, and in the meanwhile, she was endeavouring to be ready with the birds. I had no objection to this, and determined on going to Tent House the same day, requesting my wife to hasten the dinner for that purpose. She replied, that this was already her intention, as she also had a little project in her head, which I should be informed of at my return. I, for my part, had one too, which was to refresh myself after the heat and fatigue of my laborious occupations, by a plunge into the sea. I wished that Ernest, who was to accompany me, should bathe also; while Fritz was to remain at home for the protection of the family.

16
A Bathing, a Fishing, the Jumping Hare, and a Masquerade

AT the moment of departure, Fritz presented his brother and myself with a case of his own workmanship, which we stuck into our belts, and which, in reality, were well contrived for holding spoons, and knives and forks, while room was left in the middle for a little hatchet. I praised Fritz for having thus brought his idea to perfection, and for contriving to make two cases with his skin instead of one.

We had harnessed the ass and the cow to our sledge; we each took a piece of bamboo-cane in hand, to serve as a whip; and resting our guns upon our shoulders, began our journey. Flora was to accompany us, and Turk to remain behind. We bade adieu to our companions, and put our animals in motion. We took the road by the sea-shore, where the sands afforded better travelling for our vehicle, than the thick wild grass. We reached Family Bridge, on Jackal's River, and arrived at Tent House without either obstacle or adventure, and unharnessed the animals to let them graze, while we set to work to load the sledge with the cask of butter, the cask of cheese, a small barrel of gunpowder, different instruments, some ball, and some shot. These exertions had so occupied our thoughts, that it was late when we first observed that our animals, attracted by the excellent quality of the grass on the other side of the river, had repassed the bridge, and wandered so far as to be out of sight. I was in hopes they would be easily found, and directed Ernest to go with Flora and bring them back, intending in the meantime to look for a convenient place, on the other side of Tent House, to bathe in. In a short time I found myself at the extremity of Providence Bay, which ended, as I now perceived, in a marsh, producing some fine bulrushes; and further on, a chain of steep rocks, advancing somewhat into the sea, and forming a kind of creek, as if expressly contrived for bathing. The juttings of the rock even seemed like little cabinets, for separate accommodation. Enchanted with this discovery, I called out to Ernest to come and join me, and in the meantime amused myself with cutting some of the rushes, and imagining what use I could apply them to.

I desired him to fill a small bag with some of the salt he had formerly observed here, and then to empty it into the large one for the ass to carry; and to take care to fill equally on each side. 'During this time, I will take the refreshment of bathing; and then it will be your turn to bathe, and mine to take care of the animals.'

I returned to the rocks, and was not disappointed in my expectation of an enjoyment the most delicious; but I did not stay long, fearing my boy might be impatient for his share of so new a pleasure. When I had dressed myself, I returned to the place to see if his work had advanced: presently I heard his voice calling out, 'Father, father, a fish! a fish of monstrous size! Run quickly, father, I can hardly hold him! He is eating up the string of my line!' I ran to the place from which the voice proceeded, and found Ernest lying along the ground on his face, upon the extremity of a point of land, and pulling in his line, to which a large fish was hanging, and struggling to get loose. I ran hastily and snatched the rod out of his hand, for I feared the weight and activity of the fish might pull him into the water. I gave the line length, to calm the fish, and then contrived to draw him gently along, into a shallow, from which he could no longer escape, and thus he was effectually secured. We examined him thoroughly, and he appeared to weigh not less than fifteen pounds; so that our capture was magnificent, and would afford the greatest pleasure to our good steward of provisions at Falcon's Stream. 'You have now really laboured,' said I to Ernest, 'not only with your head, but with your whole body; and I would advise you to wipe the perspiration from your face, and keep a little quiet before you venture into the water. You have procured us a dish of great excellence, which will last for several days, and have conducted yourself like a true cavalier, *without fear and without reproach.*'*

'It was at least fortunate,' observed he in a modest tone, 'that I thought of bringing my fishing-rod.'

Father.—Certainly it was. But tell me how you came to see this large fish, and what made you think you could catch it?

Ernest.—I used to remark great quantities of fish in the water, just hereabout, and this made me determine to bring my fishing-tackle with me. In my way to the salt, I perceived a great number of little crabs, upon which fishes feed, near the water's brink; I thought I would try to bait my hook with one of them; so I hastened my work, and came to this spot, where I caught only a dozen little fish, which are there in my handkerchief; but I remarked, that they were chased in the water by fishes of larger size. This gave me the idea of baiting my hook with one of the small ones; but the hook was too small, and my rod too weak. I then took one of the finest of the bulrushes you had just gathered, and put a larger hook to my line, and in a short time the large fish you see there seized upon the bait, and paid his life for his voracity. However, I must confess, that if you had not come to my assistance, I must either have let go my line, or have been dragged into the water; for the fish was stronger than I.

We now examined the smaller fishes, which were mostly trout and herrings, while I felt certain that the large one was a salmon. I cut them all open, and rubbed them in the inside with salt, that they might not be injured by the heat. While I was thus employed, Ernest went to the rocks and bathed, and I had time to fill some more bags with salt, before his return. We then harnessed and loaded our animals, and then resumed the road to Falcon's Stream.

When we had proceeded about half-way, Flora, who was before us, suddenly sprang off, and by her barking gave notice that she scented some game. We soon after saw her pursuing an animal, which seemed endeavouring to escape, and made the most extraordinary jumps imaginable. The dog continuing to follow, the creature, in trying to avoid her, passed within gunshot of the place where I stood. I fired, but its flight was so rapid, that I did not hit. Ernest, who was at a small distance behind, hearing the report of my gun, prepared his own, and fired it off at the instant the singular animal was passing near him, seeking to hide itself among the tall herbage just by: he had fired so skilfully, that the animal fell dead at the same instant. I ran with extreme curiosity to ascertain what kind of quadruped it might be. It was as large as a sheep, with the tail resembling that of a tiger; both its snout and hair were like those of a mouse, and its teeth were like a hare's, but much larger; the forelegs resembled those of the squirrel, and were extremely short; but to make up for this, its hind legs were as long as a pair of stilts, and of a form strikingly singular. We examined the creature a long time in silence; I could not be sure that I had ever seen an engraving or description of it in any natural history, or book of travels. Ernest at length, clapping his hands together, joyously exclaimed, 'And have I really killed this wonderful animal? What will my mother and my brothers say? How astonished they will be! and how fortunate I am in securing so fine a prize! What do you think is its name, father? I would give all the world to know.'

Father.—And so would I, my boy; but I am as ignorant as you. One thing, however, is certain, that this is your lucky day. Let us again examine this interesting stranger, that we may be certain to what family of quadrupeds it belongs: this will perhaps throw a light upon its name.

Ernest.—I think it can hardly be named a quadruped; for the little forelegs look much more like hands, as is the case with monkeys.

Father.—They are notwithstanding legs, I can assure you. Let us look for its name among the animals who give suck; on this point we cannot be mistaken. Now let us examine its teeth.

Ernest.—Here are the four incisory teeth, like the squirrel.

Father.—Thus we see that it belongs to the order of Nibblers. Now let us look for some names of animals of this kind.

Ernest.—Besides squirrels, I recollect only mice, marmots, hares, beavers, porcupines, and jumpers.

Father.—Jumpers! That short word furnishes the necessary clue; the animal is completely formed like the gerboa or jumping hare, except that it is twice the size of those of which I have read a description. . . . Wait a moment—an idea strikes me: I will wager that our animal is one of the large jumpers, called kangaroo; it belongs properly to the genus *Didelphis* or *Philander*, because the female, who never bears more than one young one, carries it in a kind of purse placed between her hind legs. To the best of my knowledge, this animal has never been seen but on the coast of New Holland,* where it was first observed by the celebrated navigator Captain Cook.* You may then be highly flattered with your adventure of killing an animal at once so rare and so remarkable. But now let us see how we shall manage to drag him to the sledge. Ernest requested that I would rather assist him to carry it, as he was afraid of spoiling its beautiful mouse-coloured skin by dragging it on the ground. I therefore tied the forelegs of the kangaroo together; and, by means of two canes, we with considerable trouble contrived to carry it to the sledge, upon which it was securely fastened.

Having now nothing more to detain us, we continued our road towards Falcon's Stream, conversing on the subject of natural history, and on the necessity of studying it in our youth, that we might learn to class plants and animals according to their characteristic marks; and we observed, that to such a knowledge as this it was owing that we had recognized the kangaroo. Ernest entreated me to tell him all I knew about the animal. 'It is', said I, 'a most singular kind of creature. Its forelegs, as you see, have scarcely the third part of the length of the hind ones, and the most it can do, is to make them serve the purpose of walking; but the hind legs enable it to make prodigious jumps, the same as in the flea and the grasshopper. The food of the kangaroo consists of herbs and roots, which they dig up very skilfully with their forelegs. They place themselves upon their hind legs, which are doubled under them, as if on a chair, and by this means are able to look above even the tall kinds of grass; they rest too upon their tail, which is exceedingly strong, and is also of great use to them in jumping, by assisting the spring from the ground. It is said that the kangaroo, if deprived of its tail, would scarcely be able to jump at all.'

We at length arrived happily, though somewhat late, at Falcon's Stream, having heard from a great distance the salutations of our family. Our companions all ran to meet us: but it was now, on seeing the ludicrous style of the dress of the three boys, our turn for immoderate fits of laughter: one had on a sailor's shirt, which trained round him like the robe of a spectre; another was buried in a pair of pantaloons, which were fastened round his neck, and reached to the ground; and the third had a long waistcoat, which came down to the instep, and gave him the exact form of a travelling portmanteau. They all tried to jump about, but finding this impossible, from the length of their garments, they next resolved to carry off the whole with an air, by strutting slowly to and fro, in the manner of a great personage in a theatre. After some hearty laughing, I enquired of my wife what could be the cause of this masquerade, and whether she had assisted them in attempting to act a comedy for our amusement. She disclosed the mystery by informing me, that her three boys had also been bathing, and that, while thus engaged, she had washed all their clothes; but as they had not dried so soon as she expected, her little rioters had become impatient, and had fallen on the chest of sailors' clothes, and each had taken from it what article he had pleased. 'I preferred,' said she, 'that you should see them in this odd sort of a disguise, rather than quite naked, like little savages', in which opinion I assured her that I heartily joined.

It was now our turn to give an account of our journey: as we advanced in our narrative, we presented, one after another, casks, bulrushes, salt, fish, and lastly, with infinite triumph, our beautiful kangaroo. In a trice it was surrounded, examined, and admired by all, and such a variety of questions asked, that Ernest and I scarcely knew which to answer first. Fritz was the only one who was a little silent. I saw plainly by his countenance what was passing in his mind. He was jealous of the good fortune of his brother Ernest; but I also saw that he was struggling manfully against the ascendancy of so mean a passion. In a short time he had succeeded so completely, that he joined frankly and unaffectedly in our conversation and merriment. He came near the kangaroo, and examined it; then turning to his brother, he observed to him, in a kind tone, that he had had good luck, and that he must be a good shot to have killed the animal with so little difficulty.—'But, father,' said he, 'when you go again to Tent House, or on any other excursion, will it not be my turn to accompany you? For here at Falcon's Stream there is nothing new to amuse us; a few thrushes, and some pigeons; this is all we have from day to day, and I find it very tiresome.'

'I promise you cheerfully what you desire, my dear boy,' said I, 'for you have valiantly combated the jealousy and ill-humour which assailed you on witnessing your brother's success with the kangaroo. I therefore engage that you shall accompany me in my very next excursion, which will probably take place at no greater distance of time than to-morrow; and it will be another journey to the vessel. But in the meantime, let me observe to you, that the high opinion I have shown of your prudence and judgement, in leaving you here, in charge of your mother and your brothers, ought to be felt by you as more flattering than the applause you would have gained by killing a kangaroo. You have accomplished an important duty, in keeping near them all the time, and not suffering yourself to be allured by such amusements as presented themselves to your fancy; and this conduct has increased my affection and respect for you. Praise is also due to Ernest, for the moderation with which he has felt his triumph, in so extraordinary an occurrence; for he has not even told you of my humiliating failure in attempting to shoot the kangaroo. To triumph over our passions, and to have on all occasions a perfect government of our temper, is an acquisition of infinitely more value, than the showing a certain skill in firing off a gun, and happening to kill an animal. In our situation, we are forced upon the cultivation of such arts as these; but though we may practise them as necessary for our existence, we have no reason to be proud of them.'

We concluded the day with our ordinary occupations: I gave some salt to each of our animals, to whom it was an acceptable treat. We then skinned our kangaroo, and put it carefully aside till the next day, when we intended to cut it to pieces, and lay such parts in salt as we could not immediately consume. We made an excellent supper on our little fish, to which we added some potatoes; nor were our faithful companions Turk and Flora neglected. The labours of the day had more than usually disposed us all to seek repose; we therefore said our prayers at an early hour, mounted our ladder, and were soon asleep.

17
More Stores from the Wreck

I ROSE with the first crowing of the cock, descended the ladder, and set about skinning the kangaroo, taking care not to deface its beautiful smooth coat. Our dogs relished their meal on the entrails of the animal so much, that they intended themselves the pleasure of a breakfast on the carcass. Before I could descend, they had got off its head, as it hung by the hind feet, and, half friends, half foes, they were going to share their prize when I made my appearance. Recollecting our want of the means of protection against similar depredations, I thought it right to give them a slight correction for their fault. My wife, awaked by the growling they made as they slunk away to the hollow of a tree, was alarmed, and came down the ladder to see what was the matter; and now I had to perform the further task of appeasing her kind heart for what she called a cruel act. 'Kind-hearted creature,' said I, 'well I know how glad you would be if there were not a stick in the world! But I did not beat Turk and Flora through anger or revenge, but from prudence and precaution: they intended modestly only to eat up our kangaroo, which you promised yourself such pleasure in cooking; and unable as I was to acquaint them in the canine tongue, that it was not placed there for their use, it was proper to let them know this in such a way as to deter them in future; otherwise, as they are strongest, they would end by devouring all our stock.'

My wife owned I was in the right: but I observed her from a corner of my eye hovering about the hollow tree, and patting the dogs to console them. I now set about stripping my kangaroo, without injuring the skin; but I advanced so slowly in the business, that my family were assembled about us, and calling out Famine! before I had finished my work. Having at last completed it, I went to the river to wash myself thoroughly, and then to the sailors' chest to change my coat, that I might appear with decency at breakfast, and give my sons an example of that cleanliness which their mother was so eager to inculcate. Breakfast over, I ordered Fritz to get ready for Tent House, where we should prepare the boat, and proceed to the vessel.

After taking an affectionate leave of my wife, we began our journey. I left Flora with her, and entreated her not to be uneasy, and to commit herself to the care of the kind Providence who had till then so graciously watched over us, and who would again bring us back to her safe and sound, enriched with many things conducive to our welfare. But to bring her to reason on the subject of these trips to the vessel was impracticable: I left her bathed in tears, and praying God that this might be the last.

We took Ernest and Jack a little way with us, and then I sent them back with a message to their mother, which I had not the resolution to deliver myself—that we might be forced to pass the night on board the vessel, and not return till the evening of the following day. It was most essential to get out of it, if yet afloat, all that could be saved, as a moment might complete its destruction. I instructed my sons how they should soothe their mother; I exhorted them to obey and to assist her; and that their excursion might not be useless, I directed them to gather some salt, and enjoined them to be at Falcon's Stream before noon.

We got into the boat, and gaining the current, quickly cleared Safety Bay, and reached the vessel, whose open side offered us an ample space to get on board. When we had fastened our boat, our first care was to select fit materials to construct a raft, as suggested by my son Ernest. Our boat of staves had neither room nor solidity enough to carry a considerable burden; we therefore looked about, and found a sufficient number of water-casks which appeared to me proper for my new enterprise. We emptied them, replaced the bungs carefully, and threw the casks overboard, after securing them with ropes and cramps, so as to keep them together at the vessel's side: this completed, we placed a sufficient number of planks upon them to form a firm and commodious platform or deck, to which we added a gunwale of a foot in depth all round, to secure the lading. Thus we contrived a handsome raft, in which we could stow thrice as much as in our boat. This laborious task had taken up the whole day; we scarcely allowed ourselves a minute to eat some cold meat we had provided, that we might not lose any time in looking for the provisions on board the vessel. In the evening, Fritz and I were so weary, that it would have been impossible for us to row back to land; so having taken all due precautions in case of a storm, we lay down in the captain's cabin, on a good elastic mattress, which induced such sound repose, that our prudent design to watch in turn, for fear of accident, was forgot, and we both slept heavily, side by side, till broad daylight opened our eyes. We rose, and actively set to work to load our raft.

We began with stripping the cabin of its doors and windows, with their appendages; next we secured the carpenter's and gunner's chests, containing all their tools and implements: those we could remove with levers and rollers were put entire upon the raft, and we took out of the others what rendered them too heavy. One of the captain's chests was filled with costly articles, which no doubt he meant to dispose of to the opulent planters of Port Jackson, or among the savages. In the collection were several gold and silver watches, snuff-boxes of all descriptions, buckles, shirt-buttons, necklaces, rings; in short, an abundance of all the trifles of European luxury. But the discovery that delighted me most, was a chest containing some dozens of young plants of every species of European fruits, which had been carefully packed in moss for transportation. I perceived pear, plum, almond, peach, apple, apricot, chestnut trees, and vine shoots. I beheld with a feeling I cannot describe, those productions of my dear country, which once so agreeably embellished my rural dwelling, and which, I might hope, would thrive in a foreign soil. We discovered a number of bars of iron, and large pigs of lead, grinding-stones, cart-wheels ready for mounting, a complete set of farrier's instruments*, tongs, shovels, ploughshares, rolls of iron and copper wire, sacks full of maize, pease, oats, vetches, and even a little hand-mill. The vessel had been freighted with everything likely to be useful in an infant colony so distant. We found a sawmill, in a separated state, but each piece numbered, and so accurately fitted, that nothing was easier than to put it together for use.

I had now to consider what of all these treasures I should take or leave. It was impossible to carry with us in one trip such a quantity of goods; and to leave them in the vessel, was exposing ourselves to be wholly deprived of them.

We with difficulty and hard labour finished our loading, having added a large fishing-net, quite new, and the vessel's great compass. With the net, Fritz found two harpoons and a rope-windlass, such as they use in the whale-fishery. He asked me to let him place the harpoons, tied to the end of the rope, over the bow of our tub-boat, and thus be in readiness in case of seeing any large fish; and I indulged him in his fancy.

Having completely executed our undertaking, we stepped into the tub-boat, and with some small difficulty, which a little reflection and a few experiments soon enabled us to overcome, we pushed out for the current, drawing our raft triumphantly after us with a stout rope, which we had been careful to fasten securely at its head.

18
The Tortoise Harnessed

THE wind was favourable, and briskly swelled our sail. The sea was calm, and we advanced at a considerable rate. Fritz had for some time fixed his eyes on something of a large size which was floating on the water, and he now desired me to take the glass, and see what it could be. I soon perceived that it was a tortoise,* which had fallen asleep in the sun on the surface of the water. No sooner had Fritz learned this, than he entreated me to steer softly within view of so extraordinary a creature. I readily consented; but as his back was towards me, and the sail between us, I did not observe his motions, till a violent jerk of the boat, a sudden turning of the windlass, and then a second jerk, accompanied by a rapid motion of the boat, gave me the necessary explanation. 'For Heaven's sake, what are you about, Fritz?' exclaimed I, somewhat alarmed.

—'I have caught him!—I touched him!' cried Fritz, without hearing one word I had been saying.—'The tortoise is ours; it cannot escape, father! Is not this, then, a valuable prize, for it will furnish dinners for us all for many weeks?'

I soon perceived that the harpoon had caught the animal, which, feeling itself wounded, thus agitated the vessel in its endeavours to get away. I quickly pulled down the sail, and seizing a hatchet, sprung to the boat's head to cut the rope, and let the harpoon and the tortoise go; but Fritz caught hold of my arm, conjuring me to wait a moment, and not so hastily bring upon him the mortification of losing, at one stroke, the harpoon, the rope, and the tortoise: he proposed watching himself, with the hatchet in his hand, to cut the rope suddenly, should any sign of danger appear; and I yielded to his entreaties.

Thus, then, drawn along by the tortoise, we proceeded with a hazardous rapidity. I soon observed that the creature was making for the sea; I therefore again hoisted the sail: and as the wind was to the land, and very brisk, the tortoise found resistance of no avail: he accordingly fell into the track of the current, and drew us straight towards our usual place of landing, and by good fortune without striking upon any of the rocks. We, however, did not disembark without one difficult adventure. The state of the tide was such as to throw us upon a sandbank: we were at this time within a gunshot of the shore; the boat, though driven with violence, remained upright in the sand. I stepped into the water, which did not reach far above my knees, for the purpose of conferring upon our conductor his just reward for the alarm he had caused us, when he suddenly gave a plunge, and then disappeared. Following the rope, I presently saw the tortoise stretched at length at the bottom of the water, where it was so shallow that I soon found means to put an end to his pain, by cutting off his head with the hatchet, and he bled to death. Being now near Tent House, Fritz gave a halloo, and fired a gun, to apprise our relatives that we were not only arrived, but arrived in triumph. This soon produced the desired effect: the mother and her three young ones soon appeared, running towards us; upon which Fritz jumped out of the boat, placed the head of our sea-prize on the muzzle of his gun, and walked to shore, which I reached at the same moment; and all were once more received with the kindest salutations, and such questions as kindness best knows how to propose.

After some gentle reproaches from my wife, for leaving her and the boys for so long a time, the history of the tortoise was related, and excited much merriment in our auditors. The tender-hearted mother, after heaving a sigh for the hard fate of the creature, began to shudder at the thought of the danger we had been exposed to, and the escape we had effected.

Our conversation ended, I requested my wife to go with two of the younger boys to Falcon's Stream, and fetch the sledge and the beasts of burden, that we might see at least a part of our booty from the ship put safely under shelter the same evening. A tempest, or even the tide, might sweep away the whole during the night! We took every precaution in our power against the latter danger, by fixing the boat and the raft, now, at the time of its reflux, as securely as we could without an anchor. I rolled two prodigious masses of lead, with the assistance of levers, from the raft upon the shore, and then tied a rope to each, the other ends of which were fastened, one to the raft, and the other to the boat, and thus satisfied myself that they could not easily be forced away.

While we were employed on this scheme, the sledge arrived, and we placed the tortoise upon it, and also some other articles of light weight, mattresses, pieces of linen, etc.; for I reckoned that the animal itself weighed at least three quintals.* The strength of our whole party was found necessary to move it from the raft to the sledge; we therefore all set out together to unload it again at Falcon's Stream.

Our first concern, on reaching our abode, was the tortoise, which we immediately turned on his back, that we might strip off the shell, and make use of some of the flesh while it was fresh. Taking my hatchet, I separated the upper and under shell all round, which were joined together by cartilages. The upper shell of the tortoise is extremely convex; the under, on the contrary, is nearly flat. I cut away as much of the flesh of the animal as was sufficient for a meal, and laid the rest carefully on the under shell, which served as a dish, recommending to my wife to cook what I had cut off, on the other shell, with no other seasoning than a little salt, and pledged myself that she would produce a luxurious dish. 'We will then', said I, 'rub salt on what we mean to keep, and distribute the head, entrails, and feet to the dogs; for all, you know, must live.'

'Oh dear papa,' cried Francis, 'do give me the shell, it will be such a pretty plaything!'

'No, no,' bawled out another; and one and all contended for the preference. I imposed silence, declaring that the right was entirely in Fritz; 'but', continued I, 'it may be well to ask what each of you thought of doing with the shell, if he had obtained it?'

Ernest.—I should turn it into a shield to defend myself with, if the savages should come upon us.

Father.—Ah, there is my egotist again; but let us see in what way you would use it. You would fling it across your shoulders, no doubt, and take to your heels manfully. I have guessed right, my poor Ernest, have I not?—And you, Jack, what have you to say?

Jack.—I should make a nice little boat of it, which would help to amuse us all. I was thinking how cleverly we could fill it with potatoes, or the other things we want to take from Tent House to Falcon's Nest; it would glide along so nicely with the stream, and we should be saved all the fatigue we now have in carrying them.

Father.—Your scheme, I grant, is not ill-imagined; but a small raft or an old chest, would do just as well for your purpose.—And now for my little Francis; I wonder what pretty plan he had thought of?

Francis.—I thought I should build a little house, papa, and the shell would make such a nice roof to it!

Father.—Vastly well, my lads, if we had only our amusement or our ease to think of; but I want you all to form the habit of thinking and acting for the general good, rather than that of what will most gratify or accommodate his single self.—Now, then, let me ask, to what use Fritz, the only rightful claimant to the shell, had intended to apply it?

Fritz.—I thought, father, of cleaning it thoroughly, and fixing it by the side of our river, and keeping it always full of pure water for my mother's use, when she has to wash the linen, or cook our victuals.

Father.—Excellent, excellent, my boy! All honour to the founder of the *pure water-tub*! This is what I call *thinking for the general good*. And we will take care to execute the idea as soon as we can prepare some clay, as a solid foundation for its bottom.

Jack.—Hah, hah! Now then it is my turn; for I have got some clay, which I have put by to keep for use, behind those old roots yonder.

Father.—And where did you get it, boy?

Mother.—Oh, you may apply to me for this part of the information; to my cost I know where the clay was got.—This morning early, my young hero falls to digging and scrambling on the hill you see to the right, and home he comes with the news, that he has found a bed of clay; but in so dirty a condition himself, that we were obliged to think next of the washing-tub.

Jack.—And if I had minded a little dirt, mother, I should not have discovered this bed of clay, which you will see will be of great use to us. As I was returning from looking for potatoes, I thought I would take the high path along the river, just to see how rapidly it runs and forms those nice cascades: by and by I came to a large slope, watered by the river; it was so slippery, that I could not keep upon my legs; so I fell, and dirtied myself all over: on looking, I saw that the ground was all of clay, and almost liquid, so I made some of it into balls, and brought them home.

Ernest.—When the water-tub is complete, I will put some roots I have found to soak a little in it, for they are now extremely dry. I do not exactly know what they are; they look something like the radish, or horse-radish; but the plant from which I took them was almost the size of a bush: being ignorant, however, of its name or nature, I have not yet ventured to taste the roots, though I saw our sow eat heartily of them.

Father.—If my suspicion is right, you have made a beneficial discovery, which, with the assistance of our potatoes, may furnish us the means of existence as long as we may remain in this island! I think your roots are *manioc*, of which the natives of the West Indies make a sort of bread or cake which they call *cassave.* * *But we must first carry the production through a certain preparation, without which it possesses pernicious properties. Try to find the same place, and bring a sufficient quantity for our first experiment.*

We had finished unloading the sledge, and I bade the three eldest boys accompany me to fetch another load before it should be dark. We left Francis and his mother busy in preparing a refreshing meal for supper, the tortoise having presented itself most opportunely for this purpose.

Having reached the raft, we took from it as many effects as the sledge could hold, or the animals draw along. One object of my attention was to secure two chests which contained the clothes of my family, which I well knew would afford the highest gratification to my wife, who had frequently lamented that they were all compelled to wear clothes that were not their own; reminding her at every moment, she said, how much they might be wanted by their proper claimants. I reckoned also on finding in one of the chests some books on interesting subjects, and principally a large handsomely printed Bible. I added to these, four cartwheels and a hand-mill for grinding; which, now that we had discovered the manioc, I considered of signal importance. These and a few other articles completed our present load.

On our return to Falcon's Nest, we found my wife looking anxiously for our arrival, and ready with the welcome she had promised, of an ample and agreeable repast. Before she had well examined our new stores, she drew me, with one of her sweetest smiles, by the arm,—'Step this way,' said she, and leading to the shade of a tree,—'this is the work I performed in your absence,' pointing to a large cask half sunk in the ground, and the rest covered over with branches of trees. She then applied a small corkscrew to the side, and filling the shell of a coconut with the contents, presented it to me. I found the liquor equal to the best canary I had ever tasted.—'How then,' said I, 'have you performed this new miracle? I cannot believe the enchanted bag produced it.'—'Not exactly,' replied she: 'for this time it was an obliging white wave which threw it on shore. I took a little ramble in your absence yesterday, to see what I could find, and well my trouble was rewarded! The boys ran for the sledge, and had but little difficulty in getting the cask to Falcon's Stream, where we dug this place in the earth to keep it cool.'

My wife now proposed that all should be regaled with some of the delicious beverage. My own share so invigorated me, that I found myself able to complete my day's work, by drawing up the mattresses we had brought from the ship, to our chamber in the tree, by means of a pulley. When I had laid them along to advantage, they looked so inviting, that I could scarcely resist my desire of at once committing myself to the kind relief they seemed to offer to my exhausted strength.

But now the savoury smell of the tortoise laid claim to my attention. I hastened down, and we all partook heartily of the luxurious treat. We returned thanks to God, and speedily retired to taste the blessing of sound repose upon the said mattresses.

19
Another Trip to the Wreck

I ROSE before day to go to the seaside and inspect our two vessels. I gently descended the ladder without awaking my family. Above, the scene was all repose; below, everything was in life and motion. The dogs jumped about me, the cock and the hens flapped their wings and chuckled, and our goats shook their long beards as they browsed. I quickly roused and harnessed the ass, and the dogs followed without bidding. As I approached the shore, animated at different moments by hope and fear, I soon saw that the boat and raft had resisted the tide, though it had partially heaved them up. I got quickly on the raft, took a small loading, and returned to Falcon's Stream in time for breakfast; but not a single creature of its inhabitants appeared, though the sun was high above the horizon.—I gave a shout as loud as a war-whoop, which awoke my wife. 'Really, my dear,' said she, 'there must be a magic charm in the mattress you brought yesterday, that has lulled us into so sound a sleep.'—'Up, my lads,' exclaimed I, once again; 'the more we venture to parley with sloth, the longer she holds us in her chains; brave youths like you ought to awake at the first call, and leap quick and gaily out of bed.' Fritz, a little ashamed, was dressed first; Jack soon after him, and Francis next; the ever slothful Ernest was the last.—'It is so delightful', cried he, 'to lose oneself again after having been awakened! One feels sleep come on afresh so gently.'—'But it is my duty to tell you, Ernest, and that gravely, that he who indulges himself in all that flatters his senses, will end by falling a victim to them.'

After this short admonition, we all came down; and break-fast over, we returned to the seaside to complete the unloading of the raft, that it might be ready for sea on the ebbing of the tide. We were not long in taking two cargoes to Falcon's Stream. At our last trip the water was nearly up to our craft. I sent back my wife and the boys, and remained with Fritz till we were quite afloat; when observing Jack still loitering near, I guessed at his wish, and consented to his embarking with us. Shortly after, the tide was high enough for us to row off. Instead of steering for Safety Bay to moor our vessels there securely, I was tempted by a fresh sea-breeze to go out again to the wreck; but it was too late to undertake much, and I was unwilling to cause my dear partner uneasiness by passing another night on board. I therefore determined to bring away only what could be obtained with ease and speed: we searched hastily through the ship for any trifling articles that might be readily removed. Jack was up and down everywhere, at a loss what to select; and when I saw him again, he drew a wheelbarrow after him, shout-ing that he had found a vehicle for carrying our potatoes.

But Fritz next disclosed still better news, which was, that he had discovered behind the bulkhead* amid ship, a pinnace (i.e. a small craft, the forepart of which is square) taken to pieces, with all its appurtenances, and even two small guns for its defence. This intelligence so delighted me, that I quitted every thing else to run to the bulkhead, when I was convinced of the truth of the lad's assertion: but I instantly perceived, that to put it together, and launch it, would be an Herculean* task. I collected various utensils, a copper boiler, some plates of iron, tobacco-graters, two grinding-stones, a small barrel of

gunpowder, and another full of flints, which I much valued.
Jack's barrow was not forgotten; two more were afterwards
found and added, with straps belonging to them. All these
articles were hurried into the boat, and we re-embarked with
speed, to avoid the land wind that rises in the evening. As we
were drawing near to shore, we were struck with the appearance of an assemblage of small figures ranged in a long line
on the strand, that seemed to be viewing us attentively: they
were dressed in black, and all uniform, with white waistcoats
and full cravats: the arms of these beings hung down carelessly;
now and then, however, they seemed to extend them tenderly,
as if they wished to embrace or offer us a token of friendship.

'I really think,' said I to the boys, who were steadfastly
gazing at them, 'that we are in the country of the pygmies, and
that they wish to form a friendly alliance with us.'

Jack.—Oh, no! father, they are certainly Lilliputians,*
though somewhat bigger than those of whom I read the description in Gulliver's Travels.

'You then, child,' said I, 'consider those travels as true; that
there is an island of Lilliput, and inhabited by dwarfs?'

Jack.—Gulliver says so. He met also with men of an immense
stature, besides an island inhabited by horses——

'And yet I must tell you that the only reality in all his
discoveries is the rich imagination of the author, whose taste
and feeling led him to resort to allegory for the purpose of
revealing grand truths. Do you know, Jack, what an allegory is?'

'It somewhat resembles a parable, I presume.'

'Right, one is very similar to the other.'

Jack.—And the pygmies you mentioned, are any to be
found?

'No more than there are Lilliputians; they exist only in
poetical fiction, or in the erroneous account of some ancient
navigators, in which a group of monkeys has been fallaciously
described as diminutive men.'

Fritz.—Such probably are the manikins that we see now
stretching out their arms towards us.—Ah, now I begin to
perceive that they have beaks, and that their arms are short
drooping wings;—what strange birds!

'You are right, son, they are penguins or ruffs. Ernest killed
one soon after our arrival. They are excellent swimmers, but

cannot fly; and so confused are they when on land, that they run in the silliest way into danger.'

While we were talking I steered gently towards shore, to enjoy the uncommon sight the longer; but the very moment we got into shallow water, my giddy Jack leaped up to his waist into it, and was quickly on land, battering with his stick among the penguins before they were aware of his approach, so that half a dozen of them were immediately laid flat; the remainder, seeing they were so roughly accosted, plunged into the sea, dived, and disappeared.

As the sun declined, and we despaired of finishing before night set in, each of us filled a barrow, in order to take home something. I requested that the tobacco-graters and iron plates might be in the first load.

Arrived at Falcon's Stream, my wife exhibited a good store of potatoes which she had got in during our absence, and a quantity of the roots I had taken for manioc, and in which I was not mistaken; I much applauded her diligence and foresight, and gave Ernest and little Francis their share of approbation.

'But now', said I, 'for some supper and repose; and if my little workmen should be industriously inclined tomorrow, I shall reward them with the novelty of a new trade to be learned.' This did not fail to excite the curiosity of all; but I kept my word, and made them wait till the following day for the explanation I had to give.

I waked the boys very early, reminding them that I had promised to teach them a new trade. 'What is it? What is it?' exclaimed they all at once, springing suddenly out of bed and hurrying on their clothes.

Father.—It is the art of the baker, my boys. Hand me those iron plates that we brought yesterday from the vessel, and the tobacco-graters also, and we will make our experiment. Ernest, bring hither the roots found under ground: but first, my dear, I must request you to make me a small bag of a piece of strong wrapper cloth.

My wife set instantly to work to oblige me; but having no great confidence in my talents for making either bread or cakes, she first filled a copper boiler with potatoes, and put it on the fire, that we might not be without something to eat at dinner-time: in the meanwhile I spread a piece of coarse linen on the ground, and assembled my young ones round me; I gave each of the boys a grater, and showed him at the same time how to rest it on the linen, and then to grate the roots of manioc; in a short time each had produced a considerable heap of a substance somewhat resembling pollard.* The occupation, as is always the case with novelties, was amusing to them all, and they looked no further into the matter: one showed the other his heap, saying in a bantering tone: 'Will you eat a bit of nice cake made of grated radishes?'

I now informed them that the manioc was known to be the principal sustenance of whole nations of the Continent of America, and which the Europeans who inhabit those countries prefer to even our wheaten bread. I added 'there are many kinds of manioc: one of these shoots rapidly, and its roots become mature in a short time; a second sort is of more tardy growth; and there is another, the roots of which require the space of two years to be fit for use. The first two kinds have pernicious or unwholesome qualities when eaten raw, but the third may be eaten without fear: for all this, the two first are generally preferred, as being more productive, and requiring a shorter time for being fit for use.'

By this time my wife had completed the bag. I had it well filled with what we called our pollard, and she closed it securely by sewing up the end. I was now to contrive a kind of press: I cut a long, straight, stout branch, from a neighbouring tree, and stripped it of the bark; I then placed a plank across the table we had fixed between the arched roots of our tree, and which was exactly the right height for my purpose, and on this I laid the bag; I put other planks again upon the bag, and then covered all with the large branch, the thickest extremity of which I inserted under an arch, while to the other, which projected beyond the planks, I suspended all sorts of heavy substances, such as lead, our largest hammers and bars of iron, which, acting with great force as a press on the bag of manioc, caused the sap it contained to issue in streams, which flowed plentifully on the ground.

Fritz.—This machine of yours, father, though simple, is as effectual as can be desired.

Father.—Certainly. It is the simplest lever that the art of mechanism can furnish, and may be made extremely useful.

Ernest.—I thought that levers were never used but for raising heavy masses, such as blocks of stone, and things of that degree of weight; I had no notion that they were ever used for pressing.

Father.—But you see that the point at which the lever rests on the planks must always be the point of rest or compression; the point at which its extremity touches the roots of the tree would no doubt be that of the raising power, if the root was not too strong to yield to the point of the lever; but then the resistance at the point of compression or rest is still stronger, and presses effectually, as you see, the contents of the bag. The Negroes, however, have another manner of proceeding; but it would have been much too tedious in the process for us to imitate. They make tresses of the bark of a tree, and with it form a kind of basket of tolerable size; they fill it with manioc, and press it so tightly, that the baskets become shorter, and increase in breadth; they then hang the baskets to the strongest branches of trees, and fasten large stones to them, which draw the baskets again lengthways; by which action upon the manioc the sap runs out at the openings left by the tresses.

Mother.—Can one make no use of this sap?

Father.—Certainly, we may: the same Negroes use it as food, after mixing with it some pepper; and when they can procure them, some sea-crabs.

Fritz.—Father, it no longer runs a single drop; may we not now set about making the dough?

Father.—I have no objection; but as there are some poisonous kinds of manioc, it will be prudent to make only a small cake at first, by way of experiment, which we will give to the monkey and the fowls, and wait to see the effect, instead of exhausting our whole store at once.

We now opened the bag, and took out a small quantity of the pollard, which already was dry enough; we stirred the rest about with a stick, and then replaced it under the press. The next thing was to fix one of our iron plates, which was of a round form, and a little hollow, so as to rest upon two blocks of stone at a distance from each other; under this we lighted a large fire, and when the iron plate was completely heated, we placed a portion of the dough upon it with a wooden spade. As soon as the cake began to be brown underneath, it was turned, that the other side might be baked also.

Ernest.—O how nicely it smells! what a pity that we may not eat some of it immediately!

Father.—I believe you might safely venture, but it is perhaps better to wait till the evening, and run no greater risk than the loss of one or two of our fowls or of the monkey; and we may say this trial of the cake will be the first service he has rendered us.

As soon as the cake was cold, we broke some of it into crumbs, and gave it to two of the fowls, and a larger piece to the monkey, who nibbled it with a perfect relish, making all the time a thousand grimaces, while the boys stood by envying the preference he enjoyed.

Fritz.—Now tell me, father, how the savages manage to grate their manioc, for surely they have not, like us, an instrument fitted for the operation;—and tell me also, if they call their composition by the name of cake or bread, as we do?

Father.—The savages having no such article as bread in their bill of fare, have consequently no word in their language to express it. At the Antilles, the bread from the manioc is called *cassave*; the savages make a kind of grater with sharp stones, or shells; or when they can get nails, on which they set a high value, they drive them into the end of a plank, and rub the manioc upon it. But now, I pray you, good wife, give us quickly our potato dinner, and we will afterwards resume the baking trade.

The first thing after dinner was to visit our fowls. Those which had eaten the manioc were in excellent condition, and no less so the monkey.—'Now then to the bakehouse, young ones,' said I, 'as fast as you can scamper.'—The grated manioc was soon emptied out of the bag, a large fire was quickly lighted, and I placed the boys where a flat surface had been prepared for them, and gave to each a plate of iron and the quantity of a coconut-full to make a cake apiece, and they were to try who could succeed the best. They were ranged in a half-circle round me, that they might observe how I proceeded, and adopt the same method for themselves. The result was not discouraging for a first experiment, though it must be confessed we were now and then so unlucky as to burn a cake; but there was not a greater number of these than served to feed the pigeons and the fowls, which hovered round us to claim their share of the treat. My little rogues could not resist the pleasure of frequently tasting their cake, a little bit at a time, as they went on. At length the undertaking was complete; the cakes were put in a dish, and served, in company with a handsome share of milk, to each person; and with this addition they furnished us with an excellent repast: what remained we distributed among our animals and fowls.

The rest of the day was employed by the boys in making several turns with their wheelbarrows, and by myself in different arrangements in which the ass and our raft had a principal share, both being employed in drawing to Tent House the remaining articles we had brought from the ship. When all this was done we retired to rest, having first made another meal on our cakes, and concluded all with pious thanks to God for the blessings his goodness thought fit to bestow upon us.

20
The Cracker and the Pinnace

FROM the time of discovering the pinnace, my desire of returning to the vessel grew every moment more irresistible; but one thing I saw was absolutely necessary, which was, to collect all my hands to get her out from the situation where we had found her. I therefore thought of taking with me the three boys: I even wished that my wife should accompany us; but she had been seized with such a horror of the perfidious element, as she called it, the sea, that she assured me the very attempt would make her ill and useless. I had some difficulty to prevail upon her to let so many as three of the children go: she made me promise to return the same evening, and on no account to pass another night on board the wreck; and to this I was, though with regret, obliged to consent.

After breakfast then, we prepared for setting out. The boys were gay and on the alert, in the expectation of the pleasure that awaited them, particularly Ernest, who had not yet made a single voyage with us to the vessel. We took with us an ample provision of boiled potatoes and *cassave*; and in addition, arms and weapons of every kind. We reached Safety Bay without any remarkable event: here we thought it prudent to put on our cork jackets; we then scattered some food for the geese and ducks which had taken up their abode there, and soon after stepped gaily into our tub-raft, at the same time fastening the new boat by a rope to her stern, so that she could be drawn along. We put out for the current, though not without considerable fear of finding that the wreck had disappeared. We soon, however, perceived that it still remained firm between the rocks. Having got on board, all repaired, on the wings of curiosity and ardour, to that part of the vessel called the bulkhead,* which contained the enviable prize, the pinnace. On further observation, it appeared to me that the plan we had formed was subject to at least two alarming difficulties; the one was the situation of the pinnace in the ship; and the other was the size and weight it would necessarily acquire when put together. The enclosure which contained the pinnace was in the interior of the ship, and timbers of prodigious bulk and weight separated it from the breach, and in this part of the deck there was not sufficient space for us to put the pinnace together, or to give her room when done. The breach also was too narrow and too irregular to admit of her being launched from this place, as we had done with our tub-raft. In short, the separate pieces of the pinnace were too heavy for the possibility of our removing them even with the assistance of our united strength. What therefore was to be done? And how could we meet such formidable difficulties? I stood on the spot absorbed in reflection, while the boys were running from place to place, conveying everything portable they could find, on board the raft.

The cabinet which contained the pinnace was lighted by several small fissures in the timbers, which after standing in the place a few minutes to accustom the eye, enabled one to see sufficiently to distinguish objects. I discovered, with pleasure, that all the pieces of which she was composed were so accurately arranged and numbered, that without too much presumption, I might flatter myself with the hope of being able effectually to collect and put them together, if I could be allowed the necessary time, and could procure a convenient place. I therefore, in spite of every disadvantage, decided on the undertaking; and we immediately set about it. We proceeded at first so slowly as to have produced discouragement, if the desire of possessing so admirable a little vessel, quite new, perfectly safe, easy to conduct, and which might at some future day be the means of our deliverance, had not at every moment inspired us with new strength and ardour.

Evening, however, was fast approaching, and we had made but small progress; we were obliged to think of our promise to my wife; and though with reluctance, we left our occupation and re-embarked. On reaching Safety Bay, we found there our kind steward and little Francis; they had been, during the day, employed in arrangements for our living at Tent House as long as we should have occasion to continue the excursions to the vessel: this she did to shorten the length of the voyage, and that we might be always in sight of each other. In return for her kindness, I made the best display I could of two casks of salted butter, three of flour, some small bags of millet seed and of rice, and some other articles of utility and comfort for our establishment; and the whole was removed to our store-house at the rocks.

We passed an entire week in this arduous undertaking of the pinnace. I embarked every morning with my three sons, and returned every evening, and never without some small addition to our stores. We were now so accustomed to this manner of proceeding, that my wife bade us goodbye without concern, and we, on our parts, left Tent House without anxiety; she even had the courage to go several times, with no companion but her little Francis, to Falcon's Stream, to feed and take care of the poultry, and to bring back potatoes for our use. As night successively returned, we had a thousand interesting things to tell each other, and the pleasure of being together was much increased by these short separations.

At length the pinnace was completed, and in a condition to be launched: the question now was, how to manage this remaining difficulty. She was an elegant little vessel, perfect in every part: she had a small neat deck; and her mast and sails were no less exact and perfect than those of a little brig. It was probable she would sail well, from the lightness of her construction, and in consequence drawing but little water. We had pitched and towed all the seams, that nothing might be wanting for her complete appearance: we had even taken the pains of further embellishing, by mounting her with two small cannon of about a pound weight; and, in imitation of larger vessels, had fastened them to the deck with chains. But in spite of the delight we felt in contemplating a work, as it were, of our own industry; the great difficulty still remained: the said commodious, charming little vessel, still stood fast enclosed within four walls; nor could I conceive of a means of getting her out. To effect a passage through the outer side of the vessel, by means of our united industry in the use of all the utensils we had secured, seemed to present a prospect of exertions beyond the reach of man, even if not attended with dangers the most alarming. We examined if it might be practicable to cut away all intervening timbers, to which, from the nature of the breach, we had easier access; but should we even succeed in this attempt, the upper timbers being, in consequence of the inclined position of the ship, on a level with the water, our labour would be unavailing: besides, we had neither strength nor time for such a proceeding; from one moment to another, a storm might arise and engulf the ship, timbers, pinnace, ourselves, and all. Despairing, then, of being able to find a means consistent with the sober rules of art, my impatient fancy inspired the thought of a project, which could not however be tried without hazards and dangers of a tremendous nature.

I had found on board a strong iron mortar,* such as is used in kitchens. I took a thick oak plank, and nailed to different parts of it some large iron hooks: with a knife I cut a groove along the middle of the plank. I sent the boys to fetch some match-wood from the hold, and I cut a piece sufficiently long to continue burning at least two hours. I placed this train* in the groove of my plank: I filled the mortar with gunpowder, and then laid the plank, thus furnished upon it, having previously pitched* the mortar all around; and, lastly, I made the whole fast to the spot with strong chains, crossed by means of the hooks in every direction. Thus I accomplished a sort of cracker,* from which I expected to effect a happy conclusion. I hung this machine of mischief to the side of the bulkhead next the sea, having taken previous care to choose a spot in which its action could not affect the pinnace. When the whole was arranged, I set fire to the match, the end of which projected far enough beyond the plank to allow us sufficient time to escape. I now hurried on board the raft, into which I had previously sent the boys before applying a light to the match; and who, though they had assisted in forming the cracker, had no suspicion of the use for which it was intended, and believing all the while it concealed some subject of amusement for their next trip to the vessel. I confess I had purposely avoided giving them the true explanation, from the fear of the entire failure of my project, or that the vessel, pinnace, and all that it contained, might in consequence be blown up in a moment. I had naturally, therefore, some reluctance to announce myself before the time as the author of so many disasters.

On our arrival at Tent House, I immediately put the raft in a certain order, that she might be in readiness to return speedily to the wreck, when the noise produced by the cracker should have informed me that my scheme had taken effect. We set busily to work in emptying her; and during the occupation, our ears were assailed with the noise of an explosion of such violence, that my wife and the boys, who were ignorant of the cause, were so dreadfully alarmed as instantly to abandon their employment. 'What can it be?—What is the matter?—What can have happened?' cried all at once. 'It must be cannon. It is perhaps the captain and the ship's company who have found their way hither! Or can it be some vessel in distress? Can we go to its relief?'

Mother.—The sound comes in the direction of the wreck: perhaps she has blown up.—From the bottom of her heart she made this suggestion, for she desired nothing more earnestly than that the vessel should be annihilated, and thus an end be put to our repeated visits.

Father.—If this is the case, said I, we had better return immediately, and convince ourselves of the fact? Who will be of the party?

'I, I, I', cried the boys; and the three young rogues lost not a moment in jumping into their tubs, whither I soon followed them, after having whispered a few words to my wife, somewhat tending to explain, but still more to tranquillize her mind during the trip we had now to engage in.

We rowed out of the bay with more rapidity than on any former occasion; curiosity gave strength to our arms. When the vessel was in sight, I observed with pleasure that no change had taken place in the part of her which faced Tent House, and that no sign of smoke appeared: we advanced, therefore, in excellent spirits; but instead of rowing, as usual, straight to the breach, we proceeded round to the side, on the inside of which we had placed the cracker. The horrible scene of devastation we had caused now broke upon our sight. The greater part of the ship's side was shivered to pieces; innumerable splinters covered the surface of the water; the whole exhibited a scene of terrible destruction, in the midst of which presented itself our elegant pinnace, entirely free from injury! I could not refrain from the liveliest exclamations of joy, which excited the surprise of the boys, who had felt the disposition such a spectacle naturally inspired, of being dejected. They fixed their eyes upon me with the utmost astonishment.—'Now then she is ours!' cried I—'The elegant little pinnace is ours! For nothing is now more easy than to launch her. Come, boys, jump upon her deck, and let us see how quickly we can get her down upon the water.'

Fritz.—Ah! now I understand you, father, you have yourself blown up the side of the ship with that machine you contrived in our last visit, that we might be able to get out the pinnace; but how does it happen that so much of the ship is blown away?

Father.—I will explain all this to you when I have convinced myself that the pinnace is not injured, and that there is no danger of any of the fire remaining on board; let us well examine. We entered by the new breach, and had soon reason to be satisfied that the pinnace had wholly escaped from injury, and that the fire was entirely extinguished. The mortar, however, and pieces of the chain, had been driven forcibly into the opposite side of the enclosure. Having now every reason to be satisfied and tranquil, I explained to the boys the nature of a cracker, the manner of its operation, and the important service for which I was indebted to the old mortar.

I then examined the breach we had thus effected, and next the pinnace. I perceived that it would be easy, with the help of the crow and the lever, to lower her into the water. In putting her together, I had used the precaution of placing her keel on rollers, that we might not experience the same difficulty as we had formerly done in launching our tub-raft. Before letting her go, however, I fastened the end of a long thick rope to her head, and the other end to the most solid part of the wreck, for fear of her being carried out too far. We put our whole ingenuity and strength to this undertaking, and soon enjoyed the pleasure of seeing our pretty pinnace descend gracefully into the sea; the rope keeping her sufficiently near, and enabling us to draw her close to the spot where I was loading the tub-boat, and where, for that purpose, I had lodged a pulley on a projecting beam, from which I was enabled also to advance with the completing of the necessary masts and sails for our new barge. I endeavoured to recollect minutely all the information I had ever possessed on the art of equipping a vessel; and our pinnace was shortly in a condition to set sail.

On this occasion, a spirit of military affairs was awakened in the minds of my young group, which was never after extinguished. We were masters of a vessel mounted with two cannon, and furnished amply with guns and pistols! This was at once to be invincible, and in a condition for resisting and destroying the largest fleet the savages could bring upon us! In the height of exultation, it was even almost wished they might assail us! For my own part, I answered their young enthusiasm with pious prayers that we might ever escape such a calamity as the being compelled to use our firearms. Night surprised us before we had finished our work, and we accordingly prepared for our return to Tent House, after drawing the pinnace close under the vessel's side. We arrived in safety, and took great care, as had been previously agreed on, not to mention our new and invaluable booty to the good mother, till we could surprise her with the sight of it in a state of entire completeness. In answer, therefore, to her enquiries as to the noise she heard, we told her that a barrel of gunpowder had taken fire, and had shivered to pieces a small part of the ship.

Two whole days more were spent in completely equipping and loading the beautiful little barge we had now secured. When she was ready for sailing, I found it impossible to resist the earnest importunity of the boys, who, as a recompense for the industry and discretion they had employed, claimed my permission to salute their mother, on their approach to Tent House, with two discharges of cannon. These accordingly were loaded, and the two youngest placed themselves, with a lighted match in hand, close to the touch-holes, to be in readiness. Fritz stood at the mast, to manage the ropes and cables, while I took my station at the rudder. These matters being adjusted, we put off with sensations of lively joy, which was demonstrated by loud huzzas and suitable gesticulation. The wind was favourable, and so brisk, that we glided with the rapidity of a bird along the mirror of the waters; and while my young ones were transported with pleasure by the velocity of the motion, I could not myself refrain from shuddering at the thought of some possible disaster.

Our old friend the tub-raft had been deeply loaded, and fastened to the pinnace, and it now followed as an accompanying boat to a superior vessel. We took down our large sail as soon as we found ourselves at the entrance of Safety Bay, to have the greater command in steering the pinnace; and soon the smaller ones were lowered one by one, that we might the more securely avoid being thrown with violence upon the rocks so prevalent along the coast: thus, proceeding at a slower rate, we had greater facilities for managing the important affair of the discharge of the cannon. Arrived within a certain distance —'*Fire!*' cried Commander Fritz. The rocks behind Tent House returned the sound.—'*Fire!*' said Fritz again—Ernest and Jack obeyed, and the echoes again majestically replied. Fritz at the same moment had discharged his two pistols, and all joined instantly in three loud huzzas.

'Welcome! welcome! dear ones,' was the answer from the anxious mother, almost breathless with astonishment and joy! 'Welcome!' cried also little Francis, with his feeble voice, as he stood clinging to her side, and not well knowing whether he was to be sad or merry! We now tried to push to shore with our oars in a particular direction, that we might have the protection of a projecting mass of rocks, and my wife and little Francis hastened to the spot to receive us: 'Ah, dear deceitful ones!' cried she, throwing herself upon my neck, and heartily embracing me, 'what a fright have you, and your cannon, and your little ship, thrown me into! I saw it advancing rapidly towards us, and was unable to conceive from whence it could come, or what it might have on board: I stole with Francis behind the rocks, and when I heard the firing, I was near sinking to the ground with terror; if I had not the moment after heard your voices, God knows where we should have run to—but come, the cruel moment is now over, and thanks to Heaven, I have you once again in safety! But tell me where you got so unhoped-for a prize as this neat charming little vessel? In good truth, it would almost tempt me once more to venture on a sea-voyage, especially if she would promise to convey us back to our dear country! I foresee of what use she will be to us, and for her sake I think that I must try to forgive the many sins of absence you have committed against me.'

Fritz now invited his mother to get on board, and gave her his assistance. When they had all stepped upon the deck, they entreated for permission to salute, by again discharging the cannon, and at the same moment to confer on the pinnace the name of their mother—*The Elizabeth.*

My wife was particularly gratified by these our late adventures; she applauded our skill and perseverance: 'but do not', said she, 'imagine that I bestow so much commendation without the hope of some return in kind: on the contrary, it is now my turn to claim from you, for myself and little Francis, the same sort of agreeable recompense; for we have not, I assure you, remained idle while the rest were so actively employed for the common benefit.—No, not so; little Francis and his mother found means to be doing something also, though not at this moment prepared to furnish such unquestionable proofs as you, by your salutations of cannon, etc.; but wait a little, good friends, and our proofs shall hereafter be apparent in some dishes of excellent vegetables which we shall be able to regale you with.—It depends, to say the truth, only on yourselves, dear ones, to go with me and see what we have done.'

We did not hesitate to comply, and jumped briskly out of the pinnace for the purpose. Taking her little coadjutor Francis by the hand, she led the way, and we followed in the gayest mood imaginable. She conducted us up an ascent of one of our rocks, and stopping at the spot where the cascade is formed from Jackal's River, she displayed to our astonished eyes a handsome kitchen-garden, laid out properly in beds and walks, and, as she told us, everywhere sowed with the seed of useful plants.

'This', said she, 'is the pretty exploit we have been engaged in, if you will kindly think so of it. In this spot the earth is so

light, being principally composed of decayed leaves, that
Francis and I had no difficulty in working in it, and then
dividing it into different compartments: one for potatoes, one
for manioc, and other smaller shares for lettuces of various
kinds, not forgetting to leave a due proportion to receive some
plants of the sugar-cane. You, dear husband, and Fritz, will
easily find means to conduct sufficient water hither from the
cascade, by means of pipes of bamboo, to keep the whole in
health and vigour; and we shall have a double source of pleasure from the general prosperity; for both the eye and the palate
will be gratified. But you have not yet seen all: there, on the
slope of the rock, I have transplanted some plants of the
ananas.* Between these I have sowed some melon seeds, which
cannot fail to succeed, thus securely sheltered, and in so warm
a soil: here is a plot allotted to pease and beans, and this other
for all sorts of cabbage. Round each bed or plot I have sowed
seeds of maize,* on account of its tall and bushy form, to serve
as a border, which at the same time will protect my young plants
from the scorching heat of the sun.'

I stood transported, in the midst of so perfect an exhibition
of the kind zeal and persevering industry of this most amiable
of women! I could only exclaim, that I should never have
believed in the possibility of such a labour in so short a time,
and particularly with so much privacy as to leave me wholly
unsuspicious of the existence of such a project.

Mother.—To confess the truth, I scarcely myself expected to
succeed, so I resolved to be silent, to avoid being put to the
blush for my presumption. But as I found my little calculations
answer better than I expected, I was encouraged, and the hope
of surprising you so agreeably gave me new strength and
activity. I, however, was not without my suspicions that your
daily visits to the wreck were connected with some great mystery, which at a certain time you would unfold.—So, mystery
for mystery, thought I; and thus my love, it has turned out.
Though acting in different directions, one only object has been
our mutual aim—the substantial good of our beloved companions of the desert!

After a few jocose remarks, with which we closed this conversation, we moved towards Tent House. This was one of our
happiest days; for we were all satisfied with ourselves and with

each other; we had conferred and received benefits; and I led my children to observe the goodness of Providence, who renders even labour a source of enjoyment, and makes our own happiness result from that of the objects of our affection, and our pride to arise from the commendations of which those objects may be deserving.

'I had almost forgot, though,' said my wife, after a short pause, 'one little reproach I had to make you: your trips to the vessel have made you neglect the bundle of precious fruit-saplings we laid together in mould at Falcon's Stream; I fear they by this time must be dying for want of planting, though I took care to water and cover them with branches. Let us go, my love, and see about them.'

I readily consented, as many other matters required our presence at Falcon's Stream. We had now in possession the greater part of the cargo of the vessel; but almost the whole of these treasures were at present in the open air, and liable to injury from both sun and rain.

My wife prepared with alertness for our walk. We hastened to unload the boat, and to place the cargo safely under shelter along with our other stores.

The pinnace was anchored on the shore, and fastened with a rope, by her head, to a stake. When all our stores were thus disposed of, we began our journey to Falcon's Stream, but not empty-handed; we took with us everything that seemed to be absolutely wanted for comfort; and when brought together, it was really so much, that both ourselves and our beasts of burden had no easy task to perform.

21
Gymnastic Exercises; Various Discoveries; Singular Animals, etc.

I RECOMMENDED to my sons to resume the exercise of the shooting of arrows; for I had an extreme solicitude about their preserving and increasing their bodily strength and agility. Nothing tends more to the extinction of personal courage in a human being, than the consciousness of wanting that strength of limb, or that address which may be necessary to aid us in defending ourselves, or in escaping from dangers. On this

occasion, I added the exercises of running, jumping, getting
up trees, both by means of climbing by the trunk, or by a
suspended rope, as sailors are obliged to do to get to the
mast-head. We began at first by making knots in the rope, at a
foot distance from each other; then we reduced the number
of knots, and before we left off we contrived to succeed without
any. I next taught them an exercise of a different nature, which
was to be effected by means of two balls made of lead, fastened
one to each end of a string about a fathom in length. While I
was preparing this machinery, all eyes were fixed upon me.
—'What can it be intended for?' cried one: 'How can we use
it?' asked another: 'Will it soon be ready?' continued a third.

'I am endeavouring', said I, 'to imitate the arms used by a
valiant nation, remarkable for their skill in the chase, and
whom you all must have heard of: I mean the Patagonians,*
inhabitants of the most southern point of America; but, instead
of balls, which they are not able to procure, they tie two heavy
stones, one at each end of a cord, but considerably longer than
the one I am working with: every Patagonian is armed with this
simple instrument, which they use with singular dexterity. If
they desire to kill or wound an enemy, or an animal, they fling
one of the ends of this cord at him, and begin instantly to draw
it back by the other, which they keep carefully in their hand,
to be ready for another throw if necessary: but if they wish to
take an animal alive, and without hurting it, they possess the
singular art of throwing it in such a way as to make it run
several times round the neck of the prey, occasioning a perplexing tightness; they then throw the second stone, and with
so certain an aim, that they scarcely ever miss their object: the
operation of the second is, the so twisting itself about the
animal as to impede his progress, even though he were at a full
gallop. The stones continue turning, carrying with them the
cord: the poor animal is at length so entangled, that he can
neither advance nor retire, and thus falls a prey to the enemy.'

This description was heard with much interest by the boys,
who now all entreated I would that instant try the effect of my
own instrument upon a small trunk of a tree which we saw at
a certain distance. My throws entirely succeeded; and the string
with the balls at the end so completely surrounded the tree,
that the skill of the Patagonian huntsmen required no further

illustration. Each of the boys must then needs have a similar instrument; and in a short time Fritz became quite expert in the art, as indeed he was in every kind of exercise that required strength or address: he was not only the most alert of my children, but being the eldest, his muscles were more formed, and his intelligence was more developed, than could yet be expected in the other three.

The next morning as I was dressing, I remarked from my window in the tree, that the sea was violently agitated, and the waves swelled with the wind. I rejoiced to find myself in safety in my home, and that the day had not been destined for out-of-door occupation. I observed then to my wife, that I should not leave her the whole day, and therefore was ready to execute anything she found wanting in our domestic arrangement. We now fell to a more minute examination than I had hitherto had time for, of all our various possessions at Falcon's Stream. She showed me many things she had herself found means to add to them during my repeated absences from home: among these was a pair of young pigeons which had been lately hatched, and were already beginning to try their wings, while their mother was again sitting on her eggs. From these we passed to the fruit trees we had laid in earth to be planted, and which were in real need of our assistance. I immediately set myself to prevent so important an injury. I had promised the boys, the evening before, to go all together to the wood of gourds, to provide ourselves with vessels of different sizes to keep our provisions in: they were enchanted with the idea, but I bar- gained that they must first assist me to plant all the young trees; which was no sooner said than set about.

When we had finished, the evening was too far advanced for so long a walk. By sunrise the next morning all were on foot; for nothing can exceed the alertness of young persons who act in expectation of a pleasurable change of scene. The ass, harnessed to the sledge, played the principal character: his office was to carry our dinners, a bottle of Canary wine, and some powder and shot, and to bring home our service of empty gourds. Turk, according to custom, led the way as our advanced guard: next followed the three eldest boys, equipped for sporting: after them, the tender mother, leading the little one: and Flora brought up the rear, with the monkey on her back, to which the boys had given the name of Knips. On this occasion I took with me a double-barrelled gun, loaded on the one side with shot for game, and, on the other with ball, in case of meeting with an enemy.

In this manner we set out, full of good humour and high spirits, from Falcon's Stream. Turning round Flamingo Marsh, we soon reached the pleasant spot which before had so delighted us. Fritz took a direction a little further from the sea-shore; and sending Turk into the tall grass, he followed himself, and both disappeared. Soon, eager for sport, we heard Turk barking loud; a large bird sprung up, and almost at the same moment a shot from Fritz brought it down: but though wounded it was not killed; it raised itself, and got off with incredible swiftness, not by flying but by running. Turk followed, and seizing the bird, held it fast till Fritz came up. Now a different scene succeeded from that which took place at the capture of the flamingo. The legs of that bird are long and weak, and it was able to make but a poor resistance. The present captive was large in size, and strong; it struck the dogs, or whoever came near, with its legs, with so much force, that Fritz, who had received a blow or two, dared not again approach the enemy. Fortunately I reached the spot in time to give assistance, and was pleased to see that it was a female bustard* of the largest size. I had long wished to possess and to tame a bird of this species for our poultry-yard, though I foresaw that it would be somewhat difficult.

To secure the bird without injuring it, I threw my pocket handkerchief over the head of the bustard; it could not disengage itself, and its efforts only served to entangle it the more. As it could not now see me, I got near enough to pass a string with a running knot over its legs, which, for the present, I drew tight, to prevent further mischief from such powerful weapons. I gently released its wing from Turk's mouth, and tied it, with its fellow, close to the bird's body. In short, the bustard was our own! and that in a condition to promise its preservation when we should once have conveyed it to Falcon's Stream, and could administer care and kindness to compensate for the rough treatment it had experienced at our hands.

We removed the prisoner to the spot where our companions had been waiting our return. Ernest and Jack ran briskly forward, bawling out, 'Oh, what a handsome bird! And what a size! What beautiful feathers!'—'I think it is a female bustard,' said Ernest. 'And you are right,' answered I; 'its flesh is excellent, having somewhat of the flavour of the turkey, to which it also in some other respects has resemblance. Let us endeavour to tame and preserve it by all means. We have gained for our poultry-yard a bird of rare value on account of its size, which will, it may be hoped attract its mate, and thus furnish us with a brood of its species.'

I now fixed the bustard on the sledge, in a posture the most favourable to its ease. As we advanced on our way, I was frequently obliged to use the hatchet to make a free passage for the ass in the tall grass. The heat also increased, and we were all complaining of thirst, when Ernest, whose discoveries were generally of a kind to be of use, made one of a most agreeable nature. He has already been described as a lover of natural history, and now he had gathered, as he proceeded, such plants as he met with, with the view of adding to his stock of knowledge. He found a kind of hollow stalk of some height, which grew at the foot of trees, and entangled our feet in walking. He cut one of them, and was surprised to see a drop of pure fresh water issue at the place where the knife had been applied: he showed it to us, put it to his lips, and found it pure, and felt much regret that there was no more. I then fell to examining the phenomenon myself, and soon perceived that the want of air prevented a more considerable issue of water. I made some more incisions, and presently water flowed out as if from a small conduit. Ernest, and after him the other boys, quenched their thirst at this new fountain, in the completest manner. I tried the experiment of dividing the plants longways, and they soon gave out water enough to supply even the ass, the monkey, and the bustard. For my own part, touched with deep gratitude for the goodness of God towards me and my beloved family, I raised my eyes to Heaven in thankfulness.

We were still compelled to fight our way through thick bushes, till at length arrived at the wood of gourds, we were not long in finding the spot where Fritz and I had once before enjoyed so agreeable a repose. Our companions had not soon done admiring and wondering at the magnificence of the trees they now beheld, and the prodigious size of the fruit which grew upon the trunk.

Jack and Ernest employed themselves in collecting dried branches and flints, while their mother was occupied in attending to the poor bustard. She remarked to me, that it was cruel to keep her any longer blinded, and her legs tied together on the sledge. To please her, I took off the covering and loosened the string on the legs, but still left it so as to be a guard against its running away, or inflicting blows on those who might approach. I tied her by a long string to the trunk of a tree, that she might relieve herself by walking about.

My wife now gave us notice, that she should want some vessels to contain milk, a large flat spoon to cut out butter by pieces, and next, some pretty plates for serving it at table, made from the gourd rinds.

Father.—You are perfectly reasonable in your demand, dear wife, said I; and, on my part, I require some nests for the pigeons, some baskets for eggs, and some hives for bees.

All.—Oh yes, these things must all be made, we will set earnestly to work.

Jack.—But first, father, tell us how to divide one of the rinds with a string.

I made them gather or collect, till we were in possession of a sufficient number. We now began our work: some had to cut; others to saw, scoop out, and model into agreeable forms. It was a real pleasure to witness the activity exhibited in this our manufacture of porcelain: each tried what specimens he could present for the applause of his companions. For my own part, I made a pretty basket, large enough to carry eggs, with one of the gourds, leaving an arch at the top to serve as a cover. I likewise accomplished a certain number of vessels, also with covers, fit to hold our milk, and then some spoons to skim the cream. My next attempt was some bottles large enough to hold fresh water, and these occasioned me more trouble than all the rest. It was necessary to empty the gourd through the small opening of the size of one's finger, which I had cut in it; I was obliged, after loosening the contents with a stick, to get them out by friction with shot and water well shaken on the inside. Lastly, to please my wife, I undertook the labour of a set of plates for her use. Fritz and Jack engaged to make the hives for the bees and nests for the pigeons and hens. For this last object, they took the largest gourds, and cut a hole in front, the size of the animal for whose use it was intended: they had, when finished, so very pretty an appearance, that little Francis was ready to cry that he was not quite small enough to get into and live in one of them. The pigeons' nests were intended to be tied to the branches of our tree; those for the hens, the geese, and the ducks, were to be placed between its roots, or on the sea-shore, and to represent a sort of hen-coop.

Our work, added to the heat of the day, had made us all thirsty; but we found nothing on this spot like our *fountain* plants, as we had named them. The boys entreated me to go with them in different directions, and try to find some water, not daring by themselves to venture further into the wood.

Ernest with great eagerness proposed relieving me of this trouble, and putting himself in my place. It was not long before we heard him calling loudly to us, and saw him returning in great alarm. 'Run quick, father,' said he, 'here is an immense wild boar. Oh, how he frightened me! I heard him grunting quite close to me, and then he scampered away to the wood. I hear him at this very moment.'

I then cried out to the boys to call the dogs quickly.

'Halloo, here, Turk! Flora!' The dogs arrived full gallop. Ernest was our leader, and conducted us to the place where he saw the boar; but he was gone, and we saw nothing but a plot of potatoes which appeared to have been ransacked by the animal. The ardour for the chase had been somewhat checked in Jack and Ernest, when they considered that they had so formidable a creature to encounter: they stopped short, and began to dig potatoes, and left it to Fritz and me to follow the traces of the dogs. We soon heard the cry of the latter; for they had overtaken the runaway, and soon after the most hideous growling assailed our ears from the same quarter. We advanced with caution, holding our guns in readiness to fire together the instant the animal should be within the proper distance. Presently the spectacle of the two brave creatures attacking him on the right and left presented itself; each held one of his ears between their teeth. But it was not a boar, but our own sow which had run away and so long been lost! After the first surprise we could not resist a hearty laugh; and then we hastened to disencumber our old friend of the teeth of her two adversaries. Her frightful squalling resounded through the wood, and drew our young companions to the place, when a warfare of banter and accusation went round among the parties. But here the attention of all was attracted to a kind of small potato which we observed lying thick on the grass around us, and which had fallen from some trees which appeared loaded with the same production: our sow devoured them greedily, thus consoling herself for the pain and fright the dogs had occasioned her.

The fruit was of different colours, and extremely pleasing to the eye. Fritz expressed his apprehension that it was the poisonous apple called the Mancenilla;* but the sow ate them with so much eagerness, and the tree which bore them having neither the form nor foliage ascribed by naturalists to the Mancenilla, made me doubt of the truth of his idea. I desired my sons to put some of the fruit in their pockets, to make an experiment with them upon the monkey. We now again, from extreme thirst, began to recollect our want of water, and determined to seek for some in every direction. Jack sprang off, and sought among the rocks, hoping that he should discover some little stream: but scarcely had he left the wood, than he bawled to us that he had found a crocodile!

'A crocodile!' cried I with a hearty laugh, 'you have a fine imagination, my boy! Whoever saw a crocodile on such scorching rocks as these, and with not a drop of water near? Now, Jack, you are surely dreaming. . . .'

'Not so much of a dream as you may think, father,' answered Jack, trying to speak in a low voice; 'fortunately he is asleep; —he lies here on a stone at his full length;—do, father, step here and look at it; it does not stir in the least.'

We stole softly to the place where the animal lay; but instead of a crocodile, I saw before me a large sort of lizard, named by naturalists *Leguana* or *Yguana*,* an animal by nature of a mild character, and excellent as food. Instantly all were for seizing him, and presenting so rare a prize to their mother. Fritz was already taking aim with his gun; but I prevented him, observing, that the animal being protected by a coat of scales, it might be difficult to destroy him, and that he is known to be dangerous, if approached when angry. 'Let us try', said I, 'another sort of experiment; as he is asleep, we need not be in a hurry: only a little contrivance is necessary to have him safe in our power alive, and the process will afford us an amusing spectacle.'

I cut a stout stick from a bush, to the extremity of which I tied a string with a running knot. I guarded my other hand simply with a little switch, and thus with cautious steps approached the creature. When I was very near to him, I began to whistle a lively air, taking care to make the sounds low at first, and to increase in loudness till the lizard was awaked. The creature appeared entranced with pleasure as the sounds fell upon his ear; he raised his head to receive them still more distinctly, and looked round on all sides to discover from whence they came. I now advanced by a step at a time, without a moment's interval in the music, which fixed him like a statue to the place. At length I was near enough to reach him with my switch, with which I tickled him gently, still continuing to whistle, one after the other, the different airs I could recollect. The lizard was bewildered by the charms of the music; the attitudes he threw himself into were expressive of a delirious voluptuousness; he stretched himself at full length, made undulating motions with his long tail, threw his head about, raised it up, and by this sort of action, disclosed the formidable range of his sharp-pointed teeth, which were capable of tearing us to pieces if we had excited his hostility. I dexterously seized the moment of his raising his head, to throw my noose over him. When this was accomplished, the boys drew near also, and wanted instantly to draw it tight and strangle him at once; but this I positively forbade, being unwilling to cause the poor animal so unmerited a suffering. I had used the noose only to make sure of him, in case it should happen that a milder mode of killing him, which I intended to try, failed of success, in which case I should have looked to the noose for protection; but this was rendered unnecessary. Continuing to whistle my most affecting melodies, I seized a favourable moment to plunge my switch into one of his nostrils. The blood flowed in abundance, and soon deprived him of life, without his exhibiting the least appearance of being in pain: on the contrary, to the last moment, he seemed to be still listening to the music.

As soon as he was dead, I allowed the boys to come quite near, and to tighten the noose, which we now found useful to draw him to the ground from the large stone on which he lay. My sons were delighted with the means I had used for killing him without pain. We had now to consider of the best way for transporting to Falcon's Stream so large and valuable a booty. After a moment of reflection, I perceived that I had better come at once to the determination of carrying him across my shoulders; and the figure I made with so singular an animal on my back, with his tail dragging on the ground, was not the least amusing circumstance of the adventure. Fritz and Jack presented themselves as pages, contending which should support my train, as they called the tail, which, independently of the good humour inspired amongst us, considerably eased me of the weight, and gave me the air of an old Chinese emperor, habited in a superb royal mantle of many colours; for those of the lizard shone like precious stones in the eyes of the sun.*

We were proceeding in our return, when we distinguished the voice of my wife and little Francis calling loudly upon my name. Our long absence had alarmed them: we had forgot on this occasion to give them notice of our approach by firing our gun, and they had imagined some terrible disaster must have befallen us. No sooner, however, did our cheerful notes reach their ear, than their lamentations were changed to joy, and we were soon the happiest of beings, assembled under a large gourd tree, where we related every particular of the excursion we had made, and laid the lizard at her feet. We had so many things to tell, that, till reminded by my wife, we forgot to mention that we had failed of procuring any water. My sons had taken out some of the unknown apples from their pockets, and they lay on the ground by our side. Knips soon scented them, and came slily up and stole several, and fell to chewing them with great eagerness. I myself threw one or two to the bustard, who also ate them without hesitation. Being now convinced that the apples were not of a poisonous nature, I announced to the boys, who had looked on with envy all the time, that they also might begin to eat them, and I myself set the example. We found them excellent in quality, and I began to suspect that they might be the sort of fruit called *guava*,* which is much esteemed in such countries. The tree which bears them is sometimes twenty feet in height, and of so fertile a nature, that in inhabited countries they are sometimes thinned and cut down, on account of the quantity of land they would occupy. This regale of the apples had in some measure relieved our thirst; but on the other hand, they had increased our hunger; and as we had not time for preparing a portion of the lizard, we were obliged to content ourselves with the cold provisions we had brought with us.

We had scarcely finished, before my wife earnestly entreated we would begin our journey home, and it appeared to me, as the evening was so far advanced, that it would be prudent to return this once without the sledge, which was heavy laden, and the ass could have drawn it but slowly: I therefore determined to leave it on the spot till the following day, when I could return and fetch it, contenting myself with loading the ass, for the present, with the bags which contained our new sets of porcelain; the lizard, which I feared might not keep fresh so long; and our little Francis, who began to complain of being tired. I took these arrangements upon myself, and left to my wife and Fritz the care of confining the bustard in such a manner that she could walk before us without danger of escaping.

When these preparations were complete, our little caravan was put in motion, taking the direction of a straight line to Falcon's Stream. The course of our route now lay along a wood of majestic oaks, and the ground was covered with acorns. My young travellers could not refrain from tasting them, and finding them both sweet and mild to the palate, I had the pleasure of reckoning them as a new means of support. On considering, I recognized that they were a kind of oak which remains always green,* and are a common production of the woods in Florida, and that the Indians of North America extract from its fruit an excellent kind of sweet oil, which they use in cooking their rice. Numerous kinds of birds subsist upon these acorns. This we were led to remark, by the wild and discordant cries of several sorts of jays and parrots, which were skipping merrily among the foliage and the branches.

We arrived shortly at Falcon's Stream, and had time to employ ourselves in some trifling arrangements, before it was completely dark. My wife had great pleasure in taking out her service of porcelain, and using some of the articles that very evening; particularly the handsome egg-basket, and the vessels for the milk. Fritz was instructed to dig a place in the ground, to serve for a kind of cooler, the better to preserve the milk; and we covered it with boards, and put heavy stones to keep them down. Jack took the pigeons' nests, and scampered up the tree, where he nailed them to the branches; he next laid some dry moss within, and placed one of the female pigeons we had contrived to tame, and which at the time was brooding, upon it; he put the eggs carefully under the mother, who seemed to accept his services, and to coo in return, with gratitude.

We concluded the exertions of the day with a plain repast, and the contriving a comfortable bed for the bustard by the side of the flamingo, and then stretched our weary limbs upon the homely couch, rendered by fatigue luxurious, in the giant tree.

22
Excursion into Unknown Tracts

MY first thought the next morning, was to fetch the sledge from the wood. I had a double motive for leaving it there, which I had refrained from explaining to my wife, to avoid giving her uneasiness. I had formed a wish to penetrate a little farther into the land, and ascertain whether anything useful would present itself beyond the wall of rocks. I was, besides, desirous to be better acquainted with the extent, the form, and general productions of our island: I wished Fritz only, who was stronger and more courageous than his brothers, and Turk, to accompany me. We set out very early in the morning, and drove the ass before us for the purpose of drawing home the sledge.

As we were picking up some acorns, different birds of exquisite plumage flitted about us; for this once, I could not refuse Fritz the pleasure of firing upon them, that we might learn their species. He brought down three. I recognized one to be the great blue Virginia jay, and the other two were parrots. One of the two was a superb red parrot; the other was green and yellow.

While Fritz was reloading his gun, we heard a singular sort of noise, which came from a distance. At one moment it resembled a muffled drum, at another, the noise made in sharpening a saw. My first idea was of music played by savages, and we retreated quickly to hide ourselves among the bushes to listen. By degrees we advanced towards the place from whence the sound appeared to come: but perceiving nothing to alarm us, we separated some of the branches with our hands, and then discovered a handsome bird, about the size of the English cock; and, like it too, adorned with elegantly formed smooth feathers round the neck, and a comb upon his head. The animal stood erect on a decayed trunk of a tree, which was lying on the ground, and at this moment exhibited some singular gestures. His tail was spread in the form of a fan, similar to that of the turkey-cock, but shorter; the feathers round his neck and head were erect and bristling.* He sometimes agitated them with so quick a motion, as to make them appear like a vapour which suddenly enclosed him: sometimes he whirled himself round and round on the trunk of the tree; at others he moved his head and eyes in such a manner as to express a state of distraction, making, at the same time, the singular kind of noise with his voice which had alarmed us, and which was preceded and followed by a sort of explosion. This last was caused by the motion of his wing striking in a quick measure on the trunk, which was hollow and dry, and made the noise resemble a muffled drum. There were assembled around him a great number of birds of the same species, but much smaller, and of a less beautiful form. One and all fixed their eyes upon him, and seemed delighted with the pantomime. I contemplated this extraordinary spectacle, of which I had formerly read an account, with astonishment. The number of the spectators of the feathered actor increased every moment, and the performance increased in spirit also in proportion, presenting the idea of a perfect intoxication or delirium of the creature. At this moment Fritz, who stood a little behind me, put an end to the scene by firing off his gun. The actor fell from the stage, and stretching himself on the sand, breathed his last, and the spectators betook themselves suddenly to flight. I must confess, the interest I took in the exhibition was of so lively a nature, that I could not refrain from reproaching Fritz in an angry tone. 'Why', said I, 'must we be always applying the means of death and annihilation to the creatures that fall in our way? Is not nature a thousand times more exhilarating in her animated movements, which express life and enjoyment, than in the selfish scheme of destruction you seem so fond of? Some allowance should no doubt be made for the curiosity of youth, for necessities caused by our situation, and even for the taste you have acquired for sporting. I, therefore, as you well know, do not object to your killing now and then a little game, or some singular or dangerous kind of animal; but moderation is on all occasions useful, and the spectacle of this bird, employed in such earnest endeavours to draw his females around him, was at least as amusing, as to see him stretched there at length, and lifeless, who but two minutes ago exhibited such rapid and lively motions! His pretty hens too, all dispersed in terror, and deprived of the possibility of ever more admiring him, or being his companions!'

Fritz looked down, ashamed and sorry. I observed to him, the thing being done, there was now no remedy; that the fetras, or heath cock, was much esteemed as game; and that he had better take it from the ground, and carry it to his mother.

We now laid the dead cock upon the ass's back, and proceeded on our journey. We soon arrived at the guava trees, and a little after at the spot where we had left the sledge, when we found our treasures in the best possible condition: but as the morning was not far advanced, we entered upon our intended project of penetrating beyond the wall of rocks.

We pursued our way in a straight line at the foot of these massy, solid productions of nature, every moment expecting to reach their extremity, or to find some turn, or breach, or passage through them, that should conduct us into the interior of the island, if, as I presumed, it was not terminated by these rocks. We walked on, continually looking about, that nothing might escape us worthy of notice, and to anticipate and avoid such dangers as should threaten. Turk, with his usual bravery, took the lead, the ass followed with lazy steps, shaking his long ears, and Fritz and I brought up the rear.

We next entered a pretty little grove, the trees of which were unknown to us. Their branches were loaded with large quantities of berries of an extraordinary quality, being entirely covered with a wax which stuck to our fingers as we attempted to gather them. I knew of a sort of bush producing wax that grows in America, and named by botanists *Myrica cerifera;** I had no doubt that this was the plant, and the discovery gave me great pleasure. 'Let us stop here,' said I to Fritz, 'for we cannot do better than collect a great quantity of these berries as a useful present to your mother.'

A short time after, another object presented itself with equal claims to our attention; it was the singular modes of behaviour of a kind of bird scarcely larger than a chaffinch,* and clothed in feathers of a common brown colour. These birds appeared to exist as a republic, there being among them one common nest, inhabited at pleasure by all their tribes. We saw one of these nests in a tree, in a somewhat retired situation; it was formed of plaited straws and bulrushes intermixed; it enclosed great numbers of inhabitants, and was built round the trunk of the tree: it had a kind of roof formed of roots and bulrushes, carefully knit together. We observed in the sides small apertures, seemingly intended as doors and windows to each particular cell of this general receptacle; from a few of these apertures issued some small branches, which served the birds as points of rest for entering and returning: the external appearance of the whole, excited the image of an immensely large open sponge. The inhabitants were very numerous they passed in and out continually, and I estimated that it might contain at least a million. The males were somewhat larger than the females, and there was a trifling difference in their plumage: the number of the males was very small in proportion to the females: I do not know whether this had been the cause of their thus assembling together.

While we were attentively examining this interesting little colony, we perceived a very small kind of parrot hovering about the nest. Their gilded green wings, and the variety of their colours, produced a beautiful effect; they seemed to be perpetually disputing with the colonists, and not unfrequently endeavoured to prevent their entrance into the building; they attacked them fiercely, and even tried to peck at us, if we but advanced our hand to the structure. Fritz, who was well trained in the art of climbing trees, was earnestly desirous to take a nearer view of them, and to secure, if possible, a few individuals. He threw his burden down, and climbed to the nest; he then tried to introduce his hand into one of the apertures, and to seize whatever living creature it should touch, in that particular cell; what he most desired, was to find a female brooding, and to carry both her and the eggs away. Several of the cells were empty, but by perseverance he found one in the situation he wished; but he received so violent a peck from an invisible bird, that his only care was now to withdraw his hand; presently, however, he ventured a second time to pass his hand into the nest, and succeeded in seizing his prey, which he laid hold of, and, in spite of the bird's resistance, he drew it through the aperture, and squeezed it into the pocket of his waistcoat; and buttoning it securely, he slided down the tree, and reached the ground in safety. The signals of distress sent forth by the prisoner collected a multitude of birds from their cells, who all surrounded him, uttering loud cries, and attacking him with their beaks, till he had made good his retreat. He now released the prisoner, and we discovered him to be a beautiful little green parrot, which Fritz entreated he might be allowed to preserve, and make a present of to his brothers, who would make a cage to keep him in, and would then tame him and teach him to speak.

On the road home, we observed to each other, that from the circumstance of this young nestling within the structure, it appeared probable that the true right of property was in this species, and that the brown-coloured birds we at first observed, were intruders, endeavouring to deprive them of it. 'Thus we find', said I to Fritz, 'the existence of social dispositions in almost every class of the animal kingdom, which leads to the combining together for a common cause or benefit. A multitude of causes may induce animals to form a body or society, instead of living singly: among them may be supposed the deficiency of females or of males; the charge of the young; providing them with food; or as a means for their safety and protection. Who shall dare to fix limits to the instinct or to the faculties of the animal creation?'

Fritz.—I do not, however, recollect any kind of animals who live thus together in society, except the bees.

Father.—What say you then to wasps, drones, and different kinds of ants?

Fritz.—I did not indeed recollect the ants, though I have so often amused myself with looking at them: nothing can be more interesting than the ingenious little houses they construct; observing them attentively, we perceive their industry, their economy, their care of their young, in a word, all their undertakings, conducted on a plan of society and numbers.

Father.—Have you also observed with what a provident kind of instinct they bring out their eggs to be warmed by the sun, and for this end remove them from place to place till the time of their maturity?

Fritz.—Is it not probable, father, that what we take for eggs, are chrysales of ants, which, like many other insects, are thus shut up while the process of their taking wings is in the operation?

Father.—You may be right. Writers on natural history have considered the industry and frugality of these insects, as a subject not unworthy of their consideration; but if the common ant of our own country excited so much of your admiration, what will be your astonishment at the labours performed by the ants of other regions! There is a kind which build nests of 4, 6 and 8 feet in height, and large in proportion: the external walls of these structures are so thick and solid, that neither sun nor rain can penetrate them. They are houses which contain within, little streets, arched roofs, piazzas, colonnades, and particular apartments for the offices of housewifery. The ant is an animal of pilfering propensities, on the profits of which it principally lives; it is also remarkable for constancy in its designs, and remaining ever in one place: a species of them exists, however, in America, which is known by the name of the cephalate,* or visiting ant; they make their appearance in numerous troops every two or three years, and disperse themselves in every house; as soon as their visit is observed, it is customary to open all the apartments and receptacles for stores; they enter everywhere, and in a short time it is found that they have exterminated as effectually the rats, mice, bugs, kakerles* (a sort of insect that gives great annoyance in hot countries);—in a word, all the different animals offensive or injurious to man, as if sent on a special mission to remedy the evils these occasion. They do no injury to man, unless they find in him an enemy, who pursues and disturbs their quiet; in

which case they attack his shoes so violently, that they are
destroyed with in- credible rapidity. This curious species does
not build its house above ground, but digs holes, sometimes
not less than 8 feet in depth, and plasters the walls according
to the rules of the art of masonry.

Fritz.—You mentioned just now, that in each class of the
animal creation there were some individuals which formed
themselves into societies; pray tell me which they are?

Father.—I know of no instance among birds, but that we
have just been witnessing: but among quadrupeds there is at
least one striking example of the social principle:—try to
recollect it yourself.

Fritz.—It is perhaps the elephant or the sea-otter.

Father.—Neither is the one I thought of: the animals you
have named discover also a strong disposition to live in society
with their species, but they build nothing like a common house
of reception:—try again.

Fritz.—Ah, is it not the beaver, father? Is it not true, that
these animals possess an intelligence, that enables them to
contrive and place dams to such streams or rivers as obstruct
their design of building entire villages, and that by this operation they are furnished with a sort of ditch, which they use for
their purposes?

Father.—You are quite right; and, strictly speaking, the marmoset* also may be included in the number of sociable
quadrupeds; for they dig themselves a common place of abode,
a sort of cavern, in the mountains, and in these whole families
of them pass the winter comfortably, in a continual sleep.

We reached a wood, the trees of which in a small degree
resembled the wild fig-tree; at least the fruit they bore, like the
fig, was round in form, and contained a soft juicy substance
full of small grains. Their height was from 40 to 60 feet: the
bark of the trunk was scaly, like the pineapple, and wholly bare
of branches, except at the very top.

The leaves of these trees are very thick; in substance, tough,
like leather; and their upper and under surfaces are different
in colour. But what surprised us the most, was a kind of gum,
which issued in a liquid state from the trunk of the tree, and
became immediately hardened by the air. This discovery
awakened Fritz's attention: in Europe he had often made use

of the gum produced by cherry trees, either as a cement or varnish in his youthful occupations; and the thought struck him, that he could do the same with what he now saw.

As we walked, he looked frequently at his gum, which he tried to soften with his breath, but without success: he now discovered a still more singular property in the substance; that of stretching on being pulled at the extremities; and, on letting go, of reducing itself instantly, by the power of an elastic principle. He was struck with surprise, and sprang towards me, repeating the experiment before my eyes, and exclaiming, 'Look, father! if this is not the very thing we formerly used, to rub out bad strokes in our drawings.'

'Ah! what do you tell me?' cried I with joy: 'Such a discovery would be valuable indeed. The best thanks of all will be due to you, if it is the true *caoutchouc* tree* which yields the Indian rubber. Quick, hand it here, that I may examine it.'—Having satisfied myself of our good fortune, I had now to explain, that caoutchouc is a kind of milky sap, which runs from its tree, in consequence of incisions made in the bark. 'This liquor is received in vessels placed expressly for the purpose: it is afterwards made to take the form of dark-coloured bottles of different sizes, such as we have seen them, in the following manner. Before the liquor has time to coagulate, some small earthen bottles are dipped into it a sufficient number of times to form the thickness required. These vessels are then hung over smoke, which completely dries them, and gives them a dark colour. Before they are entirely dry, a knife is drawn across them, which produces the lines or figures with which you have seen them marked. The concluding part of the operation is to break the mould, and to get out the pieces by the passage of the neck, when there remains the complete form of a bottle.'

Fritz.—This process seems simple enough, and we will make some bottles of it for carrying liquids, when we go far in pursuit of game. But still I do not perceive how the discovery is of so much value to us?

Father.—Not by this use of it alone, certainly; but its quality is excellent for being made into shoes and boots without seams, if we can add the assistance of earthen moulds of the size of the leg or foot to be fitted. We must consider of some means of restoring masses of the caoutchouc to its liquid form, for spreading upon the moulds; and if we should not succeed, we must endeavour to draw it in sufficient quantities, in its liquid state, from the trees themselves.

We continued our way till we reached another wood, the skirts of which we had already seen, it being the same which stretches from the sea-shore to the top of the rocks. In this spot alone, and mixed with a quantity of coconut trees, I discovered a sort of tree of smaller growth, which I presumed must be the sago palm:* one of these had been thrown down by the wind, so that I was able to examine it thoroughly. I perceived that the trunk of it contained a large quantity of a mealy substance; I therefore, with my hatchet laid it open longways and cleared it of the whole contents; and I found on tasting, it was exactly like the sago I had often eaten in Europe. We now began to consider how much further we would go: the thick bushes of bamboo, through which it was impossible to pass, seemed to furnish a natural conclusion to our journey. We were therefore unable to ascertain whether or not we should have found a passage beyond the wall of rocks: we perceived then no better resource than to turn to the left towards Cape Disappointment, where the luxurious plantations of sugar-canes now again drew our attention. That we might not return empty-handed to Falcon's Stream, and might deserve forgiveness for so long an absence, we each took the pains to cut a large bundle of the canes, which we threw across the ass's back, not forgetting the ceremony of reserving one apiece to refresh ourselves with along the road. We soon arrived on the well-known shore of the sea, which at length afforded an open and a shorter path; we next reached the wood of gourds, where we found our sledge loaded as we had left it the night before; we took the sugar-canes from the ass, and fastened them to the sledge, and then we harnessed the ass, and the patient animal began to draw towards home.

We arrived at Falcon's Stream without any further adventure. We received at first some kind reproofs; we were next questioned, and lastly thanked, as we displayed our various treasures, but particularly the sugar-canes: each of the boys seized one and began to suck it, as did their mother also. Nothing could be more amusing than to hear Fritz relate, with unaffected interest, our new discoveries, and to see him imitate the gestures of the heath cock, as he held it up for examination:—his hearers continued to shout with laughter for many minutes. Then came the history of the colony of birds and their singular habitation, and of the green parrot, all of which was listened to with the delight excited by a fairy-tale. Fritz showed them the handsome red parrot dead, also the great blue jay, both of which they did not cease to admire; but when he took out of his pocket the little green parrot all alive, there were no bounds to their ecstasy: they jumped about like mad things, and I was obliged to interpose my authority to prevent their tearing him to pieces, in the struggle who should have him first. Francis nearly devoured the little animal with kisses, repeating a thousand times pretty little parrot! At length the bird was fastened by the leg to one of the roots of the trees, till a cage could be made for him; and was fed with acorns, which he appeared exceedingly to relish. We next gave an account of the prospect I now had of furnishing not only candles but boots and shoes. Fritz took a bit of the rubber from his pocket and drew it to its full length, and then let it go suddenly, to the great amusement of little Francis.

Soon after nightfall, we partook of a hearty meal: being much fatigued, we went earlier than usual to rest, and having carefully drawn up the ladder, we fell exhausted, into sound and peaceful slumbers.

23

Useful Occupations and Labours; Embellishments; a Painful but Natural Sentiment.

ON the following day, my wife and the boys importuned me to begin my manufactory of candles: I therefore set myself to recollect all I had read on the subject. I soon perceived that I should be at a loss for a little fat to mix with the wax I had procured from the berries, for making the light burn clearer; but I was compelled to proceed without. I put as many berries into a vessel as it would contain, and set it on a moderate fire; my wife in the meantime employed herself in making some wicks with the threads of sailcloth. When we saw an oily matter of a pleasing smell and light green colour, rise to the top of the liquid the berries had yielded, we carefully skimmed it off and put it into a separate vessel, taking care to keep it warm. We continued this process till the berries were exhausted, and had produced a considerable quantity of wax; we next dipped the wicks one by one into it, while it remained liquid, and then hung them on the bushes to harden: in a short time we dipped them again, and repeated the operation, till the candles were increased to the proper size, and they were then put in a place and kept, till sufficiently hardened for use. We, however, were all eager to judge of our success that very evening, by burning one of the candles, with which we were well satisfied. In consequence of this new treasure, we should now be able to sit up later, and consequently spend less of our time in sleep; but independently of this advantage, the mere sight of a candle, which for so long a time we had been deprived of, caused ecstasies of joy to all.

Our success in this last enterprise, encouraged us to think of another, the idea of which had long been cherished by our kind steward of provisions: it was to make fresh butter of the cream we every day skimmed from the milk, and which was frequently, to her great vexation, spoiled, and given to the animals. The utensil we stood in need of, was a churn, to turn the cream in. Having earnestly applied my thoughts, as to the best manner of conquering the difficulty, I suddenly recollected what I had heard read in a book of travels, of the method used by the Hottentots for making butter; but instead of a sheepskin sewed together at its extremities, I emptied a large gourd, washed it clean, filled it again with cream, and stopped it close with the piece I had cut from the top. I placed my vase of cream on a piece of sailcloth with four corners, and tied to each corner a stake: I placed one boy midway between each stake, and directed them to shake the cloth briskly, but with a steady measure, for a certain time. This exercise, which seemed like children's play, pleased them mightily, and they called it rocking the cradle. They performed their office singing and laughing all the time, and in an hour, on taking off the cover, we had the satisfaction of seeing some excellent butter. We heartily congratulated each other, and praised the workmen, who by their constancy of labour, had thus produced a most agreeable article for food. I had now to propose to my sons a work of a more difficult nature than we had hitherto accomplished: it was the constructing a cart, in all its forms, for the better conveyance of our effects from place to place, instead of the sledge, which caused us so much fatigue to load and draw. Many reasons induced me to confine my attempt in the first instance to a two-wheel cart, and to observe the result before I ventured on one with four wheels. I tried earnestly and long to accomplish such a machine; but it did not entirely succeed to my wishes, and I wasted in the attempt both time and timber; I however produced what from courtesy we called a cart, and it answered the purpose for which it was designed.

When I had no occasion for the boys, they with their mother engaged in other useful matters. They undertook to transplant the European fruit-trees, to place them where they would be in a better situation for growth, according to the properties of each. They planted vine shoots round the roots of the magnificent tree we inhabited, and round the trunks of some other kinds of trees which grew near; and we watched them, in the fond anticipation that they would in time ascend to a height capable of being formed into a sort of trellis, and help to cool us by their shade. Lastly, we planted two parallel lines of saplings, consisting of chestnut, cherry, and the common nut-trees, to form an avenue from Family Bridge to Falcon's Stream, which would hereafter afford us a shaded walk to Tent House. This last undertaking was not to be effected without a degree of labour and fatigue the most discouraging:—the ground was to be cleared of everything it had produced, and a certain breadth covered with sand, left higher in the middle than on the sides for the sake of being always dry. The boys fetched the sand from the seaside in their wheelbarrows.

Our next concern was to introduce, if possible, some shade and other improvements on the barren site of Tent House, and to render our occasional abode or visits there more secure. We began by planting in a quincunx* all those sorts of trees that thrive best in the sun, such as lemon, pistachio, almond, mulberry, and lime trees; lastly, some of a kind of orange tree, which attains to a prodigious size, and bears a fruit as large as the head of a child. The commoner sorts of nut-trees we placed along the shore. The better to conceal and fortify our tent, which enclosed all our stores, we formed on the accessible side, a hedge of wild orange and lemon trees, which produce an abundant prickly foliage; and to add to the agreeableness of their appearance, we here and there interspersed the pomegranate; nor did I omit to make a little arbour of the guava shrub, which is easily raised from slips, and bears a small fruit rather pleasant to the taste. We also took care to introduce at proper places a certain number of the largest sorts of trees, which in time would serve the double purpose of shading annual plants, and, with benches placed under them, of a kind of private cabinet. Should any accident or alarm compel us to retire to the fortress of Tent House, a thing of the first importance would be to find there sufficient food for our cattle. For the greater security, I formed a plantation of the thorny fig-tree, of sufficient breadth to occupy the space between our fortress and the river, thus rendering it difficult for an enemy to approach.

The curving form of the river having left some partial elevations of the soil within the enclosure, I found means to work them into slopes and angles, so as to serve as bastions to our two cannon from the pinnace and our other firearms, should we ever be attacked by savages. When this was all complete, we perceived that one thing more was wanting, which was to make such alterations in Family Bridge as would enable us to use it as a drawbridge, or to take it away entirely, this being the only point at which the passage of the river could be easily effected. But as we could not do all at once, we contented ourselves, for present safety, with taking away the first planks of the bridge at each end every time we passed it. My concluding labour was to plant some cedars along the usual landing-places, to which we might fasten our vessels.

We employed six whole weeks in effecting these laborious arrangements; but the exercise of mind and body they imposed, contributed to the physical and moral health of the boys, and to the support of cheerfulness and serenity in ourselves. The more we embellished our abode by the work of our own hands, the more it became dear to our hearts. The constant and strict observance of the Sabbath-day afforded such an interval of rest as could not fail to restore our strength, and inspire us with the desire of new exertions. The sentiment of gratitude which filled our minds towards the Supreme Being, who had saved us from destruction, and supplied us with all things needful, demanded utterance, and on Sundays we might allow ourselves the indulgence of pouring out our hearts in thankfulness.

By this time we had nearly exhausted our stock of clothes, and we were compelled once more to have recourse to the vessel, which we knew still contained some chests fit for our use. To this motive we added an earnest desire to take another look at her, and, if practicable, to bring away a few pieces of cannon, which might be fixed on the new bastions at Tent House, and thus we should be prepared for the worst.

The first fine day I assembled my three eldest sons, and put my design into execution. We reached the wreck without any striking adventure, and found her still fixed between the rocks, but somewhat more shattered than when we had last seen her. We secured the chests of clothes, and whatever remained of ammunition stores; powder, shot, and even such pieces of cannon as we could remove, while those that were too heavy we stripped of their wheels, which might be extremely useful.

But to effect our purpose, it was necessary to spend several days in visits to the vessel, returning constantly in the evening, enriched with everything of a portable nature which the wreck contained; doors, windows, locks, bolts, nothing escaped our grasp: so that the ship was now entirely emptied, with the exception of the large cannon, and three or four immense copper cauldrons. We by degrees contrived to tie the heaviest articles to two or three empty casks well pitched, which would thus be sustained above water. I supposed that the wind and tide would convey the beams and timbers ashore, and thus with little pains we should be possessed of a sufficient quantity of materials for erecting a building at some future time. When these measures were taken, I came to the resolution of blowing up the wreck, by a process similar to that with which I had so well succeeded with the pinnace. We accordingly prepared a cask of gunpowder, which we left on board for the purpose: we rolled it to the place most favourable for our views: we made a small opening in its side, and at the moment of quitting the vessel, we inserted a piece of matchwood which we lighted at the last moment, as before. We then sailed with all possible expedition for Safety Bay, where we arrived in a short time. We could not, however, withdraw our thoughts from the wreck, and from the expected explosion, for a single moment. I had cut the match a sufficient length for us to hope that she would not go to pieces before dark. I proposed to my wife to have our supper carried to a little point of land from whence we had a view of her, and here we waited for the moment of her destruction with lively impatience.

About the time of nightfall, a majestic rolling sound like thunder, accompanied by a column of fire and smoke, announced that the ship, so awfully concerned with our peculiar destiny, which had brought us to our present abode in a desert, and furnished us there with such vast supplies for general comfort, was that instant annihilated, and withdrawn forever from the face of man!—At this moment, love for the country that gave us birth, that most powerful sentiment of the human heart, sunk with a new force into ours. The ship had disappeared forever! Could we then form a hope ever to behold that country more? We had made a sort of jubilee of witnessing the spectacle: the boys had clapped their hands and skipped about in joyful expectation; but the noise was heard;—the smoke and sparks were seen!—while the sudden change which took place in our minds could be compared only to the rapidity of these effects of our concerted scheme against the vessel. We all observed a mournful silence, and all rose, as it were, by an impulse of mutual condemnation, and with our heads sinking on our bosoms, and our eyes cast upon the ground, we took the road to Tent House.

My wife was the only person who was sensible of motives for consolation in the distressing scene which had been passing; she was now relieved from all the cruel fears for our safety in our visits to a shattered wreck, that was liable to fall to pieces during the time we were on board. From this moment she conceived a stronger partiality for our island, and the modes of life we had adopted.

A night's repose had in some measure relieved the melancholy of the preceding evening, and I went rather early in the morning with the boys, to make further observations as to the effects of this remarkable event. We perceived in the water, and along the shore, abundant vestiges of the departed wreck; and amongst the rest, at a certain distance, the empty casks, cauldrons, and cannon, all tied together, and floating in a large mass upon the water. We jumped instantly into the pinnace, with the tub-boat fastened to it, and made a way towards them through the numberless pieces of timber, etc. that intervened, and in a little time reached the object of our search, which from its great weight moved slowly upon the waves. Fritz, with his accustomed readiness, flung some rope round two four-pounders,* and contrived to fasten them to our barge; after which he secured also an enormous quantity of poles, laths, and other useful articles. With this rich booty we returned to land.

We performed three more trips for the purpose of bringing away more cannon, cauldrons, fragments of masts, etc., all of which we deposited for present convenience in Safety Bay: and now began our most fatiguing operations,—the removing such numerous and heavy stores from the boats to Tent House. We separated the cannon and the cauldrons from the tub-raft, and from each other, and left them in a place which was accessible for the sledge and the beasts of burden. With the help of the crow we succeeded in getting the cauldrons upon the sledge, and in replacing the four wheels we had before taken from the cannon; and now found it easy to make the cow and the ass draw them.

The largest of the boilers or copper cauldrons we found of the most essential use. We brought out all our barrels of gunpowder, and placed them on their ends in three separate groups, at a short distance from our tent; we dug a little ditch round the whole, to draw off the moisture from the ground, and then put one of the cauldrons turned upside down upon each, which completely answered the purpose of an outhouse. The cannon were covered with sailcloth, and upon this we laid heavy branches of trees; the larger casks of gunpowder we prudently removed under a projecting piece of rock, and covered them with planks, till we should have leisure for executing the plan of an ammunition store-house, about which we had all become extremely earnest.

My wife, in taking a survey of these our labours, made the agreeable discovery, that two of our ducks and one of the geese had been brooding under a large bush, and at the time were conducting their little families to the water. The news produced general rejoicings; and the sight of the little creatures, so forcibly carried our thoughts to Falcon's Stream, that we all conceived the ardent desire of returning to the society of the numerous old friends we had left there. One sighed for his monkey, another for his flamingo; Francis for his parrot, and his mother for her poultry-yard, her various housewifery accommodations, and her comfortable bed. We therefore fixed the next day for our departure, and set about the necessary preparations.

24
A New Domain; the Troop of Buffaloes; The Vanquished Hero

ON entering our plantation of fruit-trees forming the avenue to Falcon's Stream, we observed that they had not a vigorous appearance, and that they inclined to curve a little in the stalk: we therefore resolved to support them with sticks, and I proposed to walk to the vicinity of Cape Disappointment, for the purpose of cutting some bamboos. I had no sooner pronounced the words, than the three eldest boys and their mother exclaimed, at once, that they would accompany me. Their curiosity had been excited by our accounts of the amusing objects we had met with in our visit to the spot: each found a sound and special reason why he must not fail to be of the party. Our provision of candles was nearly exhausted, and a new stock of berries must therefore be procured, for my wife now repaired our clothes by candle-light, while I employed myself in composing a journal of the events of every day: —then, the sow had again deserted us, and nothing could be so probable as that we should find her in the acorn-wood: Jack would fain gather some guavas for himself; and Francis must needs see the plantation of sugar-canes. In short, all would visit this land of Canaan.*

We accordingly fixed the following morning, and set out in full procession. For myself, I had a great desire to explore more thoroughly this part of our island. I therefore made some preparations for sleeping, should we find the day too short for all we might have to accomplish: I took the cart instead of the sledge, having fixed some planks across it for Francis and his mother to sit upon when they should be tired: I was careful to be provided with the different implements we might want;

some rope machinery I had contrived for rendering the climbing of trees more easy; and lastly, some provisions, some water in a gourd-flask, and one bottle of wine from the captain's store. When all was placed in the cart, I for this time harnessed to it both the ass and the cow, as I expected the load would be increased on our return; and we set out, taking the road of the potato and manioc plantations. Our first halt was at the tree of the colony of birds, which I now examined with more attention, and recollected to what species they belonged, by naturalists named *Loxia gregaria* (Sociable Grosbeak).

It was not without much difficulty that we conducted the cart through the thick entangled bushes, the most intricate of which I everywhere cut down, and we helped to push it along with all our strength. We succeeded tolerably well at last; and that the poor animals might have time to rest, we determined to pass several hours in this place, which furnished such a variety of agreeable and useful objects. We began by gathering a bag full of the guavas; and after regaling ourselves plentifully, we put the remainder into the cart.

We continued our way, and soon arrived at the caoutchouc, or gum-elastic trees. I thought we could not do better than to halt here, and endeavour to collect a sufficient quantity of the sap to make the different utensils, and the impenetrable boots and shoes, as I had before proposed. It was with this design that I had taken care to bring with me several of the most capacious of the gourd rinds. I made deep incisions in the trunks, and fixed some large leaves of trees, partly doubled together lengthways, to the place, to serve as a sort of channel to conduct the sap to the vessels I had kept in readiness to receive it. We had not long begun this process before we perceived the sap begin to run out as white as milk, and in large drops, so that we were not without hopes, by the time of our return, to find the vessels full, and thus to have obtained a sufficient quantity of the ingredient for a first experiment.

We left the sap running, and pursued our way, which led us to the wood of coconut trees; from thence we passed to the left, and stopped half-way between the bamboos and the sugar-canes, intending to furnish ourselves with a provision of each. We aimed our course so judiciously, that on clearing the skirts of the wood, we found ourselves in an open plain, with the sugar-cane plantations on our left and on our right those of bamboo interspersed with various kinds of palm trees, and, in front, the magnificent bay formed by Cape Disappointment, which stretched far out into the sea.

The prospect that now presented itself to our view was of such exquisite beauty, that we determined to choose it for our resting-place, and to make it the central point of every excursion we should in future make: we were even more than half disposed to desert our pretty Falcon's Stream, and transport our possessions hither: a moment's reflection, however, betrayed the folly of quitting the thousand comforts we had there with almost incredible industry assembled; and we dismissed the thought with promising ourselves to include this ravishing spot ever more in our projects for excursions. We disengaged the animals, that they might graze and refresh themselves under the shade of the palm-trees, and sat down to enjoy our own repast, and to converse on the beauty of the scene.

It was now evening; and as we had determined to pass the night in this enchanting spot, we began to think of forming some large branches of trees into a sort of hut, as is practised by the hunters in America, to shelter us from the dew and the coolness of the air. While we were thus engaged, we were suddenly roused by the loud braying of the ass, which we had left to graze at a distance but a short time before. On going to the place, we saw him throwing his head in the air, and kicking and prancing about; and while we were thinking what could be the matter, he set off on a full gallop. Unfortunately, Turk and Flora, whom we sent after him, took the fancy of entering the plantation of the sugar-canes, while the ass had preferred the direction of the bamboos on the right. We began to fear the approach of some wild beast might have frightened the creature, and to think of assembling our firearms. In a little time the dogs returned, and showed no signs, by scenting the ground or otherwise, of any pursuit. I made a turn round the hut to see that all was well, and then sallied forth with Fritz and the two dogs in the direction the ass had taken, hoping the latter might be enabled to trace him by the scent.

Fatigued, and vexed with the loss of the useful creature, I entered the hut, which I found complete, the boys having covered it with sailcloth, and strewed branches on the ground for sleeping, and collected some reeds for making a fire, which the freshness of the evening air rendered agreeable to all: it served us also for cooking our supper. When all was safe, I watched and replenished the fire till midnight, rather from habit than the fear of wild beasts, and then took possession of the little corner assigned me near my slumbering companions.

The following morning we breakfasted on some milk from the cow, some boiled potatoes, and a small portion of Dutch cheese, and formed during our meal the plan of the business for the day. It was decided that one of the boys and myself, attended by the two dogs, should seek the ass through the bamboo plantation. I took with me the agile Jack, who was almost beside himself with joy at this determination.

We soon reached the bamboo plantation, and found means to force ourselves along its intricate entanglements. After great fatigue, and when we were on the point of relinquishing all further hope, we discovered the print of the ass's hoofs on the soil, which inspired us with new ardour in the pursuit. After spending a whole hour in further endeavours, we at length, on reaching the skirts of the plantation, perceived the sea in the distance, and soon after found ourselves in an open space, which bounded the great bay. A considerable river flowed into the bay at this place, and we perceived that the ridge of rocks which we had constantly seen, extended to the shore, and terminated in a perpendicular precipice, leaving only a narrow passage between the rocks and the river, which during every flux of the tide must necessarily be under water, but which at that moment was dry and passable. The probability that the ass would prefer passing by this narrow way, to the hazard of the water, determined us to follow in the same path: we had also some curiosity to ascertain what might be found on the other side of the rocks, for as yet we were ignorant whether they formed a boundary to our island, or divided it into two portions; whether we should see there land or water. We continued to advance, and at length reached a stream which issued foaming from a large mass of rock, and fell in a cascade into the river. The bed of this stream was so deep, and its course so rapid, that we were a long time finding a part where it might be most practicable for us to cross. When we had got to the other side, we found the soil again sandy, and mixed with a fertile kind of earth: in this place we no longer saw naked rock; but the print of the ass's hoofs were again visible on the ground.

By observing closely, we saw with astonishment, the prints of the feet of other animals, much larger and different in many respects from those of the ass. Our curiosity was so strongly excited, that we resolved to follow the traces; and they conducted us to a plain at a great distance, which presented to our wondering eyes a terrestrial paradise. We ascended a hill which partly concealed from our view this delicious scene, and then, with the assistance of a glass, we beheld an extensive range of country exhibiting every kind of rural beauty, and in which a profound tranquillity had seemed to take up its abode. To our right appeared the majestic wall of rocks which divided the island. Some of these appeared to touch the heavens; others to imprint the clouds with wild fantastic forms, while mists, broken into pieces, partially concealed their tops. To the left, a chain of gently rising hills, the long green verdure of which, tinged with blue, stretched as far as the eye could discern, and were interspersed at agreeable distances with little woods of palm-trees. The river we had crossed flowed in a serpentine course through this exquisite valley, presenting the idea of a broad floating silver riband, while its banks were adorned with reeds and various aquatic plants. I could with difficulty take my eyes from this enchanting spectacle, and I seated myself on the ground to contemplate and enjoy it at my leisure. Neither on the plain nor on the hills was there the smallest trace of the abode of man, nor of any kind of cultivation; it was everywhere a virgin soil, in all its original purity; nothing endowed with life appeared to view, excepting a few birds, which flew fearlessly around us, and a quantity of brilliantly coloured butterflies, which the eye frequently confounded with the different sorts of unknown flowers, which here and there diversified the surface of the soil.

By straining our eyes, however, as far as we could see, we thought we perceived at a great distance some specks upon the land, that seemed to be in motion. We hastened towards the spot; and as we drew nearer, to our inexpressible surprise beheld a pretty numerous group of animals, which in the assemblage presented something like the outline of a troop of horses or of cows. I observed them sometimes run up to each other, and then suddenly stoop to graze. Though we had not lately met with farther traces of the ass, I was not entirely without the hope of finding him among these animals. On a nearer approach, we perceived they were wild buffaloes.* This animal is formed at first sight to inspire the beholder with terror; it is endowed with an extraordinary degree of strength, and two or three of them would have been capable of destroying us in a moment, should they attack us. My alarm was so great that I remained for a few moments fixed to the spot like a statue. By good luck, the dogs were far behind us, and the buffaloes gave no sign of fear or of displeasure at our approach: they stood perfectly still, with their large round eyes fixed upon us in vacant surprise: those which were lying down got up slowly, but not one among them seemed to have any hostile disposition towards us. The circumstance of the dogs' absence was most likely, on this occasion, the means of our safety; as it was, we had time to draw back quietly, and prepare our firearms. It was not, however, my intention to make use of them in any way but for defence, being sensible that we were unequal to the encounter, and recollecting also to have read, that the sound of a gun drives the buffalo to a state of desperation. I therefore thought only of retreating; and with my poor Jack, for whom I was more alarmed than for myself, was proceeding in this way, when unfortunately Turk and Flora ran up to us and we could see were noticed by the buffaloes. The animals instantly, and all together, set up such a roar as to make our nerves tremble; they struck their horns and their hoofs upon the ground, which they tore up by pieces and scattered in the air. I with horror foresaw the moment when, confounding us with the dogs, which no doubt they mistook for jackals, they would seize upon and tear us to pieces. Our brave Turk and Flora, fearless of danger, ran, in spite of all our efforts, into the midst of them, and, according to their manner of attacking, laid hold of the ears of a young buffalo, which happened to be standing a few paces nearer to us than the rest; and though the creature began a tremendous roar and motion with his hoofs, they held him fast, and were dragging him towards us. Thus hostilities had commenced; and unless we could resolve to abandon the cause of our valiant defenders, we were now forced upon the measure of open war, which, considering the strength and number of the enemy, wore a face of the most pressing and inevitable danger. Our every hope seemed now to be in the chance of the terror the buffaloes would feel at the noise of our musketry, which, perhaps, for the first time, would assail their organs, and most likely excite them to flight. With, I must confess, a palpitating heart, and trembling hands, we fired both at the same moment: the buffaloes, terrified by the sound and by the smoke, remained for an instant motionless, as if struck by a thunderbolt, and then one and all betook themselves to flight with such incredible rapidity, that they were soon beyond the reach of our sight. We heard their loud roaring from a considerable distance, which by degrees subsided into silence, and we were left with only one of their terrific species near us; this one, a female, was no doubt the mother of the young buffalo which the dogs had seized and still kept a prisoner; she had drawn near on hearing its cries, and had been wounded by our guns, but not killed; the creature was in a furious state: after a moment's pause, she took aim at the dogs, and with her head on the ground, as if to guide her by the scent, was advancing in her rage, and would have torn them to pieces, if I had not prevented her by firing upon her with my double-barrelled gun, and thus putting an end to her existence.

It was only now that we began to breathe. A few moments before, death, in the most horrible and inevitable form, seemed to stare us in the face! But now we might hope that every danger was over: I was enchanted with the behaviour of my boy, who, instead of giving way to fears and lamentations, as other lads of his age might have done, had stood all the time in a firm posture by my side, and had fired with a steady aim in silence. I bestowed freely on him the commendation he had so well deserved, and made him sensible how necessary it is in times of danger to preserve a presence of mind, which in many cases is of itself sufficient to effect the sought-for deliverance. The young buffalo still remained a prisoner with his ears in the mouths of the dogs, and the pain occasioned him to be so furious, that I was fearful he might do them some injury; I therefore determined to advance and give them what assistance I might find practicable. To say the truth, I scarcely knew in what way to effect this. The buffalo, though young, was strong enough to revenge himself if I were to give the dogs a sign to let go his ears. I had the power of killing him with a pistol at a stroke; but I had a great desire to preserve him alive, and to tame him, that he might be a substitute for the ass, which we had but little hope of recovering. I found myself in a perplexing state of indecision, when Jack suddenly interposed an effective means for accomplishing my wishes. He had his string with balls in his pocket; he drew it out hastily, and making a few steps backward, he threw it so skilfully as to entangle the buffalo completely, and throw him down. As I could then approach him safely, I tied his legs two and two together with a very strong cord; the dogs released his ears, and from this moment we considered the buffalo as our own. Jack was almost mad with joy. 'What a magnificent creature! How much better than the ass he will look, harnessed to the cart! How my mother and the boys will be surprised and stare at him as we draw near!' repeated he, many and many times.

The question was now, how we were to get the buffalo home: having reflected, I conceived that the best way would be to tie his two forelegs together so tight that he could not run, yet loose enough for him to walk; 'and', pursued I, 'we will next adopt the method practised in Italy; you will think it somewhat cruel, but the success will be certain; and it shall afterwards be our study to make him amends by the kindest care and treatment. Hold you the cord which confines his legs with all your strength, that he may not be able to move:—I then called Turk and Flora, and made each again take hold of the ears of the animal; I took from my pocket a sharp pointed knife, and taking hold of the snout, I made a hole in a nostril, into which I quickly inserted the string, which I immediately tied so closely to a tree, that the animal was prevented from the least motion of the head, which might have inflamed the wound and increased his pain. I drew off the dogs the moment the operation was performed. The creature, thus rendered furious, would have run away, but the stricture of the legs and the pain in the nostril prevented it. The first attempt I made to pull the cord, found him docile and ready to accommodate his motions to our designs, and I perceived that we might now begin our march. I left him for a short time to make some other preparations.

I was unwilling to leave so fine a prey as the dead buffalo behind us: I therefore, after considering what was to be done, began by cutting out the tongue, which I sprinkled with some of the salt we had in our provision-bag: I next took off the skin from the four feet, taking care not to tear it in the operation. I remembered that the Americans use these skins, which are of a soft and flexible quality, as boots and shoes, and I considered them as precious articles. I lastly cut some of the flesh of the animal with the skin on, and salted it, and abandoned the rest to the dogs, as a recompense for their behaviour. I then repaired to the river to wash myself, after which we sat down under the shade of a large tree, and ate the rest of our provisions.

As we were not disposed to leave the spot in a hurry, I desired Jack to take the saw and cut down a small quantity of the reeds, which from their enormous size might be of use to us. We set to work but I observed that he took pains to choose the smallest.—'What shall we do', said I, 'with these small-sized reeds? You are thinking, I presume, of a bagpipe, to announce a triumphal arrival to our companions?'—'You are mistaken, father,' answered Jack; 'I am thinking of some candlesticks for my mother, who will set so high a value on them!'

'This is a good thought,' said I; 'I am pleased both with the kindness and the readiness of your invention, and I will assist you to empty the reeds without breaking them: if we should not succeed, at least we know where to provide ourselves with more.'

We had so many and such heavy articles to remove, that I dismissed for that day all thoughts of looking further for the ass. I began now to think of untying the young buffalo; and on approaching him, perceived with pleasure that he was asleep, which afforded me a proof that his wound was not extremely painful. As I began to pull him gently with the string, he gave a start; but he afterwards followed me without resistance. I fastened another string to his horns, and led him on by drawing both together; and he performed the journey with so unexpected a docility, that to ease ourselves of a part of our heavy burdens, we even ventured on the measure of fastening the bundles of reeds upon his back, and upon these we laid the salted pieces of the buffalo. The creature did not seem aware that he was carrying a load; he followed in our path, as before, and thus on the first day of our acquaintance he rendered us an essential service.

In a short time we found ourselves once more at the narrow passage between the torrent and the precipice of the rocks, which I have already mentioned. I had tied the young buffalo to a tree near the cascade, without remarking of what species it might be; when I went to release him, I saw that it was a kind of small palm-tree, and on looking about me, I also observed some other palm-trees, which I had not before met with. One of the kinds, I now remarked, was from ten to twelve feet in height; its leaves were armed with thorns, and it bore a fruit resembling a small cucumber in form, but which at this time was immature, so that we could not taste it. The second, which was smaller, was also thorny; it was now in blossom, and had no fruit. I suspected that the first of these was the *little royal palm*, sometimes called *awiva*, or *Adam's Needle*; and the other, the *dwarf palm*. I resolved to avail myself of both, for further fortifying my enclosure at Tent House, and also to protect the outer side of the narrow pass immediately over the torrent of the cascade. I determined to return and plant a line of them there, as close to each other as the consideration of their growth would allow; for my intention, of course, was to effect this by means of the young shoots, which presented themselves in great abundance: we also hoped by that time to find their fruit ripe, and to ascertain their kind. We repassed the river in safety, and accompanied by the agreeable sounds of its foaming cascades, we regained the narrow pass at the turn of the rocks. We proceeded with caution, and when safe on the other side, we thought of quickening our pace to arrive the sooner at the hut.

The first solicitudes about health and safety being answered, we entered upon the narrative of our adventures; when question after question was so rapidly proposed to us, that we, on our parts, were obliged to ask for the necessary time for our replies. All agreed that our success with the buffalo was the most extraordinary of our achievements: all longed for the morning, when they might take their fill of looking at the spirited creature we had brought with us. The day concluded with supper, and sound repose.

25
The Malabar Eagle; Sago Manufactory; Bees

MY wife the next morning began the conversation. She told me that the boys had been good and diligent; that they had ascended Cape Disappointment with her, and had gathered wood, and made some torches for the night and, what seemed almost incredible had ventured to fell and bring down an immense palm-tree. It lay prostrate on the ground, and covered a space of at least seventy feet in length. To effect their purpose, Fritz had got up the tree with a long rope, which he fastened tight to the top of it. As soon as he had come down again, he and Ernest worked with the axe and saw to cut it through. When it was nearly divided, they cautiously managed its fall with the rope, and in this manner they succeeded. Fritz was in high spirits too on another account: he brought me on his wrist a young bird of prey, of the most beauteous plumage; he had taken it from the nest in one of the rocks near Cape Disappointment. Very young as the bird was, it had already all its feathers, though they had not yet received their full colouring; it answered to the description I had read of the beautiful eagle of Malabar, and I viewed it with the admiration it was entitled to:—meeting with one of these birds is thought a lucky omen; and it being neither large nor expensive in its food, I was desirous to keep it and train it like a falcon, to pursue smaller birds. Fritz had already covered its eyes and tied a string to its foot; and I advised him to hold it often, and for a length of time, on his hand, and to tame it with hunger, as falconers do.

When all the narratives were concluded, I ordered a fire to be lighted, and a quantity of green wood to be put on it, for the purpose of raising a thick smoke, over which I meant to hang the buffalo meat I had salted, to dry and preserve it for our future use. The young buffalo was beginning to browse, and we gave him also a little of the cow's milk; and in a few days we fed him with a heap of sliced potatoes, which he greedily devoured; and this led us to conclude that the pains from the wound in his nose had subsided, and that he would soon become tame.

The morning of this day was spent in again talking over our late extraordinary adventures; we left our meat suspended over the smoke of the fires during our sleep; we tied the young buffalo by the side of the cow, and were pleased to see them agree and bid fair to live in peace together. At night the dogs were set upon the watch. Fritz resolved to go to bed with his eaglet fastened on his wrist, and its eyes still bound: it remained in this state throughout the night without disturbing its master. The time of repose elapsed so calmly, that none of us awoke to keep in the torch-lights, which now for the first time the industry of the boys had supplied us with, and we did not get up till after sunrise. After a moderate breakfast, I chanted the accustomed summons for our setting out; but my young ones had some projects in their heads, and neither they nor their mother were just then in the humour to obey me.

'Let us reflect a little first,' said my wife: 'as we had so much difficulty in felling the palm-tree, would it not be a pity to lose our labour, by leaving it in this place? Ernest assures me it is a sago-tree; if so, the pith would be an excellent ingredient for our soups. Do, my dear, examine it, and let us see if in any way we can turn it to account.'

I found she was in the right: but in that case it was necessary to employ a day in the business; since, to lay open from one end to the other a tree of such a length and substance, was no trivial task. I however consented; as, independent of the use of the farinaceous* pith, I could, by emptying it, obtain two handsome and large troughs for the conveyance of water from Jackal's River to my wife's kitchen-garden at Tent House, and thence to my new plantations of trees.

Fritz.—One of the halves, father, will answer that purpose, and the other will serve as a conduit for our little stream from Falcon's Nest into my pretty basin lined with tortoiseshell; we then shall be constantly regaled with the agreeable view of a fountain close to our dwelling:—I fancy it now before my eyes, and that I see its course. 'And I, for my part,' said Ernest, 'long for a sight of the sago formed into small grains, as I have seen it in Europe.'

I now desired them to bring me the graters they had used for the manioc, and observed that they had to assist me in raising the palm-tree from the ground, which must be done, continued I, by fixing at each end two small cross pieces or props to support it; to split it open as it lies would be a work of too much labour: this done, I shall want several wooden wedges to keep the cleft open while I am sawing it, and afterwards a sufficient quantity of water. 'There is the difficulty,' said my wife; 'our Falcon's Stream is too far off, and we have not yet discovered any spring in the neighbourhood of this place.'

Ernest.—That is of no consequence, mother; I have seen hereabouts so great an abundance of the plants which contain water, that we need not be at a loss; for they will fully supply us, if I could only contrive to get vessels enough to hold it.

We now produced the enormous reeds we had brought home, which being hollow, would answer the purpose of vessels; and as some time was required to draw off the water from such small tubes, he and Francis at once set to work; they cut a number of the plants, which they placed slantingly over the brim of a vessel, and whilst that was filling, they were preparing another. The rest of us got round the tree, and with our united strength we soon succeeded in raising the heavy trunk, and the top of it was then sawed off. We next began to split it through the whole length, and this the softness of the wood enabled us to effect with little trouble. We soon reached the pith or marrow that fills up the middle of the trunk the whole of its length. When divided, we laid one half on the ground, and we pressed the pith together with our hands, so as to make temporary room for the pith of the other half to the trunk, which rested still on the props. We wished to empty it entirely, that we might employ it as a kneading-trough, leaving merely enough of the pith at both ends to prevent a running out; and then we proceeded to form our paste.

My young manufacturers fell joyfully to work: they brought water, and poured it gradually into the trough, whilst we mixed it with the flour. In a short time the paste appeared sufficiently fermented; I then made an aperture at the bottom of the grater on its outside, and pressed the paste strongly with my hand; the farinaceous parts passed with ease through the small holes of the grater, and the ligneous* parts which did not pass were thrown aside in a heap, in the hope that mushrooms, etc. might spring from them. My boys were in readiness to receive in the reed vessels what fell from the grater, and conveyed it directly to their mother, whose business was to spread out the small grains in the sun upon sailcloth, for the purpose of drying them. Thus we procured a good supply of a wholesome and pleasant food; and should have had a larger stock of it, had we not been restricted as to time; but the privilege of renewing the process at pleasure, by felling a sago-tree, added to some impatience to take home our two pretty conduits, and employ them as proposed, prompted us to expedite the business. The paste which remained was thrown upon the mushroom-bed, and watered well to promote a fermentation.

We next employed ourselves in loading the cart with our tools and the two halves of the tree. Night coming on, we retired to our hut, where we enjoyed our usual repose, and early next morning were ready to return to Falcon's Stream. Our buffalo now commenced his service, yoked with the cow; he supplied the want of the ass, and was very tractable: it is true, I led him by the cord in his nose, and thus restrained him whenever he was disposed to deviate from his duty.

We returned the same way as we came, in order to load the cart with a provision of berries, wax, and elastic gum. I sent forward Fritz and Jack as a vanguard, with one of the dogs; they were to cut an ample road through the bushes for our cart. The two water conductors, which were very long, produced numerous difficulties, and somewhat impeded our progress. We reached the wax and gum trees with tolerable speed and without any accident, and halted to place our sacks of berries in the cart. The elastic gum had not yielded as much as I expected, from the too rapid thickening caused by an ardent sun. We obtained however about a quart, which sufficed for the experiment of the impenetrable boots I had so long desired.

We set out again, still preceded by our pioneers, who cleared the way for us through the little wood of guavas. Suddenly we heard a dreadful noise, which came from our vanguard, and beheld Fritz and Jack hastening towards us. I began now to fear a tiger or panther was near at hand, or had perhaps attacked them. Turk began to bark so frightfully, and

Flora joined in so hideous a yell, that I prepared myself for a bloody conflict. I advanced at the head of my troop to the assistance of my high-mettled dogs, who ran furiously up to a thicket, where they stopped, and with their noses to the ground, and almost breathless, strove to enter it. I had no doubt some terrible animal was lurking there; and Fritz, who had seen it through the leaves, confirmed my suspicions; he said it was about the size of the young buffalo, and that his hair was black and shaggy. I was going to fire at it promiscuously in the thicket, when Jack, who had thrown himself on his face on the ground to have a better view of the animal, got up in a fit of laughter—'It is only', exclaimed he, 'our old sow, who is never tired of playing off her tricks upon us.' Half vexed, half laughing, we broke into the midst of the thicket, where in reality we found our old companion stretched supinely on the earth, but by no means in a state of dreary solitude; she had round her seven little creatures, which had been littered a few days, and were sprawling about, contending with each other for the best place near their mother for a hearty meal. This discovery gave us considerable satisfaction, and we all greeted the good matron, who seemed to recollect and welcome us with a sociable kind of grunting, while she licked her young without any ceremony or show of fear. And now a general consultation took place—should this new family be left where we found it, or conveyed to Falcon's Stream? Opinions being at variance, it was decided that for the present they should keep quiet possession of their retreat.

We then, so many adventures ended, pursued our road, and arrived at Falcon's Stream in safety, experiencing what is so generally true, that home is always dear and sacred to the heart, and anticipated with delight. All was in due order, and our animals welcomed our return in their own jargon and manner, but which did not fail to be expressive of their satisfaction in seeing us again. We threw them some of the food they were most partial to, which they greedily accepted, and then voluntarily went back to their usual stand. It was necessary to practise a measure dictated by prudence, which was to tie up the buffalo again, to inure it by degrees to confinement; and the handsome Malabar eagle shared the same fate: Fritz chose to place it near the parrot on the root of a tree; he fastened it with a piece of packthread, of sufficient length to allow it free motion, and uncovered its eyes; till then the bird had been tolerably quiet; but the instant it was restored to light it fell into a species of rage that surprised us; it proudly raised its head, its feathers became ruffled, and its eyeballs seemed to whirl in their orbits, and dart out vivid lightnings. All the poultry were terrified and fled; but the poor luckless parrot was too near the sanguinary creature to escape. Before we were aware of the danger, it was seized and mangled by the formidable hooked beak of the eagle. Fritz vented his anger in loud and passionate reproaches; he would have killed the murderer on the spot, had not Ernest ran up and entreated him to spare its life: 'Parrots', said he, 'we shall find in plenty, but never perhaps so beauteous, so magnificent a bird as this eagle, which, as father observes, we may train for hawking. You may, too, blame only yourself for the parrot's death;—why did you uncover his eyes? I could have told you that falconers keep them covered six weeks, till they are completely tamed. But now, brother, let me have the care of him; let me manage the unruly fellow; he shall soon, in consequence of the methods I shall use, be as tractable and submissive as a new-born puppy.'

Fritz refused to part with his eagle, and Ernest did not long oppose giving him the information he wanted:—'I have read,' said he, 'somewhere, that the Caribs puff tobacco smoke into the nostrils of the birds of prey and of the parrots they catch, until they are giddy and almost senseless;—this stupefaction over, they are no longer wild and untractable.'

Fritz resolved on the experiment: he took some tobacco and a pipe, of which we had plenty in the sailors' chests, and began to smoke, at the same time gradually approaching the unruly bird. As soon as it was somewhat composed, he replaced the fillet over the eyes, and smoked close to its beak and nostrils so effectually, that it became motionless on the spot, and had the exact air of a stuffed bird. Fritz thought it dead, and was inclined to be angry with his brother; but I told him it would not hold on the perch if it were lifeless, and that its head alone was affected;—and so it proved. The favourite came to itself by degrees, and made no noise when its eyes were unbound; it looked at us with an air of surprise, but void of fury, and grew tamer and calmer every day. The care of the monkey was now by all adjudged to Ernest as a reward, and he took formal possession of it, and made it lie down near him.

We next began a business which we had long determined on; it was to plant bamboos close to all the young trees, to support them in their growth. We had our cart loaded with canes in readiness, and a large pointed iron to dig holes in the ground.

We began our work at the entrance of the avenue nearest to Falcon's Stream. The walnut, chestnut, and cherry trees we had planted in a regular line and at equal distances, we found disposed to bend considerably to one side. Being the strongest, I took the task of making holes with the implement upon myself, which, as the soil was light, I easily performed. The boys selected the bamboos, cut them of equal lengths, and pointed the ends to go into the ground. When they were well fixed, we threw up the earth compactly about them, and fastened the sapling by the branches to them with some long straight tendrils of a plant which we found near the spot. In the midst of our exertions we entered into a conversation respecting the culture of trees. Till then my boys had only thought of eating fruits, without giving themselves much trouble about their production; but now their curiosity was excited, and they asked a thousand questions, which I answered as well as I could.

Towards evening, a keen appetite hastened our return to Falcon's Stream, where we found an excellent and plentiful supper prepared by our good and patient steward.

When the sharpness of hunger was appeased, a new subject was introduced, which I and my wife had been thinking of for some time: she found it difficult, and even dangerous, to ascend and descend our tree with a rope ladder: we never went there but on going to bed, and each time felt an apprehension that one of the children, who scrambled up like cats, might make a false step, and perhaps be lamed forever: bad weather might come on, and compel us for a long time together to seek an asylum in our aerial apartment, and consequently to ascend and descend oftener.

My wife had repeatedly applied to me to remedy this evil, and my own anxiety had often made me reflect if the thing were really possible. A staircase on the outside was not to be thought of; the considerable height of the tree rendered that impracticable, as I had nothing to rest it on, and should be at a loss to find beams to sustain it; but I had for some time formed the idea of constructing winding stairs within the immense trunk of the tree, if it should happen to be hollow, or I could contrive to make it so: I had heard the boys talking of a hollow in our tree, and of a swarm of bees issuing from it, and I now, therefore, went to examine whether the cavity extended to the roots, or what its circumference might be. The boys seized the idea with ardour; they sprang up, and climbed to the tops of the roots like squirrels, to strike at the trunk with axes, and to judge from the sound how far it was hollow; but they soon paid dearly for their attempt; the whole swarm of bees, alarmed at the noise made against their dwelling, issued forth, buzzing with fury, attacked the little disturbers, began to sting them, stuck to their hair and clothes, and soon put them to flight, uttering lamentable cries. My wife and I had some trouble to stop the course of their uproar, and cover their little wounds with fresh earth to allay the smart. Jack, whose temper was on all occasions rash, had struck fiercely upon the bees' nest, and was more severely attacked by them than the rest: it was necessary, so serious was the injury, to cover the whole of his face with linen. The less active Ernest got up the last, and was the first to run off when he saw the consequences, and thus avoided any further injury than a sting or two; but some hours elapsed before the other boys could open their eyes, or be in the least relieved from the acute pain that had been inflicted. When they grew a little better, the desire of being revenged of the insects that had so roughly used them had the ascendant in their minds: they teased me to hasten the measures for getting everything in readiness for obtaining possession of their honey. The bees in the meantime were still buzzing furiously round the tree. I prepared tobacco, a pipe, some clay, chisels, hammers, etc. I took the large gourd long intended for a hive, and I fitted a place for it, by nailing a piece of board on a branch of the tree; I made a straw roof for the top, to screen it from the sun and rain; and as all this took up more time than I was aware of, we deferred the attack of the fortress to the following day, and got ready for a sound sleep, which completed the cure of my wounded patients.

26
Treatment of Bees; Staircase; Training of Various Animals;
Manufactures, etc.

NEXT morning, almost before dawn, all were up and in motion; the bees had returned to their cells, and I stopped the passages with clay, leaving only a sufficient aperture for the tube of my pipe. I then smoked as much as was requisite to stupify, without killing the little warlike creatures. Not having a cap with a mask, such as bee-catchers usually wear, nor even gloves, this precaution was necessary. At first a humming was heard in the hollow of the tree, and a noise like a gathering tempest, which died away by degrees. All was become calm, and I withdrew my tube without the appearance of a single bee. Fritz had got up by me: we then began with a chisel and a small axe to cut out of the tree, under the bees' hole of entrance, a piece three feet square. Before it was entirely separated, I repeated the fumigation, lest the stupefaction produced by the first smoking should have ceased, or the noise we had been just making revived the bees. As soon as I supposed them lulled again, I separated from the trunk the piece I had cut out, producing as it were the aspect of a window, through which the inside of the tree was laid open to view; and we were filled at once with joy and astonishment on beholding the immense and wonderful work of this colony of insects. There was such a stock of wax and honey, that we feared our vessels would be insufficient to contain it. The whole interior of the tree was lined with fine honeycombs: I cut them off with care, and put them in the gourds the boys constantly supplied me with. When I had somewhat cleared the cavity, I put the upper combs, in which the bees had assembled in clusters and swarms, into the gourd which was to serve as a hive, and placed it on the plank I had purposely raised. I came down, bringing with me the rest of the honeycombs, with which I filled a small cask, previously well washed in the stream. Some I kept out for a treat at dinner; and had the barrel carefully covered with cloths and planks, that the bees, when attracted by the smell, might be unable to get at it. We assembled round the table, and regaled ourselves plentifully with the delicious treat. My wife then put by the remainder; and I proposed to my sons to go back to the tree, to prevent the bees from swarming again there on being roused from their stupor, as they would not have failed to do, but for the precaution I took of passing a board at the aperture, and burning a few handfuls of tobacco on it, the smell and smoke of which drove them back whenever they attempted to return. At length they desisted, and became gradually reconciled to their new residence, where their queen no doubt had settled herself. I took this opportunity to relate to my children all I had read, in the interesting work by Mr Huber of Geneva*, of the queen bee, this beloved and respected mother of her subjects, who take care of and guard her, work for her, nourish the rising swarms, make the cells in which they are to lodge, prepare others of a different structure, as well as nutriment for the young queens destined to lead forth the fresh colonies. These accounts highly entertained my youthful auditory, who almost regretted having molested the repose of a fine peaceable kingdom that had flourished so long without interruption in the huge trunk. I now advised that all should watch during the night, over the whole provision of honey obtained while the bees were torpid, who, when recovered, would not fail to be troublesome, and come in legions to get back to their property; and to this end we threw ourselves on our beds, in our clothes, to take an early doze: on awakening about nightfall, we found the bees quiet in the gourd, or settled in clusters upon near branches, so we went expeditiously to business. The cask of honey was emptied into a kettle, except a few prime combs, which we kept for daily consumption; the remainder, mixed with a little water, was set over a gentle fire, and reduced to a liquid consistence, strained, and squeezed through a bag, and afterwards poured back into the cask, which was left upright, and uncovered all night to cool. In the morning the wax was entirely separated, and had risen to the surface in a compact and solid cake that was easily removed; beneath was the purest, most beautiful and delicate honey that could be seen: the cask was then carefully headed again, and put into cool ground near our wine-vessels. This task accomplished, I mounted to revisit the hive, and found every thing in order; the bees going forth in swarms, and returning loaded with wax, from which I judged they were forming fresh edifices in their new dwelling-place. I had been surprised that the numbers occupying the trunk of the tree should find room in the gourd, till I perceived the clusters upon the branches, and I thence concluded a young queen was among each of them. In consequence, I procured another gourd, into which I shook them, and placed it by the former: thus I had the satisfaction of obtaining at an easy rate two fine hives of bees in activity.

We soon after these operations proceeded to examine the inside of the tree. I sounded it with a pole from the opening I had made; and a stone fastened to a string served us to sound the bottom, and thus to ascertain the height and depth of the cavity. To my great surprise, the pole penetrated without any resistance to the branches on which our dwelling rested, and the stone descended to the roots. The trunk, it appeared, had wholly lost its pith, and most of its wood internally. It seems that this species of tree, like the willow in our climates, receives nourishment through the bark; for it did not look decayed, and its far-extended branches were luxuriant and beautiful in the extreme. I determined to begin our construction in its capacious hollow that very day. The undertaking appeared at first beyond our powers; but intelligence, patience, time, and a firm resolution, vanquished all obstacles. We were not disposed to relax in any of these requisites; I was pleased to find opportunity to keep my sons in continual action, and their minds and bodies were all the better for exertion. They grew tall and strong, and were too much engaged to regret, in ignoble leisure, any of their past enjoyments in Europe.

We began to cut into the side of the tree, towards the sea, a doorway equal in dimensions to the door of the captain's cabin, which we had removed with all its frame-work and windows. We next cleared away from the cavity all the rotten wood, and rendered the interior even and smooth, leaving sufficient thickness for cutting out resting-places for the winding stairs, without injuring the bark. I then fixed in the centre the trunk of a tree about 20 feet in length, and a foot thick, completely stripped of its branches, in order to carry my winding staircase round it: on the outside of this trunk, and the inside of the cavity of our own tree, we formed grooves, so calculated as to correspond with the distances at which the boards were to be placed to form the stairs. These were continued till I had got to the height of the trunk round which they turned. I made two more apertures at suitable distances, and thus completely lighted the whole ascent. I also effected an opening near our room, that I might more conveniently finish the upper part of the staircase. A second trunk was fixed upon the first, and firmly sustained with screws and transverse beams. It was surrounded, like the other, with stairs cut slopingly; and thus we happily effected the stupendous undertaking of conducting it to the level of our bedchamber. Here I made another door directly into it. To render it more solid and agreeable, I closed the spaces between the stairs with plank. I then fastened two strong ropes, the one descending the length of the central trunk, the other along the inside of our large tree, to assist in case of slipping. I fixed the sash windows taken from the captain's cabin in the apertures we had made to give light to the stairs; and I then found I could add nothing further to my design. When the whole was complete, it was so pretty, solid, and convenient, that we were never tired of going up and coming down it. Our success was owing to the firm resolution adopted by all, to persevere in patient industry and constant efforts to the end; and it employed us many weeks. I have now to relate some occurrences that took place during the construction of our staircase.

A few days after the commencement of our undertaking, our brave Flora whelped us six young puppies, all healthy, and likely to live. The number was so alarming, that I was under the necessity of drowning all but a male and female to keep up the breed. A few days later, the two she-goats gave us two kids, and our ewes five lambs; so that we now saw ourselves in possession of a pretty flock: but lest the domestic animals should follow the example of the ass, and run away from us, I tied a bell to the neck of each. We had found a sufficient number of bells in the vessel, which had been shipped for trading with the savages; it being one of the articles they most value. We could now immediately trace a deserter by the sound, and bring it back to the fold.

Next to the winding stairs, my chief occupation was the management of the young buffalo, whose wound in the nose was quite healed, so that I could lead it at will with a cord or stick passed through the orifice, as the Caffrarians do. I preferred the stick, which answered the purpose of a bit, and I

resolved to break in this spirited beast for riding as well as
drawing. It was already used to the shafts, and very tractable in
them; but I had more trouble in inuring him to the rider, and
to wear a girth, having made one out of the old buffalo's hide.
I formed a sort of saddle with sailcloth, and tacked it to the
girth. Upon this I fixed a burden, which I increased progress-
ively. I was indefatigable in the training of the animal, and soon
brought it to carry, patiently, large bags of potatoes, salt, and
other articles, in the place of the ass. The monkey was his first
rider, who stuck so close to the saddle, that, in spite of the
plunging and kicking of the buffalo, it was not thrown. Francis
was then tried, as the lightest of the family; but throughout
his excursion I led the beast with a halter, that it might not
throw him off. Jack now showed some impatience to mount the
animal in his turn. I next passed the stick through the buffalo's
nose, and tied strong packthread at each end of it, bringing
them together over the neck of the animal, and put this new-
fangled bridle into the hands of the young rider, directing him
how to use it. For a time the lad kept his saddle, notwithstanding the unruly gestures of the creature; at length a side jolt
threw him on the sand, without his receiving much injury.
Ernest, Fritz, and lastly myself, got on successively, with more
or less effect. His trotting shook us to the very centre, the
rapidity of his gallop turned us giddy, and our lessons in horsemanship were reiterated many days before the animal was
tamed, and could be rode with either safety or pleasure. At last,
however, we succeeded without any serious accident; and the
strength and swiftness of our saddled buffalo were prodigious.
It seemed to sport with the heaviest loads. My three eldest boys
mounted it together now and then, and it ran with them with
the swiftness of lightning. By continued attentions it at length
became extremely docile: it was not in the least apt to start;
and I really felt satisfaction in being thus enabled to make my
sons expert riders, so that if they should ever have horses, they
might get on the most restive and fiery without fear:—none
could be compared to our young buffalo; and the ass, which I
had intended to employ in the same way, was far surpassed by
this new member of our family. Fritz and Jack, with my
instructions, amused themselves in training the animal as
horses are exercised in a riding-house: and by means of the

little stick through the nose, they were able to do what they pleased with him.

In the midst of all this, Fritz did not neglect his eagle: he daily shot some small birds which he gave it to eat, placing them sometimes between the buffalo's horns, sometimes on the back of one of the hens, or of the flamingo, or on a shelf, or at the end of a stick, in order to teach it to pounce like a falcon upon other birds. He taught it to perch on his wrist whenever he called or whistled to it; but some time elapsed before he could trust it to soar without securing its return by a long string, apprehending its bold and wild nature would prompt it to take a distant and farewell flight from us.

Our whole company, including even the inert Ernest, was infected with the passion of becoming instructors. Ernest tried his talents in this way with his monkey, who seldom failed to furnish him with work. It was no poor specimen of the ludicrous to see the lad; he whose movements were habitually slow and studied, now constrained to skip and jump, and play a thousand antics with his pupil during training hours, and throughout, against the grain, carrying forward the lesson the grotesque mimic was condemned to learn, of bearing small loads, climbing the coconut trees, and to fetch and bring the nuts. He and Jack made a little hamper of rushes, very light; they put three straps to it, two of which passed under the fore, and one between the hind legs of the animal, and were then fastened to a belt in front, to keep the hamper steady on the back of the mischievous urchin. This apparatus was at first intolerable to poor Knips: he gnashed his teeth, rolled on the ground, jumping like a mad creature, and did everything to get rid of it: but all in vain, for education was the standing order, and he soon found he must submit. The hamper was left on day and night; his sole food was what was thrown into it; and in a short time pug was so much accustomed to the burden, that he began to spit and growl whenever we attempted to take it off, and everything given to the creature to hold was instantly thrown into it. Knips became at length a useful member of our society; but he would only obey Ernest, whom he at once loved and feared, thus affording a proof of at least one of the great ends of all instruction.

These different occupations filled up several hours of the day; when, after working at our stairs, we assembled in the evening round our best of friends, the good mother, to rest ourselves: and forming a little circle, every individual of which was affectionate and cheerful, it was her turn to give us some agreeable and less fatiguing occupation in the domestic concerns of Falcon's Stream: such as improving our candle manufactory, by blending the berry and the bees' wax, and employing the reed-moulds invented by Jack: but having found some difficulty in taking out the candles when cold, I adopted the plan of dividing the moulds, cleaning the inside, and rubbing it over with a little butter, to prevent the wax from adhering to it; then to rejoin both halves with a band that could be loosened at pleasure, to facilitate the extraction of the tapers. The wicks gave us most trouble as we had no cotton. We tried with moderate success the fibrous threads of the karata,* and those of the algava* or flame-wood; but each had the inconvenience of becoming a sort of coal or cinder. The production which gave us the most satisfaction was the pith of a species of elder; but it did not, however, lessen our desire to discover the only appropriate ingredient, the cotton-tree.

We now began to think of manufacturing our impenetrable boots without seams, of the caoutchouc or elastic gum. I began with a pair for myself; and I encouraged my children to afford a specimen of their industry, by trying to form some flasks and cups that could not break. They began by making some clay moulds, which they covered with layers of gum, agreeably to the instructions I had given them. In the meanwhile I filled a pair of stockings with sand, and covered them with a layer of clay, which I first dried in the shade, and afterwards in the sun. I then took a sole of buffalo-leather, well beaten, and studded round with tacks, which served me to fix it under the foot of the stocking; after this I poured the liquid gum into all the interstices, which on drying produced a close adhesion between the leather and stocking sole. I next proceeded to smear the whole with a coat of resin of a tolerable thickness; and as soon as this layer was dried on, I put on another, and so on till I had applied a sufficiency with my brush. After this I emptied the sand, drew out the stocking, removed the hardened clay, shook off the dust, and thus obtained a pair of seamless boots, as finished as if made by the best English workman; being pliant, warm, soft, smooth, and completely waterproof. I hung them up directly, that they might dry without shrinking. They fitted uncommonly well; and my four lads were so highly pleased with their appearance, that they skipped about with joy, as they asked me to make each of them a pair. I refrained from any promise, because I wished to ascertain their strength previously, and to compare them with boots made out of mere buffalo-leather. Of these I at once began a pair for Fritz, with a piece of the slaughtered buffalo's skin. They gave far more trouble than those manufactured with the caoutchouc, which I used to cover the seams and render them less pervious to water. The work turned out very imperfect, and so inferior to my incomparable boots, that Fritz wore them reluctantly; and the more so, as his brother shouted with laughter at the difficulty he had to run in them.

We had also been engaged in the construction of our fountain, which afforded a perpetual source of pleasure to my wife, and indeed to all of us. In the upper part of the stream we built with stakes and stones a kind of dam, that raised the water sufficiently to convey it into the palm-tree troughs; and afterwards, by means of a gentle slope, to glide on contiguous to our habitation, where it fell into the tortoise-shell basin, which we had elevated on stones to a certain height for our convenience; and it was so contrived, that the redundant water passed off through a cane pipe fitted to it. I placed two sticks athwart each other for the gourds, that served as pails, to rest on; and we thus produced, close to our abode, an agreeable fountain, delighting with its rill, and supplying us with a pure crystal fluid, such as we frequently could not get when we drew our water from the bed of the river, which was often encumbered with the leaves and earth fallen into it, or rendered turbid by our waterfowls. The only inconvenience was, that the water flowing in this open state through the narrow channels in a slender stream, was heated, and not refreshing when it reached us. I resolved to obviate this inconvenience at my future leisure, by employing, instead of the uncovered conduits, large bamboo canes fixed deep enough in the ground to keep the water cool. In waiting the execution of this design, we felt pleasure in the new acquisition; and Fritz, who had suggested the notion, received his tribute of praise from all.

27
The Wild Ass; Difficulty in Breaking it; The Heath-Fowl's Nest

WE were scarcely up one morning, and had got to work in putting the last hand to our winding staircase, when we heard at a distance two strange kind of voices, that resembled the howlings of wild beasts, mixed with hissings and sounds of some creature at its last gasp; and I was not without uneasiness: our dogs too pricked up their ears, and seemed to whet their teeth for a sanguinary combat with a dangerous enemy.

From their looks we judged it prudent to put ourselves in a state of defence; we loaded our guns and pistols, placed them together within our castle in the tree, and prepared to repel vigorously any hostile attack from that quarter. The howlings having ceased an instant, I descended from our citadel, well armed, and put on our two faithful guardians their spiked collars and side-guards: I assembled our cattle about the tree to have them in sight, and I reascended to look around for the enemy's approach. Jack wished they might be lions—'I should like', said he, 'to have a near view of the king of beasts, and should not be in the least afraid of him.'

At this instant the howlings were renewed, and almost close to us. Fritz got as near the spot as he could, listened attentively and with eager looks, then threw down his gun, and burst into a loud laughter, exclaiming, 'Father, it is our ass! The deserter comes back to us, chanting the hymn of return: listen! do you not hear his melodious brayings in all the varieties of the gamut?' I listened, and a fresh roar, in sounds unquestionable, raised loud peals of laughter amongst us; and then followed the usual train of jests and mutual banter at the alarm we had one and all betrayed. Shortly after, we had the satisfaction of seeing among the trees our old friend Grizzle, moving towards us leisurely, and stopping now and then to browse; but to our great joy, he was accompanied by one of his own species, of very superior beauty; and when it was nearer, I knew it to be a fine onagra,* or wild ass, which I conceived a strong desire to possess, though at the same time aware of the extreme difficulty there would be in taming and rendering her subject to the use of man. Some writers, who have described it under the name of the *Œigitai*, (or long-eared horse), given it by the Tartars, affirm that the taming it has been ever found impracticable; but my mind furnished an idea on the subject, which I was resolved to act on if I got possession of the handsome creature. Without delay I descended the ladder with Fritz, desiring his brothers to keep still; and I consulted my privy-counsellor on the means of surprising and taking the stranger captive.

I got ready, as soon as possible, a long cord with a running knot, one end of which I tied fast to the root of a tree; the noose was kept open with a little stick slightly fixed in the opening, so as to fall of itself on the cord being thrown round the neck of the animal, whose efforts to escape would draw the knot closer. I also prepared a piece of bamboo about two feet long, which I split at the bottom, and tied fast at top, to serve as nippers. Fritz attentively examined my contrivance, without seeing the use of it. Prompted by the impatience of youth, he took the ball-sling, and proposed aiming at the wild ass with it, which he said was the shortest way of proceeding. I declined adopting this Patagonian method, fearing the attempt might fail, and the beautiful creature avail itself of its natural velocity to evade us beyond recovery: I therefore told him my project of catching it in the noose, which I gave him to manage, as being nimbler and more expert than myself. The two asses drew nearer and nearer to us. Fritz holding in his hand the open noose, moved softly on from behind the tree where we were concealed, and advanced as far as the length of the rope allowed him: the onagra started on perceiving a human figure; it sprang some paces backward, then stopped as if to examine the unknown form; but as Fritz now remained quite still, the animal resumed its composure, and continued to browse. Soon after he approached the old ass, hoping that the confidence that would be shown by it, would raise a similar feeling in the stranger: he held out a handful of oats mixed with salt; our ass instantly ran up to take its favourite food, and greedily devoured it; this was quickly perceived by the other. It drew near, raised its head, breathed strongly, and came up so close, that

Fritz, seizing the opportunity, succeeded in throwing the rope round its neck; but the motion and stroke so affrighted the beast, that it instantly sprang off. It was soon checked by the cord, which, in compressing the neck, almost stopped its breath: it could go no farther, and, after many exhausting efforts, it sunk panting for breath upon the ground. I hastened to loosen the cord, and prevent its being strangled. I then quickly threw our ass's halter over its head; I fixed the nose in my split cane, which I secured at the bottom with packthread. Thus I succeeded in subduing the first alarm of this wild animal, as farriers shoe a horse for the first time. I wholly removed the noose that seemed to bring the creature into a dangerous situation; I fastened the halter with two long ropes to two roots near us, on the right and left, and let the animal recover itself, noticing its actions, and devising the best way to tame it in the completest manner.

The rest of my family had by this time come down from the tree, and beheld the fine creature with admiration, its graceful shape, and well-turned limbs, which placed it so much above the ass, and nearly raised it to the noble structure of the horse. In a few moments the onagra got up again, struck furiously with its foot, and seemed resolved to free itself from all bonds: but the pain of its nose, which was grasped and violently squeezed in the bamboo, forced it to lie down again. Fritz and I now gently undid the cords, and half led, half dragged it, between two roots closely connected, to which we fastened it afresh, so as to give the least scope for motion, and thus render its escape impracticable, whilst it enabled us to approach securely, and examine the valuable capture we had made. We also guarded against master Grizzle playing truant again, and tied him fast with a new halter, confining its forelegs with a rope. I then fastened it and the wild ass side by side, and put before both plenty of good provender to solace their impatience of captivity.

We had now the additional occupation of training the onagra for our service or our pleasure, as might turn out to be most practicable: my boys exulted in the idea of riding it, and we repeatedly congratulated each other on the good fortune which had thus resulted from the flight of our ass. Yet I did not conceal that we should have many difficulties to encounter in taming it, though it seemed very young, and not even to have reached its full growth. But I was inclined to think proper means had not been hitherto adopted, and that the hunters, almost as savage as the animals themselves, had not employed sufficient art and patience, being probably unconscious of the advantages of either. I therefore determined to resort to all possible measures. I let the nippers remain on its nose, which appeared to distress him greatly, though we could plainly perceive their good effect in subduing the creature; for without them no one could have ventured to approach him. I took them off, however, at times, when I gave it food, to render eating easier, and I began, as with the buffalo, by placing a bundle of sailcloth on its back, to inure it to carry. When accustomed to the load, I strove to render the beast by degrees still more docile, by hunger and thirst; and I observed with pleasure, that when it had fasted a little and I supplied it with food, its look and actions were less wild. I also compelled the animal to keep erect on its four legs, by drawing the cords closer that fastened it to the roots, in order to subdue gradually by fatigue its natural ferocity. The children came in turns to play with it, and scratch its ears gently, which were remarkably tender; and it was on these I resolved to make my last trial, if all other endeavours failed. For a long time we despaired of success; the onagra made furious starts and leaps when any of us went near it, kicked with its hind feet, and even attempted to bite those who touched it. This obliged me to have recourse to a muzzle, which I managed with rushes, and put on when it was not feeding. To avoid being struck by its hind feet, I partially confined them, by fastening them to the forefeet with cords, which, however, I left moderately loose, that we might not encroach too much upon the motion necessary for its health. It was at length familiarized to this discipline, and was no longer in a rage when we approached, but grew less impatient daily, and bore to be handled and stroked.

At last we ventured to free it by degrees from its restraints, and to ride it as we had done with the buffalo, still keeping the forefeet tied: but notwithstanding this precaution and every preceding means, it proved as fierce and unruly as ever for the moment. The monkey, who was first put on its back, held on pretty well by clinging to its mane, from which it was suspended as often as the onagra furiously reared and plunged; it was therefore for the present impracticable for either of my sons to get upon it. The perverse beast baffled all our efforts, and the perilous task of breaking it was still to be persevered in with terror and apprehension. In the stable it seemed tolerably quiet and gentle; but the moment it was in any degree unshackled, it became wholly ferocious and unmanageable.

I was at length reduced to my last expedient, but not without much regret, as I resolved, if it did not answer, to restore the animal to full liberty. I tried to mount the onagra, and just as in the act of rearing up violently to prevent me, I seized with my teeth one of the long ears of the enraged creature, and bit it till it bled; instantly it stood almost erect on its hind feet, motionless, and as stiff as a stake; it soon lowered itself by degrees, while I still held its ear between my teeth. Fritz seized the moment, and sprang on its back; Jack, with the help of his mother, did the same, holding by his brother, who on his part clung to the girth. When both assured me they were firmly seated, I let go the ear: the onagra made a few springs less violent than the former, and checked by the cords on its feet, it gradually submitted, began to trot up and down more quietly, and ultimately grew so tractable, that riding it became one of our chief pleasures. My lads were soon expert horsemen; and their horse, though rather long-eared, was very handsome and well broken in. Thus patience on our parts conquered a serious difficulty, and gained for us a proud advantage.

I now explained to my companions that I learned this extraordinary mode of taming from a horsebreaker I met with by chance. He had lived long in America, and carried on the skin-trade with the savages, to whom he took, in exchange, various European goods. He employed in these journeys half-tamed horses of the southern provinces of that country, which are caught in snares or with nooses. They are at first unruly, and resist burdens; but as soon as the hunter bites one of their ears, they become mild and submissive, and at last so docile that anything may be done with them. The journey is continued through forests and over heaths to the dwellings of the savages; skins are given in barter for the goods brought them, with which the horses are reloaded. They set out again on their return, and are directed by the compass and stars to the European settlements, where they profitably dispose of their skins and horses.

In a few weeks the onagra was so effectually tamed, that we all could mount it without fear: I still, however, kept its two forelegs confined together with the cord, to moderate the extreme swiftness of its running. In the room of a bit, I contrived a curb, and with this and a good bite applied, as wanted, to the ear, it went to right or left at the will of the rider. Now and then I mounted it myself, and not without an emotion of pride at my success in subduing an animal that had been considered by travellers and naturalists as absolutely beyond the power of man to tame. But how superior was my gratification, on seeing Fritz spring at any time on the creature's back, drive along our avenue like lightning, and do what he pleased with it, in depicting to my fond imagination, that even on a desert unknown island, I could qualify my dear children to re-enter society, and become in such respects its ornament! In beholding their physical strength and native graces unfold themselves, and these keeping pace with the improvement of their intelligence and their judgement; and in anticipating that, buried as they were in a distant retreat, far from the tumult of the world, and all that excites the passions, their sentiments would be formed in exact conformity to the paternal feelings of my heart! I had not lost the hope that we should one day return to Europe in some vessel chance might throw on our coast, or even with the aid of our pinnace: but I felt, at the same time, and my wife still more, that we should not leave the island without a lively regret, and I determined to pursue my arrangements as if we were to close existence on a spot where all around us prospered.

During the training of our horse, which we named *Lightfoot*, a triple brood of our hens had given us a crowd of little feathered beings; forty of these at least were chirping and hopping about us, to the great satisfaction of my wife, whose zealous care of them sometimes made me smile. Some of these we kept near us, while others were sent in small colonies to feed and breed in the desert, where we could find them as they were wanted for our use.

This increase of our poultry reminded us of an undertaking

we had long thought of, and was not in prudence to be deferred any longer; this was the building, between the roots of our great tree, covered sheds for all our bipeds and quadrupeds. The rainy season, which is the winter of these countries, was drawing near, and to avoid losing most of our stock, it was requisite to shelter it.

We began by forming a kind of roof above the arched roots of our tree, and employed bamboo-canes for the purpose: the longest and strongest supported the roofing in the place of columns, the smaller more closely united and composed the roof itself. I filled up the interstices with moss and clay, and I spread over the whole a thick coat of tar. By these means I formed a compact and solid covering, capable of bearing pressure. I then made a railing round it, which gave the appearance of a pretty balcony, under which, between the roots, were various stalls sheltered from rain and sun, that could be easily shut and separated from each other by means of planks nailed upon the roots: part of them were calculated to serve as a stable and yard, part as an eating-room, a store-room, etc., and as a hayloft to keep our hay and provisions dry in. This work was soon completed; but afterwards it was necessary to fill these places with stores of every kind for our supply throughout the wet season. In this task we engaged diligently, and went daily here and there with our cart to collect every thing useful, and that might give us employment when the weather prevented our going far.

One evening, on our return from digging up potatoes, as our cart loaded with bags, drawn by the buffalo, ass, and cow, was gently rolling along, seeing still a vacant place in the vehicle, I advised my wife to go home with the two youngest boys, whilst I went round by the wood of oaks with Ernest and Fritz, to gather as many sweet acorns as we could find room for. We had still some empty sacks. Ernest was accompanied by his monkey, who seldom left him; and Fritz, horseman like, was on his dear onagra, which he had appropriated to himself, inasmuch as he had helped to take and tame it, and indeed because he knew how to manage it better than his brothers. Ernest was too lazy, and preferred walking at ease with the monkey on his shoulder, and the more so, because it spared him the trouble of gathering fruit.

When we reached the oaks, Lightfoot was tied to a bush, and we set actively to work to gather the acorns that had dropped from the trees. While all were busily employed, the monkey quitted its master's shoulder and skipped unperceived into an adjoining bush. It had been there some time when we heard on that side the loud cries of birds and flapping of wings, and this assured us a sharp conflict was going on between master Knips and the inhabitants of the bushes. I dispatched Ernest to reconnoitre. He went stoutly towards the place, and in an instant we heard him exclaim, 'Come quickly, father! A fine heath fowl's nest full of eggs; Mr Knips, as usual, wished to make a meal of them; the hen and he are fighting for it: come quick, Fritz, and take her; I am holding greedy-chops as well as I can.'

Fritz ran up directly, and in a few moments brought out alive the male and female heath fowl,* both very beautiful; the cock finely collared, similar to one he had killed on a former occasion. I was rejoiced at this discovery, and helped my son to prevent their escape, by tying their wings and feet, and holding them while he returned to the bush for the eggs. And now Ernest came forward driving the monkey before him, and carrying his hat with the utmost care: he had stuck his girdle full of narrow sharp-pointed leaves, in shape like a knife-blade, which reminded me of the production named sword-grass; but I did not pay much attention, as I was too busily engaged in our egg-hunt, and considered his decoration as childishness. On coming up to me he uncovered his hat, and gave it me in a transport of joy, crying out, 'Here, father, are some heath fowl's eggs. I found them in a nest so well concealed under these long leaves, that I should not have observed them had not the hen, in defending herself against the monkey, scattered them about. I am going to take them home, they will please my mother; and these leaves will amuse Francis, for they are like swords, and he will like them for a plaything.' I applauded Ernest's kind thought, and I encouraged him and Fritz to be thus ever considerate for the absent. The kindnesses conferred on those who are separated from us have in themselves more merit, and are more valued, than those which are personally received. It was now time to think of moving homeward: my two sons filled the bags with acorns, and put them on

Lightfoot. Fritz mounted, Ernest carried the eggs I took charge of the hen, and we proceeded to Falcon's Stream followed by our wagon-train. Our good cattle were in such complete subjection, that it was only necessary to speak to them. I remarked Ernest often applying his ear to the hat which held the eggs, as if he thought the little ones were near coming forth; I listened also and observed some shells already broken and the young protruding: we were overjoyed at our good luck, and Fritz could not refrain from trotting on briskly to bear the tidings to his mother. When arrived, our first care was to examine the eggs: the female bird was too frightened and wild to sit upon them: fortunately we had a hen that was hatching; her eggs were immediately removed, and the new ones put in their place: the female heath fowl was put into the parrot's cage, and hung up in the room, to accustom it to our society. In less than three days all the chickens were hatched; they kept close to their foster-mother, and ate greedily a mixture of sweet acorns bruised in milk, such as we gave our tame poultry: as they grew up I plucked out the large feathers of their wings, lest they should naturally take flight: but they and their real parent gradually became so domesticated, that they daily accompanied our feathered stock in search of food, and regularly came back at night to the roost I had prepared for them, and in which this little new colony of feathered beings seemed to delight.

28
Flax; and the Rainy Season

FRANCIS for a short time was highly amused with his sword-leaves, and then, like all children, who are soon tired of their toys, he grew weary of them, and they were thrown aside. Fritz picked up some of them that were quite soft and withered; holding up one which was pliable as a riband in his hand: 'Francis,' said he, 'you can make whips of your sword-grass, and they will be of use in driving your goats and sheep.' It had been lately decided that it should be the business of Francis to lead these to pasture. Fritz accordingly sat down to help him divide the leaves, and afterwards plait them into whip-cords. As they were working, I saw with pleasure the flexibility and strength of the bands; I examined them more closely, and found they

were composed of long fibres or filaments; and this discovery
led me to surmise that this supposed sword-grass might be a
very different thing, and not improbably the flax-plant,* of
New Zealand, called by naturalists *Chlomidia*, and by others
Phormion. This was a valuable discovery in our situation: I knew
how much my wife wished for the production, and that it was
the article she felt most the want of: I therefore hastened to
communicate the intelligence to her, and she expressed the
liveliest joy: 'This', said she, 'is the most useful thing you have
found; lose not a moment in searching for more of these
leaves, and bring me the most you can of them; I will make you
stockings, shirts, clothes, thread, ropes—In short, give me flax,
looms, and frames, and I shall be at no loss in the employment
of it.' I could not help smiling at the scope she gave to her
imagination, on the bare mention of flax, though so much was
to be done between the gathering the leaves, and having the
cloth she was already sewing in idea. Fritz whispered a word in
Jack's ear; both went to the stable, and without asking my leave,
one mounted Lightfoot, the other the buffalo, and galloped
off towards the wood so fast that I had no time to call them
back: they were already out of sight: their eagerness to oblige
their mother in this instance pleaded their forgiveness, and I
suffered them to go on without following them, purposing to
proceed and bring them back if they did not soon return. In
waiting for them I conversed with my wife, who pointed out to
me, with all the animation and spirit of useful enterprise so
natural to her character, the various machinery I must contrive
for spinning and weaving her flax for the manufactory of cloths,
with which she said she should be able to equip us from head to
foot; in speaking of which, her eyes sparkled with doing good, the
love of the purest kind of joy, and I promised her all she desired
of me.

In a quarter of an hour our deserters came back: like true
hussars, they had foraged the woods, and heavily loaded their
cattle with the precious plant, which they threw at their
mother's feet with joyful shouts. It was next proposed that all
should assist her in preparations for the work she was to engage
in, and previously in steeping the flax.

Fritz.—How is flax prepared, father, and what is meant by
steeping it?

Father.—Steeping flax, or hemp, is exposing it in the open air, by spreading it on the ground to receive the rain, the wind, and the dew, in order in a certain degree to liquify the plant; by this means the ligneous parts of the flax are separated with more ease from the fibrous; a kind of vegetable glue that binds them is dissolved, and it can then be perfectly cleaned with great ease, and the parts selected which are fit for spinning.

Fritz.—But may not the natural texture of this part be destroyed by exposing it so long to wet?

Father.—That certainly may happen when the process is managed injudiciously, and the flax not duly turned; the risk, however, is not great, the fibrous part has a peculiar tenacity, which enables it to resist longer the action of humidity: flax may be even steeped altogether in water without injury. Many think this the best and quickest method, and I am of their opinion.

My wife coincided with me, especially in the sultry climate we inhabited: she therefore proposed to soak the flax in Flamingo Marsh, and to begin by making up the leaves in bundles, as they do hemp in Europe. We agreed to her proposal, and joined in this previous and necessary preparation of the flax during the rest of the day.

Next morning the ass was put to the small light car, loaded with bundles of leaves; Francis and the monkey sat on them, and the remainder of the family gaily followed with shovels and pick-axes. We stopped at the marsh, divided our large bundles into smaller, which we placed in the water, pressing them down with stones, and leaving them in this state till it was time to remove and set them in the sun to dry, and thus render the stems soft and easy to peel. In the course of this work we noticed with admiration the instinct of the flamingos, in building their cone-shaped nests above the level of the marsh, each nest having a recess in the upper part, in which the eggs are securely deposited, while the contrivance enables the female to sit with her legs in the water: the nest is of clay closely cemented, so as to resist all danger from the element till the young can swim.

In a fortnight we took the flax out of the water, and spread it on the grass in the sun, where it dried so rapidly that we were able to load it on our cart the same evening, and carry it to

Falcon's Stream, where it was put by till we had time to make the beetles,* wheels, reels, carding-combs, etc. required by our chief for the manufacture. It was thought best to reserve this task for the rainy season, and to employ the present time in collecting a competent stock of provisions for ourselves and for all the animals. Occasional slight showers, the harbingers of winter, had already come on: the temperature, which hitherto had been warm and serene, became gloomy and variable; the sky was often darkened with clouds, the stormy winds were heard, and warned us to avail ourselves of the favourable moment to get all that might be wanted ready.

Our first care was to dig up a full supply of potatoes and yams for bread, with plenty of coconuts, and some bags of sweet acorns. It occurred to us while digging, that the ground being thus opened and manured with the leaves of plants, we might sow in it to advantage, the remainder of our European corn. Notwithstanding all the delicacies this stranger land afforded us, the force of habit still caused us to long for the bread we had been fed with from childhood: we had not yet laid ourselves out for regular tillage, and I was inclined to attempt the construction of a plough of some sort as soon as we had a sufficient stock of corn for sowing. For this time, therefore, we committed it to the earth with little preparation: the season, however, was proper for sowing and planting, as the ensuing rain would moisten and swell the embryo grain, which otherwise would perish in an arid, burning soil. We accordingly expedited the planting of the various palm-trees we had discovered in our excursions, at Tent House, carefully selecting the smallest and the youngest. In the environs we formed a large handsome plantation of sugar-canes, so as to have hereafter everything useful and agreeable around us, and thus be dispensed from the usual toil and loss of time in procuring them.

These different occupations kept us several weeks in unremitted activity of mind and body; our cart was incessantly in motion, conveying home our winter stock: time was so precious, that we did not even make regular meals, and limited ourselves to bread, cheese, and fruits, in order to shorten them, to return quickly to our work, and dispatch it before the bad season should set in.

Unfortunately, the weather changed sooner than we had expected, and than, with all our care, we could be prepared for: before we had completed our winter establishment, the rain fell in such heavy torrents, that I could not refrain from painful apprehension in surmising how we should resist such a body of water, that seemed to change the whole face of the country into a lake.

The first thing to be done, was to remove our aerial abode, and to fix our residence at the bottom of the tree, between the roots and under the tarred roof I had erected; for it was no longer possible to remain above, on account of the furious winds that threatened to bear us away, and deluged our beds with rain through the large opening in front, our only protection here being a piece of sailcloth, which was soon dripping wet and rent to pieces. In this condition we were forced to take down our hammocks, mattresses, and every article that could be injured by the rain; and most fortunate did we deem ourselves in having made the winding stairs, which sheltered us during the operation of the removal. The stairs served afterwards for a kind of lumber-room; we kept all in it we could dispense with, and most of our culinary vessels, which my wife fetched as she happened to want them. Our little sheds between the roots, constructed for the poultry and the cattle, could scarcely contain us all; and the first days we passed in this manner were painfully embarrassing, crowded all together, and hardly able to move in these almost dark recesses, which the fetid smell from the close adjoining animals rendered almost insupportable: in addition, we were half stifled with smoke whenever we kindled a fire, and drenched with rain when we opened the doors. For the first time since our disaster, we sighed for the comfortable houses of our dear country: —but what was to be done! we were not there, and losing our courage and our temper would only increase the evil. I strove to raise the spirits of my companions, and obviate some of the inconveniences. The now doubly precious winding stair was, as I have said, every way useful to us; the upper part of it was filled with numerous articles that gave us room below; and as it was lighted and sheltered by windows, my wife often worked there, seated on a stair, with her little Francis at her feet. We confined our livestock to a smaller number, and gave them a freer current of air, dismissing from the stalls those animals that, from their properties, and being natives of the country, would be at no loss in providing for themselves. That we might not lose them altogether, we tied bells round their necks; Fritz and I sought and drove them in every evening that they did not spontaneously return. We generally got wet to the skin and chilled with cold, during the employment, which induced my wife to contrive for us a kind of clothing more suitable to the occasion; she took two seamen's shirts from the chest, and with some pieces of old coats, she made us a kind of cloth hoods joined together at the back, and well formed for covering the head entirely: we melted some elastic gum, which we spread over the shirts and hoods; and the articles thus prepared answered every purpose of waterproof overalls, that were of essential use and comfort to us. Our young rogues were ready with their scornful jokes the first time they saw us in them: but afterwards they would have been rejoiced to have had the same: this, however, the reduced state of our gum did not allow, and we contented ourselves with wearing them in turn, when compelled to work in the rain, from the bad effects of which they effectually preserved us.

As to the smoke, our only remedy was to open the door when we made a fire; and we did without as much as we could, living on milk and cheese, and never making a fire but to bake our cakes: we then used the occasion to boil a quantity of potatoes, and salt meat enough to last us a number of days. Our dry wood was also nearly expended, and we thanked Heaven the weather was not very cold; for had this been the case, our other trials would have much increased. A more serious concern was, our not having provided sufficient hay and leaves for our European cattle, which we kept housed to avoid losing them; the cow, the ass, the sheep, and the goat, the two last of which were increased in number, required a large quantity of provender, so that we were ere long forced to give them our potatoes and sweet acorns, which by the by, they found very palatable, and we remarked that they imparted a delicate flavour to their milk;—the cow, the goats, and even the sheep, amply supplied us with that precious article: milking, cleaning the animals, and preparing their food, occupied us most of the morning, after which we were usually employed

in making flour of the manioc root, with which we filled the
large gourds, previously placed in rows. The gloom of the
atmosphere and our low windowless habitation, sensibly
abridged our daylight; fortunately, we had laid in a huge store
of candles, and felt no want of that article: when darkness
obliged us to light up, we got round the table, where a large
taper fixed on a gourd gave us an excellent light, which
enabled my wife to pursue her occupation with the needle,
while I on my part was forming a journal, and recording what
the reader has perused of the narrative of our shipwreck and
residence in this island, assisted from time to time by my sons
and their admirable mother, who did not cease to remind me
of various incidents belonging to the story. To Ernest, who
wrote a fine hand, was entrusted the care of writing off my
pages in a clear legible character: Fritz and Jack amused themselves by drawing from memory, the plants and animals which
had most struck their observation; while one and all contributed to teach little Francis to read and write: we concluded
the day with a devotional reading in the Holy Bible, performed
by each in turn, and we then retired to rest, happy in ourselves,
and in the innocent and peaceful course of our existence. Our
kind and faithful steward often surprised us agreeably on our
return from looking after the cattle, by lighting a fagot of dried
bamboo, and quickly roasting by the clear and fervent heat it
produced, a chicken, pigeon, or duck, from our poultry-yard,
or some of the thrushes we had preserved in butter, which were
excellent, and welcomed as a treat to reward extraordinary toil.
Every four or five days the kind creature made us new fresh
butter in the gourd-churn; and this, with some fragrant honey
spread on our manioc cakes, formed a collation that would
have raised the envy of European epicures. These unexpected
regales represented to our grateful hearts so many little
festivals, the generous intention of which made us forget our
bad accommodations and confinement.

The fragments of our meals belonged in right to our
domestic animals, as part of the family. We had now four dogs,
the eagle, and the monkey, to feed; they relied with just confidence on the kindness of their respective masters, who
certainly would have deprived themselves to supply the wants
of their helpless dependants. Francis had taken under his

mighty protection the two puppies; my wife, Flora; and I, the brave Turk:—thus each had his attendant, of which he took care, and no one was dispensed from the offices of tenderness and vigilance. If the buffalo, the onagra, and pig, had not found sustenance abroad, they must have been killed or starved, and that would have given us much pain. In the course of these discomforts, it was unanimously resolved on, that we would not pass another rainy season exposed to the same evils; even my gentle-tempered and most beloved consort, was a little ruffled now and then with our inconvenient situation, and insisted more than any of us on the plan of building elsewhere a more spacious winter residence: she wished, however, to return to our castle in the tree every summer, and we all joined with her in that desire. The choice of a fresh abode now engrossed our attention, and Fritz in the midst of consultation came forward triumphantly with a book he had found in the bottom of our clothes' chest. 'Here', said he, 'is our best counsellor and model, *Robinson Crusoe* ; since Heaven has destined us to a similar fate, whom better can we consult? As far as I remember, he cut himself a habitation out of the solid rock: let us see how he proceeded; we will do the same, and with greater ease, for he was alone; we are six in number, and four of us able to work.' This idea of Fritz was hailed by all. We assembled, and read the famous history with an ardent interest; it seemed, though so familiar, quite new to us: we entered earnestly into every detail, and derived considerable information from it, and never failed to feel lively gratitude towards God, who had rescued us all together, and not permitted one only of us to be cast, a solitary being, on the island. The occurrence of this thought produced on overwhelming sense of affection among us; we could not refrain from throwing ourselves into each other's arms, embracing repeatedly, and the pathetic scene ended in mutual congratulations.

Francis expressed his wish to have a *Man Friday* ; Fritz thought it better to be without such a companion, and to have no savages to contend with. Jack was for the savages, warfare, and encounters. The final result of our deliberations was to go and survey the rocks round Tent House, and to examine whether any of them could be excavated for our purpose.

Our last job for the winter, undertaken at my wife's solicitation, was a beetle for her flax, and some carding-combs. I filed large nails till they were even, round, and pointed; I fixed them at equal distances in a sheet of tin, and raised the sides of it like a box; I then poured melted lead between the nails and the sides, to give firmness to their points, which came out 4 inches. I nailed this tin on a board, and the machine was fit for work. My wife was impatient to use it; and the drying, peeling, and spinning her flax, became from this time a source of inexhaustible delight.

29
Spring; Spinning; Salt Mine

I CAN hardly describe our joy, when, after many tedious and gloomy weeks of rain, the sky began to brighten, the sun to dart its benign rays on the humid earth, the winds to be lulled, and the state of the air became mild and serene. We issued from our dreary hovels with joyful shouts, and walked round our habitation breathing the enlivening balmy ether, while our eyes were regaled with the beauteous verdure beginning to shoot forth on every side. Reviving nature opened her arms, every creature seemed reanimated, and we felt the genial influence of that glorious luminary which had been so long concealed from our sight, and now returned, like a friend who had been absent, to bring us back blessings and delight. We rapidly forgot in new sensations the embarrassments and weary hours of the wet season, and with jocund, hopeful hearts, looked forward to the toils of summer as enviable amusements.

The vegetation of our plantation of trees was rapidly advancing; the seed we had thrown into the ground was sprouting in slender blades that waved luxuriantly; a pleasing tender foliage adorned the trees; the earth was enamelled with an infinite variety of flowers, whose agreeable tints diversified the verdure of the meadows. Odorous exhalations were diffused through the atmosphere; the song of birds was heard around; they were seen between the leaves, joyfully fluttering from branch to branch; their various forms and brilliant plumage heightened this delightful picture of spring, and we were at once struck with wonder and penetrated with gratitude towards the Creator of so many beauties. Under these impressions we celebrated the ensuing Sunday in the open air, and, if possible, with stronger emotions of piety than heretofore. The blessings which surrounded us were ample compensation for some uneasy moments which had occasionally intervened; and our hearts, filled with fresh zeal, were resolved to be resigned, if it should be the will of God, to pass the residue of our days in this solitude with serenity of soul. The force of paternal feelings, no doubt, made me sometimes form other wishes for my children; but these I buried in my own breast, for fear of disturbing their tranquillity; but if I secretly indulged a desire for some event that might prolong and even increase their happiness, I nevertheless wholly submitted all to the Divine will.

Our summer occupations commenced by arranging and thoroughly cleaning Falcon's Nest, the order and neatness of which the rain and dead leaves blown by the wind had disturbed: in other respects, however, it was not injured, and in a few days we rendered it fit for our reception; the stairs were cleared, the rooms between the roots reoccupied, and we were left with leisure to proceed to other employments. My wife lost not a moment in resuming the process of her flax. Our sons hastened to lead the cattle to the fresh pastures; whilst it was my task to carry the bundles of flax into the open air, where, by heaping stones together, I contrived an oven sufficiently commodious to dry it well. The same evening we all set to work to peel, and afterwards to beat it and strip off the bark; and lastly to comb it with my carding machine, which fully answered the purpose. I took this laborious task on myself, and drew out such distaffs full of long soft flax ready for spinning, that my enraptured wife ran to embrace me, to express her thankfulness, requesting me to make her a wheel without delay, that she might enter upon her favourite work.

At an earlier period of my life I had practised turnery* for my amusement; now, however, I was unfortunately destitute of the requisite utensils; but as I had not forgotten the arrangement and component parts of a spinning-wheel* and reel, I by repeated endeavours found means to accomplish those two machines to her satisfaction; and she fell so eagerly to spinning, as to allow herself no leisure even for a walk, and scarcely time to dress our dinners: nothing so much delighted her as to be left with her little boy, whom she employed to reel as fast as she could spin, and sometimes the other three were also engaged in turns at the wheel, to forward her business whilst she was occupied in culinary offices; but not one of them was found so tractable as the cool-tempered, quiet Ernest, who preferred this to more laborious exertions. Our first visit was to Tent House, and here we found the ravages of winter more considerable than even at Falcon's Stream: the tempest and rain had beaten down the tent, carried away a part of the sailcloth, and made such havoc amongst our provisions, that by far the largest portion was spotted with mildew, and the remainder could be only saved by drying them instantly. Luckily, our handsome pinnace had been for the most part spared; it was still at anchor, ready to serve us in case of need; but our tub-boat was in too shattered a state to be of any further service.

In looking over the stores, we were grieved to find the gunpowder, of which I had left three barrels in the tent, the most damaged. The contents of two were rendered wholly useless. I thought myself fortunate on finding the remaining one in tolerable condition, and derived from this great and irreparable loss, a cogent motive to fix upon winter quarters, where our stores, our only wealth, would not be exposed to such cruel dilapidations.

Fritz and Jack were constant in their endeavours to make me undertake the excavation in the rock, but I had no hopes of success. Robinson Crusoe* found a spacious cavern that merely required arrangement; no such cavity was apparent in our rock, which bore the aspect of extreme solidity and impenetrableness; so that, with our limited powers, three or four summers would scarcely suffice to execute the design. Still the earnest desire of a more substantial habitation, to defend us from the elements, perplexed me incessantly, and I resolved to make at least the attempt of cutting out a recess that should protect the gunpowder, the most valuable of all our treasures. I accordingly set off one day, accompanied by my two boys, leaving their mother at her spinning with Ernest and Francis. We took with us pickaxes, chisels, hammers, and iron levers, to try what impression we could make on the rock. I chose a part nearly perpendicular, and much better situated than our tent: the view from it was enchanting; for it embraced the whole range of Safety Bay, the banks of Jackal's Stream, and Family Bridge, and many of the picturesque projections of the rocks. I marked out with charcoal the opening we wished to make, and we began the heavy toil of piercing the quarry. We made so little progress the first day, that, in spite of our courage, we were tempted to relinquish the undertaking; we persevered, however, and my hope was somewhat revived as I perceived the stone was of a softer texture as we penetrated deeper: I concluded from this, that the ardent rays of the sun striking upon the rock had hardened the external layer, and that the stone within would increase in softness as we advanced; and it occurred to me, that the substance might be a species of calcareous stone.* When I had cut about a foot in depth, we could loosen it with a spade like dried mud; this determined me to proceed with double ardour, and my boys assisted me with a spirit and zeal beyond their years.

After a few days of assiduous labour, we measured the opening, and found we had already advanced seven feet into the rock. Fritz removed the fragments in a barrow, and discharged them in a line before the place, to form a sort of terrace; I applied my own labour to the upper part, to enlarge the aperture; Jack, the smallest of the three, was able to get in and cut away below. He had with him a long iron bar sharpened at the end, which he drove in with a hammer, to loosen a piece at a time; suddenly he bawled out: 'It is pierced through, father! Fritz, I have pierced it through!'

'Hah, hah, master Jack at his jokes again!—But let us hear, what have you pierced? Is it the mountain? Not peradventure your hand or foot, Jack?' cried I.

Jack.—No, no, it is the mountain (the rocks resounding with his usual shout of joy); huzza, huzza! I have pierced the mountain!

Fritz now ran to him. 'Come, let us see then: it is no doubt the globe at least you have pierced,' said he, in a bantering tone: 'you should have pushed on your tool boldly, till you reached Europe, which they say is under our feet; I should have been glad to peep into that hole.'

Jack.—Well, then, peep you may, but I hardly know what you will see; come and look how far the iron is gone in, and tell me if it is all my boasting.

'Come hither, father,' said Fritz, 'this is really extraordinary; his iron bar seems to have got to a hollow place; see, it can be moved in every direction.' I approached, thinking the incident worth attention: I took hold of the bar, which was still in the rock, and working it about, I made a sufficient aperture for one of my sons to pass, and I observed that in reality the rubbish fell within the cavity, which I judged, from the falling of the stones, was not much deeper than the part we stood on. My two lads offered to go in together and examine it: this, however, I forbade. I even made them remove from the opening, as I smelled the mephitic* air, that issued abundantly from it, and began myself to feel giddiness in consequence of having gone too near; so that I was compelled to withdraw quickly, and inhale a purer air. 'Beware, my dear children,' said I, in terror, 'of entering such places, for the loss of life might be the consequence.'

Jack.—How can that be, father?

Father.—Because the air is mephitic, that is, foul, and therefore unfit for breathing in.

Jack.—How does air become mephitic?

Father.—In different ways: for example, when it is replete with noxious vapours, or when it contains too many igneous or inflammable particles, or when it is too heavy or dense, as fixed air is; but in general, when it merely loses its elasticity, it no longer passes freely into the lungs; respiration is then stopped, and suffocation speedily ensues, because air is indispensable to life and the circulation of the blood.

Jack.—Then all to be done is, to be off quickly when one feels a stoppage of breath.

Father.—This is certainly the natural course, when it can be taken; but the attack usually begins by a vertigo or dizziness of the head, so violent as to intercept motion, which is followed by an insurmountable oppression; efforts are made to breathe, fainting follows, and, without speedy help, a sudden death takes place.

Fritz.—What assistance can be administered?

Father.—The first thing to be done is to remove the person so affected to pure fresh air, and to throw cold water over his body; he must then be well dried, and afterwards rubbed with warm cloths; vital air must be infused, or tobacco-smoke thrown up;—in short, he must be treated like a drowned person till signs of reanimation appear, which is not always the result.

Fritz.—But why do you think, father, the air in this cavern is mephitic, as you term it, or dangerous to breathe in?

Father.—All air confined and wholly separated from that of the atmosphere, gradually loses its elasticity, and can no longer pass through the lungs: in this state it generates injurious qualities that interrupt the process of respiration. It is in this act that the atmospheric air diffused around us, unites intimately with the blood, to which it communicates one of its most essential parts, called vital air, for without it life cannot be supported. This air failing, respiration ceases, and death succeeds in a few minutes: the consequence is similar when this air is impregnated too abundantly with injurious parts.

Fritz.—And by what is good air known? How judge that one may respire freely at a few paces from this mephitic cave?

Father.—This becomes evident when inspiration and expiration are performed with ease; besides, there is an infallible test: fire does not burn in foul air, yet it is made the means of correcting it. We must light a fire of sufficient strength in this hole to purify the air within, and render it friendly to respiration; at first the bad air will extinguish the fire, but by degrees the fire in its turn will expel the bad air and burn freely.

Fritz.—Oh! that will be an easy matter.

The boys now hastened to gather some dry moss, which they made into bundles; they then struck a light and set fire to them, and threw the moss blazing into the opening; but, as I had described, the fire was extinguished at the very entrance, thus proving that the air within was highly mephitic. I now saw that it was to be rarefied by another and more effectual method; I recollected that we had brought from the vessel a chest that was full of grenades, rockets, and other fireworks, which had been shipped for the purpose of making signals, as well as for amusement. I sought it hastily, and took some of these, together with an iron mortar for throwing; out of it I laid a train of gunpowder, and set fire to the end which reached to where we stood: a general explosion took place, and an awful report reverberated through the dark recess; the lighted grenades flew about on all sides like brilliant meteors, rebounding and bursting with a terrific sound. We then sent in the rockets, which had also a full effect. They hissed in the cavity like flying dragons, disclosing to our astonished view its vast extent. We beheld too, as we thought, numerous dazzling bodies, that sparkled suddenly, as if by magic, and disappeared with the rapidity of lightning, leaving the place in total darkness. A squib* bursting in the form of a star, presented a spectacle we wished to be prolonged. On its separating, a crowd of little winged genii came forth, each holding a small lighted lamp, and the whole fluttering in every direction with a thousand varied reverberations: everything in the cavern shone brilliantly, and offered instantly a truly enchanting sight; but they dropped in succession, fell to the ground without noise, and vanished like ethereal spirits.

After having played off our fireworks, I tried lighted straw: to our great satisfaction, the bundles thrown in were entirely consumed; we could then reasonably hope nothing was to be feared from the air; but there still remained the danger of plunging into some abyss, or of meeting with a body of water. From these considerations, I deemed it more prudent to defer our entrance into this unknown recess, till we had lights to guide us through it. I dispatched Jack on the buffalo to Falcon's Stream, to tell his mother and brothers of our discovery, directing him to return with them, and bring all the tapers* that were left: my intention was to tie them together to the end of a stick, and proceed with it lighted to examine the cavity. I had not sent Jack on his embassy without a meaning; the boy possessed from nature a lively imagination: I knew he would tell his mother such wonders of the enchanted grotto, of the fireworks, and all they had brought to our view, that he would induce her to accompany him without delay, and bring us lights to penetrate the obscure sanctuary.

Jack, overjoyed, sprang on the buffalo, gaily smacked his whip, and set off so boldly, that I almost trembled for his safety. The intrepid boy was unencumbered by fear, and made a complete racehorse of his horned Bucephalus.*

In three or four hours we saw them coming up in our car of state, which was now drawn by the cow and the ass, and conducted by Ernest. Francis too played his part in the cavalcade, and contended with his brother for the ropes that served as reins. Jack, mounted on his buffalo, came prancing before them; blew through his closed hand, in imitation of the French horn,* and now and then whipped the ass and cow to quicken their motion. When they had crossed Family Bridge, he came forward on the gallop; and when he got to us, jumped off the beast, shook himself, took a spring or two from the ground, and thus refreshed, ran up to the car to hand his mother out, like a true and gallant knight.

I immediately lighted some of the tapers; but not together, as I had intended; I preferred each taking one in his right hand, an implement in his left, another taper in his pocket, flint and steel; and thus we entered the rock in solemn procession. I took the lead, my sons followed me, and their beloved mother, with the youngest, brought up the rear. The interest and curiosity she felt were not unalloyed with tender apprehensions; and indeed I felt myself that sort of fear which an unknown object is apt to excite; even our dogs that accompanied us betrayed some timidity, and did not run before as usual; but we had scarcely advanced four paces within the grotto, when all was changed to more than admiration and surprise. The most beautiful and magnificent spectacle presented itself. The sides of the cavern sparkled like diamonds, the light from our six tapers was reflected from all parts, and had the effect of a grand illumination. Innumerable crystals of every length and shape hung from the top of the vault; which, uniting with those of the sides, formed pillars, altars, entablatures, and a variety of other figures, composing the most splendid masses. We might have fancied ourselves in the palace of a fairy, or an illumined temple. In some places, all the colours of the prism were emitted from the angles of the crystals, and gave them the appearance of the finest precious stones. The waving of the lights, their bright coruscations, dark points here and there intervening, the dazzling lustre of others—the whole, in short, delighted and enchanted the sight and the fancy.

The astonishment of my family was so great as to be almost ludicrous; they were all in a kind of dumb stupor, half imagining it was a dream. For my own part, I had seen stalactites, and read the description of the famous grotto of Antiparos;* my sensations, therefore, were not the same. The bottom was level, covered with a white and very fine sand, as if purposely strewed, and so dry, that I could not see the least mark of humidity anywhere. All this led me to hope the spot would be healthy, convenient, and eligible for our proposed residence. I now formed a particular conjecture as to the nature of the crystallizations shooting out on all sides, and especially from the arch-roof. They could scarcely be of that species of rock-crystals produced by the slow filtering of water falling in drops and coagulating in succession, and seldom found in excavations exhibiting so dry a nature, nor ever with so many of the crystals perpendicular and perfectly smooth. I was impatient to evince the truth or falsehood of this idea by an experiment, and discovered with great joy, on breaking a portion of one of them, that I was in a grotto of *sal gem*, that is, fossil or rock salt, found in the earth in solid crystallized masses, generally above a bed of spar or gypsum, and surrounded by layers of fossils or rock. The discovery of this fact, which no longer admitted a doubt, pleased us all exceedingly. The shape of the crystals, their little solidity, and finally their saline taste, were decisive evidences.

How highly advantageous to us and our cattle was this superabundance of salt, pure and ready to be shovelled out for use, and preferable in all respects to what we collected on the shore, which required to be refined! As we advanced in the grotto, remarkable figures formed by the saline matter everywhere presented themselves; columns reaching from the bottom to the top of the vault appeared to sustain it, and some even had cornices and capitals: here and there undulating masses which at certain distances resembled the sea. From the variegated and whimsical forms we beheld, fancy might make a thousand creations at its pleasure: windows, large open cupboards, benches, church ornaments, grotesque figures of men and animals; some like polished crystals or diamonds, others like blocks of alabaster.

We viewed with unwearied curiosity this repository of wonders, and we had all lighted our second taper, when I observed on the ground in some places a number of crystal fragments that seemed to have fallen off from the upper part. Such a separation might recur, and expose us to danger; a piece falling on any of our heads might prove instantly fatal. But on closer inspection, I was convinced they had not dropped of themselves spontaneously; the whole mass was too solid for fragments of that size to have been so detached from it; and had dampness loosened them, they would have dissolved gradually: I concluded they were broken off by the concussion caused by the explosion of our artillery and fireworks, and I thought it prudent to retire, as other loosened pieces might unexpectedly fall on us. I directed my wife and three of the children to place themselves in the entrance, while Fritz and I carefully examined every part that threatened danger. We loaded our guns with ball, and fired them into the centre of the cavern, to be more fully assured of what produced the separation of the former pieces; one or two more fell; the rest remained immovable, though we went round with long poles, and struck all we could reach. We at length felt confident, that in point of solidity there was nothing to fear, and that we might proceed without dread of accident. Loud exclamations, projects, consultations, now succeeded to our mute astonishment! Many schemes were formed for converting this magnificent grotto into a convenient and agreeable mansion for our abode. We had possession of the most eligible premises; the sole business was to turn them to the best account; and how to effect this was our unceasing theme: some voted for our immediate establishment there, but they were opposed by more sagacious counsel, and it was resolved that Falcon's Stream should still be our headquarters till the end of the year.

30
House in the Salt Rock; Herring Fishery

THE lucky discovery of a previously existing cavern in the rock, had, as must be supposed, considerably lessened our labour: excavation was no longer requisite: I had more room than was wanted for the construction of our dwelling; to render it habitable was the present object, and to do this did not seem a difficult task. The upper bed of the rock, in front of the cavern, through which my little Jack had dug so easily, was of a soft nature, and to be worked with moderate effort. I hoped also that, being now exposed to the air and heat of the sun, it would become by degrees as hard and compact as the first layer that had given me so much trouble. From this consideration I began, while it retained its soft state, to make openings for the doors and windows of the front. This I regulated by the measurement of those I had fixed in my winding staircase, which I had removed for the purpose of placing them in our winter tenement. Intending Falcon's Nest in future as a rural retreat for the hottest days of summer, the windows of the staircase became unnecessary; and as to the door, I preferred making one of bark similar to that of the tree itself, as it would the better conceal our abode, should we at any time experience invasion from savages or other enemies: the door and windows were therefore taken to Tent House, and to be hereafter fixed in the rock. I had previously marked out the openings to be cut for the frames, which were received into grooves for greater convenience and solidity. I took care not to break the stone taken from the apertures, or at least to preserve it in large pieces, and these I cut with the saw and chisel into oblongs an inch and half in thickness, to serve as tiles. I laid them in the sun, and was gratified in seeing they hardened quickly; I then removed them, and my sons placed them in order against the side of the rock, till they were wanted for our internal arrangements.

When I could enter the cavern freely with a good doorway, and it was sufficiently lighted by the windows, I erected a partition, for the distribution of our apartments and other conveniences. The extent of the place afforded ample room for my design, and even allowed me to leave several spaces in which salt and other articles could be stored. At the request of my children, I was cautious to injure as little as possible the natural embellishments of this new family mansion; but with all my care, I could not avoid demolishing them in the division allotted to the stables: cattle are fond of salt, and would not have failed to eat away these ornaments, and perhaps in a prejudicial quantity: however, to gratify and reward my obedient children, I preserved the finest of the pillars, and the most beautiful pieces to decorate our saloon. The large ones served us for chairs and tables; the brilliant pilasters,* at once enlivened and adorned the apartment, and at night multiplied the reflection of the lights. I laid out the interior in the following manner: a very considerable space was first partitioned off in two divisions; the one on the right was appropriated to our residence; that on the left was to contain the kitchen, stables, and work-room. At the end of the second division, where windows could not be placed, the cellar and store-room were to be formed; the whole separated by partition-boards, with doors of communication, so as to give us a pleasant and comfortable abode. Favoured so unexpectedly by what nature had already effected of the necessary labour, we were far from repining ungratefully at what remained to be done, and entertained full hope of completing the undertaking, or at least the chief parts, before winter.

The side we designed to lodge in, was divided into three apartments; the first, next the door, was the bedroom for my wife and me, the second a dining-parlour, and the last a bedroom for the boys: as we had only three windows, we put one in each sleeping-room; the third was fixed in the kitchen, where my wife would often be. A grating for the present fell to the lot of our dining-room, which, when too cold, was to be exchanged for one of the other apartments. I contrived a good fireplace in the kitchen, near the window; I pierced the rock a little above, and four planks nailed together, and passing through this opening, answered the purpose of a chimney. We made the work-room near the kitchen, of sufficient dimensions for the performance of undertakings of some magnitude; it served also to keep our cart and sledge in: lastly the stables, which were formed into four compartments, to separate the different species of animals, occupied all the bottom of the cavern on this side; on the other were the cellar and magazine.

It is readily imagined, that a plan of this extent was not to be executed as if by enchantment, and that we satisfied ourselves in the first instance with doing what was most urgent, reserving the residue for winter; yet every day forwarded the business more than we had been aware of. On every excursion, we brought something from Falcon's Stream, that found its place in the new house, where we deposited likewise, in safety, the remaining provisions from the tent.

The long stay we made at Tent House during these employments, furnished us an opportunity of perceiving several advantages we had not reckoned upon. Immense turtles were often seen on the shore, where they deposited their eggs in the sand and they regaled us with a rich treat; but, extending our wishes, we thought of getting possession of the turtles themselves for livestock, and of feasting on them whenever we pleased. As soon as we saw one on the sands, one of my boys was dispatched to cut off its retreat; meanwhile we approached the animal, and quietly, without doing it any injury, turned it on its back, then passed a long cord through the shell, and tied the end of it to a stake, which we fixed close to the edge of the water. This done, we set the prisoner on his legs again; it hastened into the sea, but could not go beyond the end of the cord; apparently it was all the happier, finding food with more facility along shore than out at sea; and we enjoyed the idea of being able to take it when wanted. I say nothing of sea lobsters, oysters, and many other small fishes, which we could catch in any number. The large lobsters, whose flesh was tough and coarse, were given to the dogs, who preferred them to potatoes; but we shortly after became possessors of another excellent winter provision, which chance unexpectedly procured us.

One morning, when near Safety Bay, a singular sort of spectacle presented itself. At some distance from the shore an extensive surface of the water seemed in a state of ebullition, as if heated by a subterraneous fire: it swelled, subsided, foamed, like boiling water: a large number of aquatic birds hovered over it, sometimes they darted along the surface of the water, sometimes rose in the air, flying in a circle, pursuing each other in every direction; we were at a loss to judge whether sportiveness, pleasure, or warfare produced their motions.

My wife and the boys stood for a long time admiring this phenomenon and indulged themselves in various ludicrous conjectures concerning it, till I at last informed them that the movable bank before us was neither more nor less than a shoal of herrings about to enter Safety Bay, and fall into our hands. All now had questions without end to propose about their appearing in such numbers, and I answered, that at a certain season, herrings leave the Frozen Sea together in a heap. They swim so close to each other, and occupy such a space, as to appear like a bank or island of sand, several leagues in breadth, some fathoms deep, and sometimes above a hundred thousand long. They afterwards divide into bodies, directing their course to the coasts and bays, where they spawn, that is, leave their eggs among the stones and sea-plants, and to these spots fishermen from all parts go to catch them. The herrings appear eager to reach those parts where the tide is lowest, to escape the voracity of the large fishes which pursue them, by getting into shallow water; but in doing this, they become an easier prey to the birds and to man. Exposed to destruction in so many ways, one might wonder the species is not extinct, if nature had not provided against these accidents by their astonishing fecundity: 68,656 eggs have been found in a moderatesized female: thus they continue undiminished, notwithstanding the vast numbers which are destroyed.

By this time the shoal of herrings had reached the entrance of our bay. They made a loud rustling noise in the water, leaping over each other, and displaying their scales of silver hue. This accounted for the luminous sparks we had seen emitted from the sea, and which we could not previously explain. We had no time for further contemplation, but hastened to unharness our team, and supply the want of nets with our hands in catching the herrings: the boys used the largest gourds in lieu of pails, which were no sooner dipped in than filled; and we should have been at a loss where to stow them, had I not thought of employing the condemned boat of tubs. It was accordingly drawn to the water's edge by the buffalo, and placed on rollers. My wife and the two youngest lads cleaned it, whilst the other two went to the cavern for salt, and I quickly fitted up a sort of tent of sailcloth on the strand, so as to keep off the rays of the sun while we were busied in salting. We then all engaged in the task, and I allotted to each a share adequate to his strength and skill. Fritz took his station in the water, to bring us the herrings as fast as caught. Ernest and Jack cleaned them with knives; their mother pounded the salt; Francis helped all, and I placed them in the tubs as I had seen done in Europe; while a joyous shout declared the general activity. I put a layer of salt at the bottom of the barrel, then of fish, the heads towards the staves, proceeding thus till my tubs were nearly full: I spread over the last layer of salt large palm-tree leaves, on these a piece of sailcloth, and fitted in two half-rounded planks for a heading which I pressed down with stones. This effected, I put the buffalo and the ass to the cart again, and conveyed it to our cool cellar in the rock. In a few days, when the herrings were sunk, I closed the barrels more accurately by means of a coating of clay and flax over the cloth, which kept out air and moisture completely, and secured us an excellent food for winter.

Scarcely had we finished our salting, when another novelty occurred: a number of sea-dogs* came into the bay and river, that had followed the herrings with the utmost greediness, sporting in the water along shore, without evincing any fear of us. The fish presented no attraction to the palate, but its skin, tanned and dressed, makes excellent leather. I was in great need of it for straps and harness, to make saddles for Fritz and Jack to ride the onagra and buffalo, and in short for our own use to cut up into soles, belts, and pantaloons, of which articles we much wanted a fresh supply: besides, I knew the fat yielded good lamp oil, that might be substituted for tapers in the long evenings of winter; and that it would be further useful in tanning and rendering the leather pliant.

We had the good fortune to be again successful, and in a short time we secured a sufficient number of them, and carefully preserved the fat, of which we collected a large quantity; it was first put into a copper, melted and cleansed properly, then poured into casks, and kept for the tan-house and lamp. When time should allow, I purposed making soap with it, and this design excited my wife's zeal in the unpleasant though ultimately useful task we were engaged in. We also took care of the bladders, which are very large, for the purpose of holding liquids; the remaining parts that could not be turned to account were thrown into the river.

At this time I likewise made some improvements in our sledge, to facilitate the carrying of stores from Falcon's Stream to our dwelling in the rock at Tent House. I raised it on two beams, on axle-trees, at the extremities of which I put on the four gun-carriage wheels I had taken off the cannon from the vessel; by this alteration I obtained a light and convenient vehicle, of moderate height, on which boxes and casks could be placed. Pleased with the operations of the week, we set out all together with cheerful hearts for Falcon's Stream, to pass our Sunday there, and once more offer our pious thanks to the Almighty, for all the benefits he had bestowed upon his defenceless creatures.

31

New Fishery; New Experiments; New Discoveries, and House

THE enterprise of our dwelling went on, sometimes as a principal, sometimes as an intermediate occupation, according to the greater or less importance of other concerns; but though we advanced thus with moderate rapidity, the progress was such as to afford the hope of our being settled within it by the time of the rainy season.

From the moment I discovered gypsum to be the basis of the crystal salt in our grotto, I foresaw some great advantages I should derive from it; but to avoid enlarging the dimensions of our house by digging further, I tried to find a place in the continuation of the rock, which I might be able to blow up: I had soon the good fortune to meet with a narrow slip between the projections of the rock, which I could easily, by the means I proposed, convert into a passage that should terminate in our work-room. I found also on the ground a quantity of fragments of gypsum, and removed a great number of them to the kitchen, where we did not fail to bake a few of the pieces at a time when we made a fire for cooking, which, thus calcined, rubbed into a powder when cold: we obtained a considerable quantity of it, which I put carefully into casks for use, when the time should come for finishing the interior of our dwelling. My notion was, to form the walls for separating the apartments, of the squares of stone I had already provided, and to unite them together with a cement of this new ingredient, which would be the means, both of sparing the timber, and increasing the beauty and solidity of the work.

It is almost incredible the immense quantity of plaster we had in a short time amassed; the boys were in a constant state of wonder as they looked at the heap. I seized the opportunity of imprinting on their minds the value of a firm and steady perseverance in an object once engaged in, the reward of which they now so agreeably experienced. 'When we first cast our eyes', continued I, 'on this rock, how little did we conceive it possible to transform it into a comfortable dwelling-place; yet we have not only in our persons sufficed for carpenters and masons, but even plasterers too; and so effectually, that, if we had it much at heart, we might adorn our walls with stucco, as is the mode in Europe: we possess both the materials and the intelligence; and with the addition of patience and industry, there is scarcely any thing, even what at first should seem impossible, too difficult for our performance.'

The first use I made of the plaster was to complete some covers I had begun with other materials for my herring tubs, four of which I stopped down to render them impenetrable to the air; the rest of the herrings we intended to dry and smoke. For this purpose we erected a little sort of hut of reeds and branches, as is practised in Holland and America by the fishermen: we placed rows of sticks, reaching from side to side, across the hut, laid the herrings upon them, and then lighted a heap composed of moss and fresh cut branches of trees, to produce a stronger and more effective vapour for the purpose: we made the door tight, and had soon the pleasure of adding a large stock of exquisitely flavoured dried herrings to our former store for the ensuing winter.

About a month after the singular visit of the herrings, which had now entirely left our shores, we received another, and not a less profitable one, from a fish of a different species: we observed Safety Bay to be filled with large fishes, which seemed eager to push to the shore, for the purpose of depositing their eggs among the stones in fresh water.

As Jack and I were walking near the mouth of Jackal's River we perceived immense quantities of a large fish moving slowly towards the banks. As they came nearer, I distinguished the largest to be sturgeons by the pointed snout, while the smallest I pronounced to be salmon. Jack now strutted about in ecstasies. 'What say you now, father?' said he; 'this is nothing like your little paltry herrings! A single fish of this troop would fill a tub!'—'No doubt,' answered I: and with great gravity I added,—'Prithee, Jack, step into the river, and fling them to me one by one, that I may take them home to salt and dry.'

He looked at me for a moment with a sort of vacant doubt if I could possibly be in earnest; then seizing suddenly a new idea—'Wait a moment, father,' cried he, 'and I will do so': and he sprang off like lightning towards the cavern, from whence he soon returned loaded with a bow and arrows, the bladders of the sea-dogs, and a ball of string to catch, as he assured me, every one of the fishes. I looked on with interest and curiosity to mark what was next to happen, while the animation of his countenance, the promptitude and boyish gracefulness of his motions, and the firm determination of his manner, afforded me the highest amusement. He tied the bladders round at certain distances with a long piece of string, to the end of which he fastened an arrow and a small iron hook; he placed the large ball of string in a hole in the ground, at a sufficient distance from the water's edge, and then he shot off an arrow, which the next instant stuck in one of the largest fishes. My young sportsman uttered a shout of joy. At the same moment Fritz joined us, and witnessed this unexpected feat without the least symptom of jealousy. 'Well done, brother Jack,' cried he, 'but let me too have my turn.'—Saying this, he ran back and fetched the harpoon and the windlass, and returned to us accompanied by Ernest. We were well pleased with their opportune arrival, for the salmon Jack had pierced struggled so fiercely, that all our endeavours to hold the string were insufficient, and we dreaded at every throw to see it break, and the animal make good its escape. By degrees, however, its strength was exhausted, and aided by Fritz and Ernest, we succeeded in drawing it to a bank, where I put an end to its existence.

This fortunate beginning of a plan for a fishery inspired us all with hope and emulation. Fritz eagerly seized his harpoon and windlass;* I, for my part, like Neptune,* wielded a trident; Ernest prepared the large fishing-rod; and Jack his arrow with the same apparatus as before, not forgetting the bladders, which were so effectual in preventing the fish from sinking when struck. We were now more than ever sensible of our loss in the destruction of the tub-boat, with which we could have pursued the creature in the water, and have been spared much pains and difficulty; but, on the other hand, such numbers of fishes presented themselves at the mouth of the river, that we had only to choose among them. Jack's arrow, after missing twice, struck the third time a large sturgeon, which was so untractable that we had great difficulty in securing him. I too had caught two of the same fish, and had been obliged to go up to the middle in the water to manage my booty. Ernest, with his rod and line and a hook, had also taken two smaller ones. Fritz, with his harpoon, had struck a sturgeon at least eight feet in length, and the skill and strength of our whole company were found necessary to conduct him safe to shore, where we harnessed the buffalo to him with strong cords to draw him to Tent House.

Our first concern was to clean our fish thoroughly inside, to preserve them fresh the longer. I separated the eggs I found in them, and which could not be less than thirty pounds, and put them aside to make a dish called caviar, greatly relished by the Russians and the Dutch. I took care also of the bladders, thinking it might be possible to make a glue from them, which would be useful for so many purposes. I advised my wife to boil some individuals of the salmon in oil, similar to the manner of preparing tunny fish* in the Mediterranean: and while she was engaged in this process, I was at work upon the caviar and the glue. For the first, I washed the berries in several waters, and then pressed them closely in gourd-rinds in which a certain number of holes had been bored. When the water had run off, the berries were taken out in a substance like cheese, which was then conveyed to the hut to be dried and smoked. For the second, we cut the bladders into strips, which we fastened firmly by one end to a stake, and taking hold of the other with a pair of pincers, we turned them round and round till the strip was reduced to a kind of knot, and these were then placed in the sun to harden; this being the simple and only preparation necessary for obtaining glue from the ingredient. When thoroughly dry, a small quantity is put on a slow fire to melt. We succeeded so well, and our glue was of so transparent a quality, that I could not help feeling the desire to manufacture some pieces large enough for panes to a window-frame.

When these various concerns were complete, we began to meditate a plan for constructing a small boat as a substitute for the tub-raft, to come close into shore. I had a great desire to make it, as the savages do, of the rind of a tree;* but the difficulty was to fix on one of sufficient bulk for my purpose; for though many were to be found in our vicinity, yet each was on some account or other of too much value to be spared. We therefore resolved to make a little excursion in pursuit of a tree of capacious dimensions, and in a situation where it was not likely to yield us fruit, to refresh us with its shade, or to adorn the landscape round our dwelling.

In this expedition, we as usual aimed at more than one object: eager as we were for new discoveries, we yet allowed ourselves the time to visit our different plantations and stores at Falcon's Stream. We were also desirous to secure a new supply of the wax berry, of gourds, and of elastic gum. Our kitchen-garden at Tent House was in a flourishing condition; nothing could exceed the luxuriance of the vegetation, and, almost without the trouble of cultivation, we had excellent roots and plants in abundance, which came in succession, and promised a rich supply of pease, beans of all sorts, lettuces, etc.; our principal labour was to give them water freely, that they might be fresh and succulent for use. We had besides, melons and cucumbers in great plenty, which, during the hottest weather, we valued more than all the rest. We reaped a considerable quantity of Turkey wheat from the seed we had sown, and some of the ears were a foot in length. Our sugarcanes were also in the most prosperous condition, and one plantation of pineapples on the high ground was also in progress to reward our labour with abundance of that delicious fruit.

This state of general prosperity at Tent House gave us the most flattering expectations from our nurseries at Falcon's Stream. Full of these hopes, we one day set out altogether for our somewhat neglected former abode.

We arrived at Falcon's Stream, where we intended to pass the night. We visited the ground my wife had so plentifully sowed with grain, which had sprung up with an almost incredible rapidity and luxuriance, and was now nearly ready for reaping. We cut down what was fairly ripe, bound it together in bundles, and conveyed it to a place where it would be secure from the attacks of more expert grain consumers than ourselves, of which thousands hovered round the booty. We reaped barley, wheat, rye, oats, pease, millet, lentils,—only a small quantity of each, it is true, but sufficient to enable us to sow again plentifully at the proper season. The plant that had yielded the most was maize, a proof that it best loved the soil. It had already shown itself in abundance in our garden at Tent House; but here there was a surface of land, the size of an ordinary field, entirely covered with its splendid golden ears, which still more than the other plants attracted the voracity of the feathered race. The moment we drew near, a dozen at least of large bustards sprang up with a loud rustling noise which awakened the attention of the dogs; they plunged into the thickest parts, and routed numerous flocks of birds of all kinds and sizes, who took hastily to flight: among the fugitives were some quails, who escaped by running; and lastly some kangaroos, whose prodigious leaps enabled them to elude the pursuit of the dogs.

We were so overcome by the surprise such an assemblage of living creatures occasioned, as to forget the resource we had in our guns; we stood as it were stupid with amazement during the first moments, and before we came to ourselves, the prey was beyond our reach, and for the most part out of sight. Fritz was the first to perceive and to feel with indignation the silly part we had been playing, and to consider in what way we could repair the mischief. Without further loss of time, he took the bandage from his eagle's eyes (for the bird always accompanied him perched upon his game-bag), and showed him with his hand the bustards still flying, and at no great distance. The eagle took a rapid flight. Fritz jumped like lightning on the back of his onagra, and galloped over everything that intervened, in the direction the bird had taken, and we soon lost sight of him.

We now beheld a spectacle which in the highest degree excited our curiosity and interest: the eagle had soon his prey in view; he mounted above one of the bustards in a direct line, without losing sight of it for an instant, and then darted suddenly down; the bustards flew about in utter confusion, now seeking shelter in the bushes, then crossing each other in every direction, in the attempt to evade the common enemy; but the eagle remained steady in pursuit of the bird he had fixed upon for his prey, and disregarded all the rest: he alighted on the unlucky bustard, fixed his claws and his beak in its back, till Fritz, arriving full gallop, got down from the onagra, replaced the bandage on the eagle's eyes, seated him once more upon the game-bag, and having relieved the poor bustard from his persecutor, he shouted to us to come and witness his triumph. We ran speedily to the place.

At the conclusion of this adventure, we hastened forward to Falcon's Stream, and dressed the wounds of the bustard. We perceived with pleasure that it was a male, and foresaw the advantage of giving him for a companion to our solitary female of the same species, which was completely tamed. I threw a few more bundles of maize into the cart, and without further delay we arrived at our tree, one and all sinking with faintness from hunger, thirst, and fatigue. It was on such occasions that my exemplary partner evinced the superior fortitude and generosity of her temper: though necessarily more a sufferer than the rest, her first thought was always what she could administer to relieve us in the shortest time. On this occasion, as we had consumed our little store of wine, and could not soon and easily procure milk from the cow, she contrived to bruise some of the maize between two large stones, and then put it in a linen cloth, and with all her strength squeezed out the sap; she then added some juice from the sugar-canes, and in a few minutes presented us with a draught of a cool refreshing liquid, invitingly white to the eye, and agreeable to the taste, which we received at her hands with feelings of grateful emotion.

The rest of the day was employed in picking the grains of the different sorts of corn from the stalks: we put what we wished to keep for sowing, into some gourd shells, and the Turkey wheat was laid carefully aside in sheaves till we should have time to beat and separate it. Fritz observed that we should also want to grind it; and I reminded him of the handmill we had secured from our departed ally, the wrecked vessel.

Fritz.—But, father, the handmill is so small, and so subject to be put out of order:—why should we not contrive a watermill, as they do in Europe? We have surely rapid streams of water in abundance.

Father.—This is true; but such a mechanism is more difficult than you imagine. The wheel alone, I conceive, would be an undertaking far beyond our strength or our capacity. I am, however, well pleased with the activity and zeal which prompted your idea; and we will hereafter consider whether it may be worth while to bestow upon it further attention. We have abundance of time before us, for we shall not want a watermill till our harvests are such as to produce plentiful crops of corn. In the meantime, let us be thinking of our proposed excursion for tomorrow; for we should set out, at least, by sunrise.

We began our preparations accordingly. My wife chose some hens and two fine cocks, with the intention of taking them with us, and leaving them at large to produce a colony of their species at a considerable distance from our dwelling-places: I, with the same view, visited our beasts, and selected four young pigs, four sheep, two kids, and one male of each species; our numbers having so much increased, that we could well afford to spare these individuals for the experiment. If we succeeded in thus accustoming them to the natural temperature and productions of our island, we should have eased ourselves of the burden of their support, and should always be able to find them at pleasure.

The next morning, after loading the cart with all things necessary, not forgetting the rope ladder and the portable tent, we quitted Falcon's Stream. The animals, with their legs tied, were all stationed in the vehicle. We left abundance of food for those that remained behind; the cow, the ass, and the buffalo, were harnessed to the cart; and Fritz, mounted on his favourite, the onagra, pranced along before us, to ascertain the best and smoothest path for the cavalcade.

We took this time a new direction, which was straight forward between the rocks and the shore, that we might make ourselves acquainted with everything contained in the island we seemed destined forever to inhabit. In effect, the line proceeding from Falcon's Stream to Safety Bay, might be said to be the extent of our dominions: for as to the adjacent exquisite country of the buffaloes, Fritz and I had discovered, that the passage to it by the end of the rocks was so dangerous, and at so great a distance, that we could not hope to domiciliate ourselves upon its soil, as we had done on our side of the rocks. We found, as usual, much difficulty in pushing through the tall tough grass, and alternately through the thick prickly bushes which everywhere obtruded themselves. We were often obliged to turn aside, while I cut a passage with my hatchet: but these accidents seldom failed to reward my toil by the discovery of different small additions to our general comforts; among others, some roots of trees curved by nature to serve both for saddles and yokes for our beasts of burden. I took care to secure several, and put them in the cart.

In about an hour we found ourselves at the extremity of the wood, and a most singular phenomenon presented itself to our view: a small plain, or rather a grove of low bushes, to appearance almost covered with flakes of snow, lay extended before us. Little Francis was the first to call our attention to it, he being seated in the cart. 'Look, father,' cried he, 'here is a place full of snow; let me get down, and make some snowballs.' I could not resist a hearty laugh, though myself completely at a loss to explain the nature of what in colour and appearance bore so near a resemblance to it. Suddenly, however, a suspicion crossed my mind, and was soon confirmed by Fritz, who had darted forward on his onagra, and now returned with one hand filled with tufts of a most excellent species of cotton, so that the whole surface of low bushes was in reality a plantation of that valuable article. This most useful of almost the whole range of vegetable productions bestowed by Providence on man, which, with the cost of only a little labour, supplies him with apparel, and commodious beds for the repose of his limbs, is found in such abundance in islands, that I had been surprised at not meeting with any before. The pods had burst from ripeness, and the winds had scattered around their flaky contents; the ground was strewed with them, they had gathered in tufts on the bushes, and they floated gently in the air.

The joy of this discovery was almost too great for utterance, and was shared by all but Francis, who was sorry to lose his pretty snowballs; and his mother, to soothe his regret, made the cotton into balls for him to play with, and promised him some new shirts and dresses; then turning to me, she poured out her kind heart in descriptions of all the comfortable things she should make for us, could I but construct a spinning-wheel, and then a loom* for weaving.—We ended with collecting as much cotton as our bags would hold, and my wife filled her pockets with the seed, to raise it in our garden at Tent House.

It was now time to proceed; and we took a direction towards a point of land which skirted the wood of gourds, and, being high, commanded a view of the adjacent country. I conceived a wish to remove our establishment to the vicinity of the cotton plantation and the gourd wood, which furnished so many of the utensils for daily use throughout the family. I pleased myself in idea, with the view of the different colonies of animals I had imagined, both winged and quadruped; and in this elevation of my fancy, I even thought it might be practicable to erect a sort of farmhouse on the soil, which we might visit occasionally, and be welcomed by the agreeable sounds of the cackling of our feathered subjects, which would so forcibly remind us of the customs of our forsaken but ever-cherished country.

We accordingly soon reached the high ground, which I found in all respects favourable to my design; behind, a thick forest gradually rose above us, which sheltered us from the north wind, and insensibly declined towards the south, ending in a plain clothed luxuriantly with grass, shrubs, and plants, and watered by a refreshing rivulet, which was an incalculable advantage for our animals of every kind, as well as for ourselves.

My plan for a building was approved by all, and we lost no time in pitching our tent, and forming temporary accommodations for cooking our victuals. When we had refreshed ourselves with a meal, I, for my part, resolved to look about in all directions, that I might completely understand what we should have to depend upon in this place, in point of safety, salubrity, and general accommodation. I had also to find a tree that would suit for the proposed construction of a boat: and lastly, to meet, if possible, with a group of trees, at such fit distances from each other as would assist me in my plan of erecting a farmhouse. I was fortunate enough in no long time to find in this last respect exactly what I wanted, and quite near to the spot we on many accounts had felt to be so enviable. I returned to my companions, whom I found busily employed in preparing excellent beds of the cotton, upon which, at an earlier hour than usual, we all retired to rest.

32

Completion of two Farmhouses; a Lake; the Beast with a Bill; a Boat and a Bull

THE trees that I had chosen for the construction of my farm embellishments, were for the most part 1 foot in diameter in the trunk; they presented the form of a tolerably regular parallelogram, with its longest side to the sea, the length 24 feet, and the breadth 16. I cut little hollow places or mortices in the trunks, at the distance of ten feet, one above the other, to form two stories. The upper one I made a few inches shorter before than behind, that the roof might be in some degree shelving; I then inserted beams 5 inches in diameter respectively in the mortices, and thus formed the skeleton of my building. We next nailed some laths from tree to tree, at equal distances from each other, to form the roof, and placed on them, in mathematical order, a covering composed of pieces of the bark of trees, cut into the shape of tiles, and in a sloping position, for the rain to run off in the wet season. As we had no great provision of iron nails, we used for the purpose the strong pointed thorn of the acacia,* which we had discovered the day before. We cut down a quantity of them, and laid them in the sun to dry, when they became as hard as iron, and were of essential service to our undertaking. We found great difficulty in peeling off a sufficient quantity of bark from trees to cover our roof. I began with cutting the bark entirely round at distances of about two feet all the length of the trunk; I next divided the intervals perpendicularly into two parts, which I separated from the tree by sliding a wedge under the corners, to raise the bark by degrees; I next placed the pieces on the ground, with stones laid on them to prevent their curving, to dry in the sun; and lastly, I nailed them on the roof, where they had the appearance of fishes' scales—an effect that was not only pleasing to the eye, but reminded us of the roofs of our native land.

On this occasion we made another agreeable discovery: my wife took up the remaining chips of the bark for lighting a fire, supposing they would burn easily; we were surprised by a delicious aromatic odour, which perfumed the air. On examining the half-consumed substance, we found some of the pieces to contain turpentine, and others gum-mastic,* so that we might rely on a supply of these ingredients from the trees which had furnished the bark. It was less with a view to the gratifying our sense of smelling, than with the hope of being able to secure these valuable drugs for making a sort of pitch to complete our meditated boat, that we indulged our earnestness in the pursuit. The instinct of our goats, or the acuteness of their smell, discovered for us another pleasing acquisition: we observed with surprise, that they ran from a distance to roll themselves on some chips of a particular bark which lay on the ground, and which they began to chew and eat greedily. Jack seized a piece also, to find out what could be the reason of so marked a preference as the goats had shown. My wife and I then followed his example, and we were all convinced that the chips were cinnamon, though not so fine a sort as that from the isle of Ceylon.*

This new commodity was certainly of no great importance to us; but we regarded it with pleasure, as it might assist to distinguish some day of rejoicing. The tree from which we had taken our bark was old, and the cinnamon was the coarser flavoured on this account: I remembered to have read, that young trees produce this spice in much greater perfection.

After our next meal we resumed with ardour our under-taking of the farm, which we continued without interruption for several days. We formed the walls with matted reeds inter-woven with pliant laths to the height of 6 feet; the remaining space to the roof was enclosed with only a simple grating, that the air and light might be admitted. A door was placed in the middle of the front. We next arranged the interior with as much convenience as the shortness of the time and our reluctance to use all our timber would allow; we divided it half-way up by a partition wall into two unequal parts; the largest was intended for the sheep and goats, and the smallest for ourselves, when we should wish to pass a few days here. At the further end of the stable we fixed a house for the fowls, and above it a sort of hayloft for the forage. Before the door of entrance we placed two benches, contrived as well as we could of laths and odd pieces of wood, that we might rest ourselves under the shade of the trees, and enjoy the exquisite prospect which presented itself on all sides. Our own apartment was provided with a couple of the best bedsteads we could make of twigs of trees, raised upon four legs, two feet from the ground, and these were destined to receive our cotton mattresses. Our aim was to content ourselves for the present with these slight hints of a dwelling, and to consider hereafter what additions either of convenience or ornament could be made, such as plastering, etc. etc. All we were now anxious about, was to provide a shelter for our animal colonists, which should encourage and fix them in the habit of assembling every evening in one place. For several days, at first, we took care to fill their troughs with their favourite food, mixed with salt, and we agreed that we would return frequently to repeat this indirect mode of invitation for their society, till they should be entirely fixed in their expectation of finding it.

I had imagined we could accomplish what we wished at the farm in three or four days; but we found in the experiment that a whole week was necessary, and our victuals fell short before our work was done. We began to consider what remedy we could apply to so embarrassing a circumstance; I could not prevail upon myself to return to Falcon's Stream, before I had completed my intentions at the farm, and the other objects of my journey. I had even come to the determination of erecting another building upon the site of Cape Disappointment; I therefore decided, that on this trying occasion I would invest Fritz and Jack with the important mission. They were accordingly dispatched to Falcon's Stream, and to Tent House, to fetch new supplies of cheese, ham, potatoes, dried fish, manioc bread, for our subsistence, and also to distribute fresh food to the numerous animals we had left there. I directed one to mount the onagra, and the other the buffalo. My two knights-errant, proud of their embassy, set off with a brisk trot; they at my desire took with them the old ass, to bring the load of provisions. Fritz was to lead him with a bridle, while Jack smacked a whip near his ears to quicken his motions; and certainly, whether from the influence of climate, or the example of his companion the onagra, he had lost much of his accustomed inactivity; and this was the more important, as I intended to make a saddle for my wife to get on his back, and relieve herself occasionally from the fatigue of walking.

During the absence of our purveyors, I rambled with Ernest about the neighbouring soil, to make what new discoveries I could, and to procure, if possible, additions to our store of provisions. We followed the winding of a river towards the middle of the wall of rocks; our course was interrupted by a marsh which bordered a small lake, the aspect of which was enchantingly picturesque. I perceived, with joyful surprise that the whole surface of the swampy soil was covered with a kind of wild rice, ripe on the stalk, and which attracted the voracity of large flocks of birds. As we approached, a loud rustling was heard, and we distinguished on the wing, bustards, Canada heath fowl,* and great numbers of smaller birds. We succeeded in bringing down five or six of them, and I was pleased to remark in Ernest a justness of aim that promised well for the future. The habits of his mind discovered themselves on this as on many previous occasions; he betrayed no ardour, he did everything with a slowness that seemed to imply dislike; yet the cool deliberation and constancy he applied to every attempt so effectually assisted his judgement, that he was sure to arrive at a more perfect execution than the other boys. He had practised but little in the study of how to fire a gun to the best advantage; but Ernest was a silent inquirer and observer, and accordingly his first essays were generally crowned with success.

Presently we saw Master Knips jump from Flora's back, and smell along the ground among some thick growing plants, then pluck off something with his two paws, and eat of it voraciously. We ran to the spot to see what it could be, when, to the relief of our parched palates, we found he had discovered there the largest and finest kind of strawberry, which is called in Europe the *Chili*, or *pine strawberry*.*—On this occasion, the proud creature, man, generously condescended to be the imitator of a monkey: we threw ourselves upon the ground, as near to Knips as we could creep, and ate as fast as we could swallow, till we felt refreshed. Many of these strawberries were of an enormous size, and Ernest, after devouring an immense quantity, recollected his absent friends, and filled a small gourd-shell with the finest fruit, then covered them with leaves, and tied them down with a tendril from a neighbouring plant, to present them in perfection to his mother. I, on my part, gathered a specimen of the rice to offer, that she might inform us if it was fit for culinary purposes.

Pursuing our way a little further along the marsh, we reached the lake, which we had described with so much pleasure from a distance, and whose banks, being overgrown with thick underwood, were necessarily concealed from the momentary view we had leisure to take of surrounding objects, particularly as the lake was situated in a deep and abrupt valley. No one, who is not a native of Switzerland, can conceive the emotion which trembled at my heart, as I contemplated this limpid, azure, undulating body of water, the faithful miniature of so many grand originals, which I had probably lost sight of forever! My eyes swam with tears! Alas! a single glance upon the surrounding picture, the different characters of the trees, the vast ocean in the distance, destroyed the momentary illusion, and brought back my ideas to the painful reality, that I and mine were—strangers in a desert island!

Another sort of object now presented itself to confirm the certainty that we were no longer inhabitants of Europe; it was the appearance of a quantity of swans gliding over the surface of the lake; but their colour, instead of white, like those of our country, was a jetty black,* and their plumage had so high a gloss as to produce, reflected on the water, the most astonishing effect. The six large feathers of the wings of this bird are white, exhibiting a singular contrast to the rest of the body; in other respects these birds were remarkable, like those of Europe, for the haughty gracefulness of their motions, and the voluptuous ease of their nature. We remained a long time in silent admiration of them: some of the swans pursued their course magnificently on the bosom of the blue water; others stopped and seemed to hold deliberations with their companions, or to admire themselves, or caress each other; many young ones followed in the train of the parent bird, who frequently turned half round, in execution of her watchful and matronly office. This was a spectacle which I could not allow to be interrupted by bloodshed, though Ernest, rendered a little vain by his success and my encomiums, would have been ready to fire upon the swans, if I had not absolutely forbidden the attempt; at the same time I consoled him with the promise, that we would endeavour to obtain a pair of the interesting creatures for our establishment at Falcon's Stream.

Flora at this moment dragged out of the water a creature she had killed. It was somewhat in shape like an otter, and like the tribe of water birds, web-footed: its tail was long and erect, and covered with a soft kind of hair; the head was very small, and the ears and eyes were almost invisible; to these more ordinary characters was added, a long flat bill, like that of a duck, which protruded from its snout, and produced so ludicrous an effect that we could not resist a hearty laugh. All the science of the learned Ernest, joined with my own, was insufficient to ascertain the name and nature of this animal.

We had no resource but to remain ignorant; in the meantime we christened it by the name of *Beast with a Bill*,* and decided that it should be carefully stuffed and preserved.

We now began to look for the shortest path for returning to the farm, which we reached at the same time with Fritz and Jack, who had well performed the object of their journey. We, on our parts, produced our offering of strawberries and our specimen of rice, which were welcomed with shouts of pleasure and surprise.

The beast with a bill was next examined with eager curiosity, and then laid aside for the plan I had formed. My wife proceeded to pluck and salt the birds we had killed, reserving one fresh for our supper, which we partook of together upon the benches before the door of our new habitation. We filled the stable with forage, laid a large provision of grain for the fowls within their house, and began arrangements for our departure.

The following day we took a silent leave of our animals, and directed our course towards the eminence in the vicinity of Cape Disappointment; we ascended it, and found it in every respect adapted to our wishes. From this eminence we had a view over the country which surrounded Falcon's Stream in one direction, and in others of a richly diversified extent of landscape, comprehending sea, land, and rocks. When we had paused for a short time upon the exhaustless beauties of the scene, we agreed with one voice, that it should be on this spot we would build our second cottage. A spring of the clearest water issued from the soil near the summit, and flowed over its sloping side, forming agreeable cascades in its rapid course; in short, every feature of the picture contributed to form a landscape worthy the homage of a taste the most delicate and refined. I presented my children with an appropriate word. —'Let us build here,' exclaimed I, 'and call the spot— *Arcadia*';* to which my wife and all agreed.

We lost no time in again setting to work; our experience at the farm enabled us to proceed with incredible rapidity, and our success was in every respect more complete. The building contained a dining-room, two bedchambers, two stables, and a store-room for preserving all kinds of provisions for man and beast. We formed the roof square, with four sloped sides, and the whole had really the appearance of a European cottage, and was finished in the short space of six days. What now remained to be done, was to fix on a tree fit for my project of a boat. After much search, I at length found one of prodigious size, and in most respects suitable to my views.

It was, however, no very encouraging prospect I had before me, being nothing less than the stripping off a piece of the bark that should be 18 feet in length, and 5 in diameter; and now I found my rope ladder of signal service; we fastened it by one end to the nearest branches, and it enabled us to work with the saw, as might be necessary, at any height from the ground. Accordingly, we cut quite round the trunk in two places, and then took a perpendicular slip from the whole length between the circles; by this means we could introduce the proper utensils for raising the rest by degrees, till it was entirely separated. We toiled with increasing anxiety, at every moment dreading that we should not be able to preserve it from breaking, or uninjured by our tools. When we had loosened about half, we supported it by means of cords and pulleys; and when all was at length detached, we let it down gently, and with joy beheld it lying safe on the grass. Our business was next to mould it to our purpose, while the substance continued moist and flexible.

The boys observed that we had now nothing more to do, than to nail a plank at each end, and our boat would be as complete as those used by the savages; but, for my own part, I could not be contented with a mere roll of bark for a boat; and when I reminded them of the paltry figure it would make, following the pinnace, I heard not another word about the further pains and trouble, and they asked eagerly for my instructions. I made them assist me to saw the bark in the middle of the two ends, the length of several feet; these two parts I folded over till they ended in a point; I kept them in this form by the help of the strong glue I had before made from fish-bladders, and pieces of wood nailed fast over the whole: this operation tended to widen the boat in the middle, and thus render it of too flat a form; but this we counteracted by straining a cord all round, which again reduced it to the due proportion, and in this state we put it in the sun, to harden and fix. Many things were still wanting to the completion of my undertaking, but I had not with me proper utensils: I therefore dispatched the boys to Tent House, to fetch the sledge, and convey it there for our better convenience in finishing.

Before our departure for Tent House, we collected several new plants for our kitchen-garden; and lastly, we made another trip to the narrow strait at the end of the wall of rocks, resolved, as I before mentioned, to plant there a sort of fortification of trees, which should produce the double effect of discouraging the invasion of savages, and allowing us to keep our pigs on the other side, and thus secure our different plantations from the chance of injury. We accomplished all these intentions to our entire satisfaction, and in addition, we placed a slight drawbridge across the river beyond the narrow pass, which we could let down or take up at pleasure on our side. We now hastened our return to Arcadia, and after a night's repose we loaded the sledge with the boat and other matters, and returned to Tent House.

As soon as we had dispatched some necessary affairs, we resumed the completion of the boat: in two days she had received the addition of a keel, a neat lining of wood, a small flat floor, benches, a small mast and triangular sail, a rudder, and a thick coat of pitch on the outside, so that the first time we saw her in the water, we were all in ecstasies at the charming appearance she made.

Our cow in the meantime had brought forth a young calf, a male; I had pierced its nose, as I had the buffalo's, so that I could manage it more easily; and as soon as the wound was healed I commenced a course of training, by accustoming it to carry the bridle and saddle of linen which had been used for the buffalo, his father.

'What shall we do with our bull?' said Fritz one day to me. 'My advice is that we make a fighting-bull of him, after the manner of the Hottentots.'

The word 'fighting' frightened my wife; and, recalling to her mind all that she had heard of Spanish bull-fights, she exclaimed, 'What! would you train that poor animal to those ferocious amusements, where blood flows in torrents, to amuse an indolent and half-barbarous population?'

'Do not fear,' I answered; 'we had no thoughts of "matadores and picadors", nor of those bloody bull-fights that so often take place at Madrid and Toledo. The fighting-bull of the

Hottentots is a useful animal, a safeguard against danger, and I think that the best thing we can do with our little bull will be to train him for that purpose. The Hottentots', continued I, 'inhabit a country that is infested by savage beasts, divided into tribes; they live almost exclusively by the productions of their herds, which, pasturing in the open country, are continually exposed to the voracity of lions, tigers, etc. It is to remedy this danger that the Hottentots train up bulls for fighting. When the bull thus educated perceives, by instinct, the approach of peril, he makes it known to the cows, who range themselves in a circle, the calves in the middle, and the entire troop offer to the aggressor a complete circle of horned heads, while the fighting-bull advances alone in front to the attack. If he be well trained he will rush immediately upon him and pierce him with his long horns; but should the foe be one of those species of animals who never recoil—for instance, the lion—the brave bull must sacrifice his own life to give the flock time to escape.'

This explanation reconciled my wife to our plan for the bull's education, and we all were sure that he would make a brave champion in the defence of our domestic animals.

The next question was to know to whom this charge should be entrusted: each one had his favourite animal whom he protected and took care of. Little Francis alone was just then without a companion, for the two dogs which had been given him had grown more rapidly than himself, and there was nothing more for him to do. Besides, I thought it useful to stimulate the activity of the little boy, and prevent his spirits from flagging.

'What do you say, Frank,' said I, suddenly, 'I have chosen you as preceptor to our bull?'

At these words his handsome blue eyes sparkled, and he answered, 'I accept the charge, papa. Have you not related to me how a very strong man, who, I think, was called Milo,* had commenced by carrying a little calf every day, and he became so strong by this exercise, that, although the calf grew up to be a bull, Milo was still able to carry him. Besides, if I am little, I know enough to make my pupil obey me, and I will treat him so well that he will be sure to love me.'

It was then arranged that the bull should be abandoned to the direction of Francis. We asked him what name he would give him; he chose that of 'Broumm', by analogy to the powerful bellowings of the animal. Jack profited by the circumstance to give an official sanction to the name of 'Storm', which he had given to his buffalo.

'How well it will sound,' said he, 'when I am dashing along at full speed, to hear you saying "Here comes Jack, riding on the Storm."'

Frank commenced his course of instruction immediately, and he was so kind and attentive to the wants of his new pupil, that the animal became firmly attached to him.

We had still two months in prospect before the rainy season, and we employed them for completing our abode in the grotto, with the exception of such ornaments as we might have time to think of during the long days of winter. We made the internal divisions of planks, and that which separated us from the stables, of stone, to protect us from the offensive smell occasioned by the animals. Our task was difficult, but from habit it became easier every day. We took care to collect or manufacture a sufficient quantity of all sorts of materials, such as beams and planks, reeds and twigs for matting, pieces of gypsum for plaster, etc. etc. At length the time of the rainy season was near at hand, and we thought of it with pleasure, as it would put us in possession of the enjoyments we had procured by such unremitting industry and fatigue. We had an inexpressible longing to find ourselves domiciliated, and at leisure to converse together on the subject of all the wondrous benefits bestowed upon us by an ever watchful and beneficent Providence!

We plastered over the walls of the principal apartments on each side with the greatest care, finishing them by pressure with a flat smooth board, and lastly a wash of size,* in the manner of the plasterers in Europe. This ornamental portion of our work amused us all so much, that we began to think we might venture a step further in European luxury, and agreed that we would attempt to make some carpets with the hair of our goats. To this effect we smoothed the ground in the rooms we intended to distinguish, with great care; then spread over it some sailcloth, which my wife had joined in breadths, and fitted exactly; we next strewed the goats' hair, mixed with wool obtained from the sheep, over the whole; on this surface we threw some hot water, in which a strong cement had been dissolved; the whole was then rolled up, and was beaten for a considerable time with hard sticks; the sailcloth was now unrolled, and the inside again sprinkled, rolled, and beaten as before; and this process was continued till the substance had become a sort of felt, which could be separated from the sailcloth, and was lastly put in the sun to harden. We thus produced a very tolerable substitute for that enviable article of European comfort, a carpet: of these we completed two; one for our parlour, and the other for our drawing-room, as we jocosely named them; both of which were completely fit for our reception by the time the rains had set in.

All we had suffered during this season in the preceding year doubled the value of the comforts and conveniences with which we were now surrounded. We were never tired of admiring our warm and well-arranged apartments, lighted with windows, and well secured with doors from wind and rain, and our granary filled with more than a sufficient winter supply of food for ourselves and for our cattle. In the morning, our first care was to feed and give them drink; and both these were now constantly at hand, without the pains of fetching or preparing: after this we assembled in the parlour, where prayers were read, and breakfast immediately served: we then adjourned to the common-room, where all sorts of industry went forward, and which contained the spinning-wheel and loom I had, though with indifferent success, constructed to gratify my wife. Here all united in the business of producing different kinds of substances, which she afterwards made into apparel. I had also contrived to construct a turning-machine, having used for the purpose one of the small cannon wheels, with the help of which the boys and I managed to produce some neat utensils for general use. After dinner, our work was resumed till night, when we lighted candles; and as they cost no more than our own trouble in collecting and manufacturing the materials, we did not refuse ourselves the pleasure of using many at a time, to admire their lights splendidly reflected by the crystals every-where pendent. We had formed a convenient portion of our dwelling into a small chapel, in which we left the crystals as produced by nature; and they exhibited a wondrous assemblage of colonnades, porticoes, altars, which, when the place was lighted to supply the want of a window, presented a truly enchanting spectacle. Divine service was performed in it regularly every Sunday. I had raised a sort of pulpit, from which I pronounced such discourses as I had framed for the instruction of my affectionate group of auditors. Jack and Francis had a natural inclination for music. I did the most I could in making a flageolet* apiece for them of two reeds, on which they so frequently practised as to attain a tolerable proficiency: they accompanied their mother, who had a sweet-toned voice, the volume of which was doubled by the echoes of the grottos, and they produced together a very pleasing little concert.

Thus, as will be perceived, we had made the first steps towards a condition of civilization: separated from society, condemned, perhaps, to pass the remainder of life in this desert island, we yet possessed the means of happiness; we had abundance of all the necessaries, and many of the comforts, desired by human beings! We had fixed habits of activity and industry; we were in ourselves serene and contented; our bodily health and strength increased from day to day; the sentiment of tender attachment was perfect in every heart; we every day acquired some new and still improving channel for the exertion of our physical and moral faculties; we everywhere beheld, and at all times acknowledged, marks of the divine wisdom and goodness; our minds were penetrated with love, gratitude, and veneration for the Providence who had so miraculously rescued and preserved us, and conducted us to the true destination of man—that of providing by his labour for the wants of his offspring! I trusted in the same goodness for restoring us once more to the society of our fellow men, or for bestowing upon us the means of founding in this desert a happy and flourishing colony of human beings, and waiting in silence for the further manifestation of his holy will, we passed our days in a course of industry, innocent pleasures, and reciprocal affection. Nearly two years have elapsed without our perceiving the smallest trace of civilized or savage man; without the appearance of a single vessel or canoe upon the vast sea by which we are surrounded. Ought we then to indulge a hope that we shall once again behold the face of a fellow creature?—We encourage serenity and thankfulness in each other, and wait with resignation the event!

33
Excursion to the Farmhouse

WE all remembered the bountiful provision we had derived from the blackbirds and ortolans* that had settled upon our giant tree at Falcon's Nest the preceding year. The time had now arrived for their reappearance, and we resolved to leave the grotto, which had become our established residence, and remove nearer to the spot, where I intended to secure as many as possible of this delicious provision for the coming winter.

My boys were all eager to start, their minds animated with the most warlike intentions. Fritz, the marksman, and Jack rejoiced at the good shots they would have; but I did not partake in their enthusiasm, for I remembered the prodigious quantity of powder that had been consumed the preceding year, and I was resolved to conduct the affair more economically.

I remembered to have read, in a book of voyages, that the inhabitants of the Pelew islands* capture birds a great deal larger and stronger than ortolans by means of limed twigs; and I intended to make, with gum-elastic and oil, a sort of glue that would save us a great deal of powder.

The provision of India-rubber which we had collected on our last excursion was exhausted; we had made waterproof boots of it, and, before I set out, I wished to give them a new coat of it. I sent Fritz and Jack to the wood of India-rubber trees, where I thought they would find, ready drawn, a sufficient quantity of the gum, as we had made large incisions in the trees, and placed calabashes under them to receive the gum; and as experience had taught us that the sun hardens it immediately, we had protected our calabashes from its rays, by surrounding them with green branches.

Our two messengers were lost to our view when my wife suddenly exclaimed, 'Stupid woman that I am, I have forgotten to give the boys a calabash in which they might put the gum, for they can not bring it home in the flat dishes we put there. I mean to go directly and see whether my gourds are ripe.'

I tranquillized my good wife by assuring her that they would not be at a loss to find something; and then, returning to the last word she had spoken, asked her what she meant by saying 'my gourds'.

She then informed me that she possessed a superb plantation of gourds, the seeds of which she had found among our
European grains, and which she had planted in her kitchen-
garden. She led the way there, and we found, among many
other plants, a quantity of those bottle-shaped gourds that the
peasants in our country carry to the field. Some were ripe,
some just formed, and others in full bloom. We selected the
ripest, and those the form of which could be useful to us, and
we commenced to empty them out. We made bottles, and
plates, and saucers, using alternately the knife and the saw. But
Ernest, my aid and companion had very little taste for such
work, and he could scarcely contain his joy when he heard me
say we had done enough. He threw down his knife and ran to
his gun, intending to give the ortolans and the jays, who were
perched in the tree, a discharge of small shot; but I soon
stopped him, for I was afraid the discharge would frighten away
the peaceable inhabitants against whom I had projected a more
quiet mode of warfare.

We now anxiously expected our young messengers back, for
the sun had already began to decline. Ernest kept a good
look-out on the side his brothers were expected from, and he
soon perceived them rapidly approaching, the one mounted
on the onagra, and the other on the buffalo.

'Well,' said I, 'have you made out well?'

'Oh yes, very well,' said Fritz, in a singular tone, as they
leaped from their coursers, and showed us what they had
brought, which consisted of a root of anise,* that Jack had
brought in his buffalo-pouch; a root, wrapped up in leaves,
which they called 'monkey root',* two calabashes of India-
rubber, and another one, half full of turpentine; a sack full of
wax-berries, and a crane, which Fritz's eagle had killed. But
while they were exhibiting their treasures, they talked so fast
and so rambling that I was obliged to make them preserve a
little order in their recital.

Jack then commenced telling us how he had obtained the
anise and the turpentine. Of these two things, one was, at least,
superfluous; but the other might be of some advantage, as I
could use it, instead of oil, in making my snares for the birds.
I then asked them concerning the 'monkey root' they had
brought; Fritz answered as follows:

'I do not know of what importance this root may be to us; but I can assure you that it far surpasses manioc, both in smell and savour. We discovered it close by the farmhouse, where a company of monkeys were regaling themselves on it. You would have laughed to see the manner in which the ugly animals pulled out the roots. They made use of a process which the labourers of Europe have no idea of—they pulled them up by turning somersaults.'

'By turning somersaults!' cried we; 'why, how wonderful!'

'Yes, somersaults,' replied Fritz. 'Every monkey, after having buried his teeth as far as possible in the root, turns himself violently over, backward, and repeats the exercise until his reiterated efforts have drawn the precious root from the ground. We looked on for a considerable time at the grimaces and contortions of the hideous animals; but, curious to judge for ourselves, of the merit of an article they seemed so very fond of, we resolved to disperse them. The report of a gun would have scattered the whole flock; but I remembered your instructions not to waste powder, and we contented ourselves by driving at full gallop among the affrighted troop, who fled in all directions. We soon tasted the root, and, finding it very delicious, I wrapped some pieces up in leaves, and brought them to you to see whether you know any other name than "monkey root" for it.'

Here Fritz stopped, and I took up the root; and, after having tasted it, I told my sons that their discovery was really a treasure, for I believed that it was the 'ginseng', the sacred root of China, which popular superstition had made a sort of universal panacea, and which the emperor alone could gather. Sentinels are placed in the spots where the 'ginseng' grows; but, notwithstanding their vigilance, immense quantities are smuggled by the Americans.

'Blessed be the monkeys, then,' said Ernest, 'that they have made known to us the existence of so precious an article.'

'Bless them, if you please,' answered Fritz; 'as for my part, I heartily abhor them. After having gathered up the roots which we have brought home,' he continued, 'we directed our course toward the gum-trees; the calabashes were full; we emptied them into others easier to carry, and, as the sun was yet very high, we thought we would go to the farmhouse and see how our little colonists got along. But imagine what was our astonishment at seeing the farmhouse overturned, the walls torn down, and the planks scattered, here and there, over the ground; the chickens were strangled, and the goats wandered through the scene of devastation. Our beautiful establishment had been sacked from top to bottom by a troop of mischievous animals, without doubt those villainous monkeys. Oh, how sorry I felt that I had not shot into the group and punished them for their audacity. We collected our poor beasts, who ran toward us on hearing our voices. We repaired their enclosure as much as we were able to; and, instead of reposing ourselves under the shade of that roof we had hoped to find, we turned, with saddened hearts, toward the "Lake of Swans", where it was that my eagle captured the crane. We then hastened our return, happy in having found so delicious a root as the ginseng, but overwhelmed with grief on beholding the ruins of our farmhouse, and thinking of your sorrow when you should hear the news.'

Fritz finished his recital. The news he had brought grieved us deeply, and I vowed vengeance against the malicious monkeys, who would soon render our island desolate. I consoled my sons by telling them I intended very soon to organize a 'monkey hunt', where they might display their courage and address.

Supper was now announced; the ginseng made its appearance, and was pronounced excellent; but as its aromatic nature made it more of a medicine than an article of food, I forbade its frequent use, while I enjoined my wife to plant a few roots in our garden. The sorrowful impression which the malice of the monkeys had caused gradually dissipated, and we separated, after the evening prayer, having decided that the first work on the morrow should be the manufacture of the bird-snares.

34
The Bird-Snares; the Monkey War; and the new Pigeons

THE next morning, after we had finished our everyday occupation, such as prayer, breakfast, and attending to our beasts, my young family reminded me of my promise of last night. All were impatient to see the snares in operation, and all prophesied marvels from this new invention.

I took a certain quantity of the liquid India-rubber, which I mixed with the turpentine, and placed the mixture over the fire; and, while the glue was thickening, I sent the boys into the copse to gather a quantity of little twigs which I needed. They soon brought me a large quantity, which I made them dip in the glue and fasten to the branches of the fig-trees, the fruit of which I observed was very much liked by the ortolans, thrushes, and beccaficos,* who frequented the place. I discovered that we had but seen the last of the season the preceding year, as at present, the birds were so numerous, that a blind man firing into the tree could not have failed to bring down a large number of them. The abundance of game suggested another idea to my mind; I thought that if the ortolans were so numerous during the day they would not be less so at night, and I resolved to try, in imitation of the Americans in Virginia, the experiment of a hunt with torches, persuaded that it would be more expeditious and successful than taking the birds by snares.

But my little boys, while employed at making the snares, had been taken in their own trap. Hands, faces, and clothes were all covered with glue, and one could not touch them without getting besmeared. They were all in great consternation, and their good mother, also, for she had but very little clean linen to spare them. I calmed their fears by assuring them that some ashes and water would remedy all the disorder, and wash out all the stains.

I rallied them a little on their awkwardness. 'I knew very well,' said I 'that my glue would trap the birds; but I had no idea it would catch little boys.'

I then taught them how to avoid the inconvenience of gluing their fingers, by plunging a packet of five or six twigs, by the aid of a pair of pincers, into the glue, instead of dipping them in singly. They adopted the plan, which succeeded perfectly. When I had made a sufficient quantity, Jack and Fritz climbed into the tree, and placed the branches of fig-trees, covered with the snares, among the limbs of the tree; and it was not long before we saw the unfortunate ortolans falling to the ground in numbers, their legs and wings stuck fast in the glue. They fell so fast that Francis, Ernest, and my wife were scarcely sufficient to gather up the game and kill them, while the two other boys again fixed in the branches the snares that had fallen with the birds, and which served three or four times. But, although the fowling was so abundant, the labour was very fatiguing, for the branches to which Fritz and Jack had to climb were as much as 60 or 70 feet from the ground. I placed a great deal of confidence in my torches, and I arranged the materials for making them, in which turpentine was a powerful auxiliary.

While I was thus occupied, Jack brought me a beautiful bird, much larger than an ortolan, which had been taken in the snares.

'I am very certain,' said Ernest, who had approached, and who, with his observing eye, had already recognized the bird, 'I am very sure that it is one of our European pigeons, one of the young ones from those who built their nest last year in the branches of the tree.'

I took the bird from Jack's hands, and recognized, with pleasure, that Ernest's conjecture was true. I rubbed the ends of his wings and his feet with ashes, to clean them from the glue, and I put him in a cage with the intention of adding a dove-cot to our domestic property. We captured others, and at night we had in our possession two fine pairs of wood-pigeons. Fritz thought that our grotto would be the best place to fix their habitation. I approved of the plan, and I promised to commence the work as soon as we had finished our present occupation.

But, notwithstanding our hard labour during the day, we were not able to fill more than one barrel. I enjoined my sons to take notice of the trees on which the ortolans roosted during the night. The bark of two or three of the fig-trees which were covered with the excrements of the birds decided the matter; and, after supper, and a few minutes of rest, I commenced my preparations. These were few in number, and consisted of two or three long bamboo-canes, two bags, torches of resin,* and some sugar-canes. Fritz, my grand huntsman, regarded me with a look of ironical incredulity. He could not understand how, with these strange instruments, I could realize the prodigies I prophesied.

We set out; and the night—which succeeds immediately to the day in these climates—soon overshadowed the earth.

Arrived at the foot of the trees that we had chosen, I lighted up my torches, and scarcely had the flame begun to burn, than a cloud of ortolans fell down around us, and began to fly wildly around the flickering flames.

'Well, gentlemen,' said I to my sons, 'you see that my strata-gem has proved not to be a bad idea. Now is your time: I have placed the game within your reach; you have but to extend your hand, and you are masters.'

I then armed each one with a bamboo-cane, and set them an example by striking right and left among the mass of orto-lans. They fell as thick and fast as rain, and we soon filled two large bags. Our flambeaus, however, would only last long enough to light us back to Falcon's Nest; and as the sacks were too heavy for me alone to carry, we placed them crosswise on the bamboo poles, and thus carried them very easily. The obscurity of the night, the torches which lighted us, and the sacks on poles, gave to our march the aspect of the funeral processions we read of in romances.

We arrived safely at Falcon's Nest, and, before we retired to rest, looked over our game, and terminated the sufferings of those poor birds that had not been killed by the blows. The next day, everyone put his hand to the work of cleaning and preparing our game—a very disagreeable, although necessary task. We filled two barrels with ortolans, half roasted, and packed down in butter.

I had not forgotten, in the midst of these culinary occupa-tions, the expedition I intended to make against the monkeys, and I fixed its commencement for the following day. We rose early; my wife gave us provisions for two days, and we set out. Fritz mounted the onagra, I took the ass, Jack and Ernest were seated together on the back of the buffalo, whom we had loaded with our provisions, our tent, and everything we had need of. Three of our dogs accompanied us. I made known to my sons that the war to which I conducted them would be a war of death, and that I was resolved to exterminate the whole race of the mischievous animals. The reason that I left Francis and their mother at Falcon's Nest was to spare them the painful sight.

The idea of death made a deep impression upon the youth-ful minds of my children, and it was not without pleasure that

I listened to their objections to spill any blood; but I persisted in my project, and forced myself to correct their ideas on this point.

'It is', said I, 'but a question of life and death between us and the monkeys. If they do not succumb, we must; it is an affair of preservation. Without doubt the useless effusion of blood is horrible, but there are circumstances that render it necessary.' But I could not convince them, nor overcome their aversion to shedding blood.

While disputing, we arrived at the shore of the lake. I selected a good place for our encampment, and we commenced immediate operations. We put up our tent and fastened hobbles to the legs of our animals to prevent their escaping, and we set off in quest of the enemy. The farmhouse was deserted, and the sight of the disorder which reigned there—the overturned walls and scattered ruins—did but confirm me in my intention to use the utmost degree of severity toward the depredators.

Fritz set off in advance, and, in a few moments, returned to announce to us that he had discovered the whole horde at a little distance tranquilly amusing themselves, and utterly unconscious of the danger near. We then drove slightly into the ground, all around the farmhouse, little stakes of wood, and we stretched between them long and flexible ropes; in different places we placed coconuts, and little vessels of cooked rice or maize with some fruit, and even some palm-wine, of which I knew by experience the monkeys were very fond. All these things we smeared over with the glue I had used to capture the birds with; we also spread it over the trunks of the trees and on everything, so that it was impossible to venture in the labyrinth we had constructed, or to touch any of the vessels, without getting stuck fast. We then retired to our tent, to give the enemy a fair chance to approach. The day passed, however, and yet nothing appeared; and I began to fear that the cunning animals had perceived our motions and suspected something wrong was going on. At night we laid down, after having done ample justice to the cold provisions we had brought with us; but nothing troubled our repose.

The next morning we awoke very early; but the monkeys had been beforehand with us, and the first sight that met our eyes was the whole company marching toward our traps. Nothing could be more laughable than to see the hideous procession; some walked on four legs, while others held themselves upright and marched majestically along, as if they were human beings. We remained still in the tent, for fear of frightening them; and we soon saw them busily engaged at the feast we had prepared for them. That which I had predicted soon happened; and immediately the whole troop formed but one solid block: they were all stuck fast together by the glued lines and the stakes, which attached themselves closely to their fur. It was a strange and truly laughable sight to see the attempts they made to extricate themselves: but their efforts were useless; and soon the whole troop set up a cry of fury and of rage. Never have I seen grimaces more hideous, or contortions more frightful. Those whom their gluttony had conducted toward the calabashes of rice or wine, carried away these vessels stuck fast to their mouths, while others dragged about the long sticks, which completely hindered them from escaping. When I thought that the disorder was complete, I let loose our dogs; they threw themselves furiously upon the troop of monkeys, tore them in pieces with their teeth, while we followed, striking right and left with our sticks. The unhappy monkeys uttered most lamentable cries, and rolled over at our feet as if to implore our pity; but I commanded that no quarter should be given, and we did not rest until the extermination was complete. Our sticks were dyed with blood, and the scene presented the appearance of a field of battle. My sons were horrified at what I had done, and they expressed themselves in so energetic a manner that I could not blame them. They declared that they would never engage in such another slaughter. The monkeys resembled man too much: their cries, their supplicating gestures rendered the whole affair most horrible.

We dug a ditch three or four feet in depth, and buried the dead bodies; and I surrounded the spot with a palisade to keep out our domestic animals.

We took a little repose, and I tried, by conversation, to dispel the sad thoughts that this bloody execution had infused into the minds of my sons. We endeavoured to repair the disorder at the farmhouse, collecting our frightened animals, and established as much order as a few hours' stay would permit; and then striking our tent, we set off for home. Before we went, however, we made a new conquest: it was that of two beautiful birds, larger than the ordinary pigeon, whom I recognized as pigeons of Molucca:* their plumage was an agreeable mixture of blue, green, yellow, purple, and violet. This capture was the work of Fritz, who had placed a little cup of rice, covered with glue, on a palm-tree, and the two birds had been taken by it. After having cleaned their wings, we fastened their feet together, and carried them with us, intending to admit them members of our future dove-cot at Tent House.

35
The Dove-Cot

WE hastened to return to Falcon's Nest, where we were gladly received by our dear ones. My wife took care of my pigeons, and approved of the plan of a dove-cot; consequently the wagon was immediately loaded with provisions, and all that was necessary for an excursion of some days, and we set out for the grotto. As soon as we arrived, I chose that part of the rock next our grotto as the situation of our dove-cot; and as the rock, after the outside layer was pierced, became softer, we soon made an excavation ten feet high, and large enough to contain twenty pairs of pigeons; two perches ran through the whole length, and, projecting out in front, with a board nailed across, formed a platform, which we protected by a slight roof; a door with a hole to admit light closed the front, and a rope ladder suspended from one of the perches enabled us to mount up, and look after the inhabitants. It cost us several weeks of constant labour to finish the construction, fix the boards strongly in their places; cover the inside with a coat of plaster, to prevent humidity, and arrange the perches, the nests, etc.: on the whole, it was a new trial of those qualities that enabled us to overcome all obstacles, patience and courage. My little workmen had learned the efficacy of these qualities, and they applied themselves to their work with a perseverance and zeal far above their years.

'There is the edifice,' said I to Fritz; 'but where are the inhabitants? We must call into action all our knowledge to find a way to force our wild stranger pigeons to dwell in the new habitation we have provided for them; and, besides, they must not only remain themselves, but must bring their companions with them.'

'It appears to me, father, that nothing short of sorcery will do it.'

'Sorcery or not, difficult as it appears, I am going to try it; and I have strong hopes of succeeding, with the assistance you can afford me.'

'Command me, then; I am ready, and impatient to know by what means you intend to effect your object.'

'It is to a pigeon-merchant that I owe the secret which I am about to put in practice. I will not warrant its success, for I have never tried it; but it consists in perfuming the new dove-cot with anise. The pigeons, it is said, are so fond of the odour of this plant that they will return, themselves, every night to respire its perfume; and it is in this manner that they insensibly change their country life for that of the pigeon-house.'

'Nothing can be easier,' replied Fritz. 'The plant of anise that Jack brought will do the business. We can break the seeds on a stone; and if the oil is not as pure as that of the chemists, it will not be less useful or less aromatic.'

'I think as you do,' I answered; 'and I am very glad that I permitted Jack to plant a root that appeared to me to be so valueless.'

We then proceeded to make the oil of anise. I rubbed the door of the dove-cot, the perches, and every place where the pigeons could touch either feet or wings, with it. I then mixed a sort of dough, with anise, salt, and clay, and, after having placed it in the middle of the dove-cot, we put in the pigeons, which we had kept in willow-baskets while their habitation was building. We shut them up, with provision for two days, and then left them to enjoy at their leisure the odour of the anise.

When our little boys had returned from our kitchen-garden at the end of that time, we formally announced to them that the pigeons had taken possession of their new abode. In a moment they flew to the ladder, in eager haste to get a sight of the new inhabitants. The two windows of isinglass* which I had placed in the door were raised by the curious, and I saw with pleasure that, instead of being frightened at the new objects that surrounded them, our prisoners appeared to have become quite tame; and when I entered, they took no more notice of me than a domestic pigeon would have done.

Two days more passed away, and I became curious myself to know the result of my *sorcery*.

On the morning of the third day I awoke Fritz very early, and commanded him to rub anew with anise the door, which was made to rise up and down by means of a pulley. He did it, and we then went, without saying anything about our preparations, to awake the still sleeping family. I then announced that the day of liberty for our prisoners had arrived, and now they were to be free.

Everybody now took their stations. I gave the cord of the door into Jack's hands; and, scarcely able to keep my countenance, I described a magic circle with a wand, and, after having murmured a pretended conjuration, I ordered Jack to pull up the cord.

The pigeons poked their heads cautiously out of the hole, then advanced on the platform, and suddenly soared up to such a height that they were lost to our sight. But in a few moments they again flew down, and settled, tranquilly, upon the platform they had just quitted.

This incident, which I did not expect, gave new proofs of my dealings in magic, and I cried out, in the most serious manner, 'I knew very well, when they flew up in the clouds, that they were not lost.'

'How could you possibly know that?' said Ernest.

'Because my charms have attached them to the dove-cot,' was my answer.

'Charms!' cried Jack; 'are you, then, a sorcerer, papa?'

'Simpleton!' replied Ernest; 'who ever *heard* of sorcerers?'

At that moment the pigeons, who had been quietly picking on the ground, attracted our attention. The two Molucca pigeons suddenly quitted their European brothers, and flew off in the direction of Falcon's Nest, with such rapidity that soon they were lost to our view.

'Adieu, gentlemen,' cried Jack, as they darted away, taking his hat off and making a thousand faces; 'adieu, a pleasant trip to you.'

My wife and Francis commenced to deplore the loss of our two handsome pigeons, while I, preserving as serious a look as possible, stretched out my hands, and, turning to the direction in which the pigeons flew, I murmured, half aloud, the following words:

'Fly, little ones, fly far, far away; till tomorrow you may stay; but then, return with your companions.'

I then turned toward my family, who stood stupefied with astonishment, not knowing what to make out of my serious address to the departed pigeons.

As for the other pigeons, they did not seem disposed to follow their companions, but appeared completely tamed: they had found the dove-cot of Europe with its shelter, and there they gladly remained.

'These two, at least,' cried Jack, 'are not so foolish; they prefer a good house and good food to the wind and the storm that the others must encounter.'

'Wait a little while,' answered Fritz: 'have you not heard father speak of his familiar spirit* that is going to bring them back?'

'Familiar spirit!' cried Ernest, shrugging his shoulders. 'Do not talk such nonsense here.'

'Not so fast, not so fast, Mr Ernest,' said I; 'it is by the fruits of my work that you must judge whether magic has been employed, and I feel very sure that my efforts will be crowned with success.'

We passed the rest of the day in the neighbourhood of the dove-cot, conversing on sorcery and the question of the pigeons: we often strained our eyes in the direction of Falcon's Nest, but nothing appeared. The evening came, and the European pigeons slept alone in their palace. We supped gayly, and retired to rest in anxious expectation of the morrow, which must establish either my defeat or my triumph.

We renewed, the next day, our habitual occupations; and though I felt a little doubtful about the return of the birds I said nothing, but anxiously awaited the evening; when, about noon, we saw Jack running furiously toward us, clapping his hands, and screaming out, 'He has returned! he has returned.'

'Who? who?' was eagerly asked.

'The blue pigeon!' he answered, 'the blue pigeon! Quick! quick! come and see him!'

'Bah!' replied the incredulous Ernest, 'a poor joke; it is not worth the trouble to run there and find the dove-cot empty.'

'Who knows?' said I to the philosopher. 'I rely on my science, and the return of the second pigeon will not astonish me more than the return of the first. Fritz asked Jack if the male pigeon had brought a female with him. But the stupid fellow could tell nothing about it. We then ran to the dove-cot, and, besides the blue pigeon, we found with him, on one of the exterior perches of the house, his mate, whom he was endeavouring to persuade to venture into the interior. He would put in his head, and then return to her, until at last he prevailed, and we had the satisfaction of seeing her enter the pigeon-house.

My sons would have immediately closed the door, but I prevented them, saying that some time or other it must be opened; 'and besides,' I added, 'how are the other pigeons to enter if we close the door?'

'I begin to think,' said my wife, at last, 'that there is something extraordinary in this; and, unless you have used some enchantment, I can not comprehend it.'

'It is chance—pure chance', interrupted Ernest.

'Chance!' replied I, laughing; 'that will do very well for one time, but when the other pigeon returns this evening, with his mate, will you think *that* chance?'

'Impossible!' answered he; 'the same phenomenon could not happen twice a day.'

While we were thus speaking, Fritz suddenly interrupted us: his eagle eyes had perceived the birds we were expecting.

'Behold them—behold them!' he cried; and really we soon saw the other pigeon and his mate alight down at our feet.

The joy which greeted their return was so loudly expressed, that I was obliged to impose silence; for the noise would have frightened the pigeons so that all the anise in the world would not have retained them. My little boys kept still; and, ere long, the newcomers entered their habitation.

'What do you say now, my little doctor?' said I to Ernest: 'Both pairs of pigeons have now returned?'

'I do not know what to say,' he answered, seriously; 'it certainly appears very extraordinary; but as for any sorcery or magic being employed, I will not believe it.'

'It gives me pleasure to see that you are not credulous; but if a third pair of Molucca pigeons should visit us to-day, would you call that chance also?'

Ernest did not answer; but his silence showed that he was far from being convinced.

We returned to our occupations, leaving Francis and his mother to provide a dinner for us. We had worked about two hours, when we saw our little Francis come running toward us. When he came near he drew up his little form, and, bowing haughtily, commenced the following speech:

'Most high and mighty lords, I am here to invite you, on the part of my good mother, to come and behold the prince of pigeons, who, with his noble spouse, has come to take possession of the magnificent palace you have provided for him.'

'You are welcome, for your good news, Mr Messenger,' was the universal answer.

We hastened to the dove-cot, where my wife, after cautioning us to make no noise, pointed out to us two superb birds, whom those in the interior were endeavouring to persuade to enter.

'I give up,' said Ernest, at last; 'my little knowledge can not comprehend it. I beg of you, papa, to explain all.'

Little Francis, who had heard the words 'magic' and 'sorcery' reiterated so many times, now begged me to explain their meaning.

'They are very nearly the same thing,' I answered. 'Magic is a Persian word, and is derived from "mages",* the name by which the wise men of Persia are known; the two words are often confounded, and mean a knowledge, more or less profound, concerning certain secrets of nature of which the vulgar are ignorant, and which may serve either for good or evil; from which comes also good and bad magic.'

Francis then asked if the magic lantern* was good or bad magic. I laughingly assured him that none could be better; and, after some questions on magic and sorcery in general, Master Ernest adroitly asked me what means of magical power I had made use of to bring back the birds.

I did not wish to prolong the embarrassment of my little doctor, and I explained to him, in detail, all that we had done. Jack laughed heartily on hearing that his plant of anise had been the charm which had so puzzled them; and I tried to persuade him to follow the example of Ernest, and not believe everything so readily.

The following days were devoted to bringing our dove-cot as near as possible to perfection; and we saw, with joy, that the new inhabitants were permanently settled, and already began to construct their nests. I observed, among the articles they gathered for that purpose, a sort of long, grey moss,* which I had seen hanging from the branches of old trees. I recognized it as being the same thing as is exported from India as a substitute for horse-hair, in the manufacture of matresses. The Spaniards make cords also of it, which are so light, that a piece twenty feet in length, if suspended from a pole, will float, like a flag, in the air.

I made this discovery known to my good wife, and one can easily imagine my news was well received; for it added another treasure to our domestic riches, and afforded promise of some fine matresses.

We found, from time to time, in the soil of the dove-cot, nutmegs, which, doubtless, the pigeons of Molucca had brought over. We washed them, and, although they were deprived of their silky covering, we committed them to the earth, without much hope of their ever germinating.

36
Jack's Misfortune; the Onagra; the Fountain

WE were occupied fifteen days in working at our dove-cot and other objects which called for our attention; and I may as well say, in this place, that the native pigeons multiplied so fast and increased so abundantly that I was obliged to reduce the number, as, in time, they would have driven out our beautiful pigeons of Molucca; and as the great multiplication was increased by swarms of emigrants from Falcon's Nest, I determined to reduce the number of European pigeons to five pairs; and, to prevent any newcomers, snares of glue were placed around the dove-cot every morning before it was opened, and they furnished us many a good fat pigeon, and relieved the eagle of Fritz from much hard duty.

An adventure of which Master Jack was the hero diverted the monotony of our existence, divided as it was between new constructions and provisioning our habitation for winter.

Jack had one day set off on an expedition, the intent and purpose of which nobody but himself had any knowledge of; but his absence was not long, for we soon saw him returning, covered from head to foot with a thick, black mud, and dragging after him a bundle of Spanish rushes, likewise covered with mud. The poor boy's eyes were filled with tears, and his irregular walk proved that he had lost one of his shoes.

At this tragi-comic sight we burst into a fit of laughter; his mother alone did not partake in the general gaiety, and she received the poor fellow rather coldly.

'Really,' said she, 'I do not think there is your equal in the world. Do you think that you have a whole wardrobe at your service, to supply you with clean clothes? What a pretty condition you are in.'

'Ha! ha!' laughed Fritz. 'He looks like a Barbary* goose.'

'Yes,' replied Ernest; 'or like the god Neptune, emerging from his watery empire.'

'Laugh away, gentleman,' cried poor Jack; 'if you had been as near death as I have been, you wouldn't laugh so heartily.'

These words arrested my attention. I reproached the boys with their unbrotherly conduct, and begged our hero to relate his story.

'Where have you been,' said I, 'to dirty yourself so?'

'In Flamingo Marsh.'

'Why, what, in the name of Heaven, were you doing there?'

'Alas!' answered the poor boy, as he heaved a deep sigh, 'I wanted to get some Spanish osiers* to make nests for the pigeons.'

'A praiseworthy intention,' said I; 'and it is doubly unjust to reward you with sarcasm. It was not your fault that the enterprise did not succeed well.'

'Oh, no; and if it had not been for these bundles of rushes I should certainly have lost my life. I wanted some thin, flexible rushes; those on the borders of the swamp were too large, and I advanced farther into the marsh, jumping from hummock to hummock, until I came to a spot where the only footing was a mass of soft, black mud; my feet slipped in, and I found myself up to my knees in the compound; and, gradually sinking deeper, I commenced screaming at the top of my voice; but nobody heard me, except my jackal,* who came running up to me, and tried to assist me by howling with all his might.'

'But why', said Ernest, 'did you not try to swim? You excel all of us in swimming.'

'Fine advice, truly; I would like to see you swimming in a swamp, up to your neck in mud, and surrounded by a thick forest of willows. When I perceived that neither my cries nor those of the jackal produced any good, I endeavoured to draw myself out; for I was sinking fast, and had no time to lose. I took my knife from my pocket, and cut, from the willows that surrounded me, two large bundles, and, placing one under each arm, they served me as a sort of hold. I then exerted all my strength, and, by moving my body, my arms, and my legs, I managed to raise myself up a little. All this time my jackal stood on the edge of the marsh, howling with all his might. I whistled him to me, and, grasping hold of his tail at last, with great difficulty, I reached terra firma.'

'God be praised, my poor child,' said I, 'that you have been preserved to us; but the risk was great, and you may thank your jackal that you are alive.'

'Who knows,' said Fritz, 'whether either of us would have thought of such a plan.'

'For my part,' added Ernest, 'I really can not tell what I would have done.'

'Oh, your inventive genius would have suggested some means of escape,' replied Jack; 'but necessity is the best master.'

'But', said his mother, 'you have forgotten one thing that should always accompany necessity—I mean prayer; for without the help of God we can do nothing.'

'Yes, dear mother,' answered Jack, 'and in the middle of my danger I repeated all the prayers I knew. I remembered our shipwreck, and how, when we prayed, God succoured us; and I also prayed that he might succour me.'

'You did well, my son,' said I; 'and God indeed heard you. It was he who gave strength to your arm, and who inspired you with the fortunate idea that accomplished your deliverance; then glory to God; let us thank him, and praise him, for all his goodness!'

His mother hastened to wash and clean the poor adventurer; his entire suit was put to soak in the Jackal's River, and we also washed the rushes, as I intended to make use of them. In their present condition they were too long and hard, and we were obliged to cut them in several strips. My sons knew nothing about basket-making, and their inexperience and awkwardness was a constant source of contention among them. I endeavoured to reconcile them as much as possible, and to impress upon their minds the importance of brotherly love and kindness in our present situation.

I profited by the willows that Jack had brought to commence the construction of a machine that my wife had long ago expected of me, viz., a weaving-machine.

Two rushes, split lengthwise, and wound round with pack-thread, so that they would dry without bending, formed four bars to make that part of the machine called the 'combs'. I made my sons cut me a quantity of little pieces of wood, to make the teeth for the combs; and when I had procured these first materials for my construction, I put them aside saying nothing to any body concerning their destination, as I wished the machine to be a surprise to my wife, and I proved insensible to all the ridicule showered upon my little sticks which Ernest facetiously called *tooth-picks*.

'What are you going to make with all those sticks?' asked my wife, with womanish curiosity.

'Oh, nothing but a whim of mine,' I answered, laughing. 'I intend to make you a superb instrument of music, such as the Hottentots* have called a *gom-gom*.* Let me alone, and, I promise you, you shall be the first to dance to its melodious sounds.'

'Dance! I assure you that I have other things to do; dancing never will occupy much of my time, I am afraid.'

'If you are really going to make a *gom-gom*,' said Ernest, 'your little sticks are useless; for *gom-goms* are made by simply stretching some cords over a piece of calabash, and the music is produced by drawing a quill across the cords.'

'A fine instrument, I declare,' cried Jack; 'why, it would frighten a jackal.'

'No matter what it will do, I have given you a correct description; for I have seen one, and the tone it emitted sounded exactly like *gom-gom*.'

Everyone was now anxious to know what I was going to do; but I waived all questions, telling my wife that she would be the first one to rejoice at its completion.

About this time, our onagra gave birth to a beautiful little ass of its species. It was received with pleasure, for it not only added to our number of useful animals, but also afforded us a courser, that in future time would make quite a figure in our cavalcades. I gave it the name of 'Rapid', as I designed him particularly for the saddle; and we saw with pleasure that his limbs were all beautifully proportioned.

The approach of the rainy season and the remembrance of the trouble we had had in collecting our animals last year, induced us to invent a method to render the service less painful; it was to accustom them to return to their homes at the sound of a conch, in which I had placed a bit of wood, like a flute. The pigs were the only ones that we could not manage. They were unruly and loved their liberty too much to be confined, we willingly abandoned them, as the dogs could easily bring them together if desired.

Among the embellishments and comforts with which we had surrounded our winter habitation, we yet wanted a reservoir of pure water, which we were obliged to bring from the Jackal's River. The distance was too great in winter, and I wished to remedy the inconvenience before the rains came on. I conceived the idea of bringing a stream of water from the river to the grotto, and to establish a fountain, as we had done at Falcon's Nest. Bamboo-canes, fitted into one another, served us for canals;* we rested them on crotches of wood, and a barrel sunk in the ground performed the office of a basin.

We proposed, when time permitted us, to give to this construction the elegance and perfection it wanted. But such as it was, it answered our purpose; and my wife assured me she was just as contented with the little fountain as if it had been built of marble, and surrounded by dolphins, and naiads,* spouting water from their mouths.

37
The Second Winter; our Studies

THE season of rains was fast approaching, and we used all possible expedition to get in everything necessary. The grain, the fruits of all sorts which surrounded our habitation, potatoes, rice, guavas, sweet acorns, pineapples, anise, manioc, bananas, nothing in short was forgotten. We sowed our seeds as we had done the year before, hoping that the European sorts would sprout quicker, and more easily on account of the moisture of the atmosphere.

My wife made us sacks of canvas which we filled, and, by the aid of our patient beasts carried to the magazines, where they were emptied into large hogsheads prepared for them. But these labours were not accomplished without much trouble; for, as we had planted our corn and wheat at different times, we were obliged to choose out the ripe stalks from a whole field—a work of no small difficulty. I resolved to devise some plan for a more regular cultivation the next year. We had a pair of buffaloes for all the labour that would have to be done; and all that was required, in addition to our present stock of harness, was a double yoke, which I intended to make during our winter seclusion. It was necessary now for us to become labourers, as we had successively been carpenters, coal-burners, and basket-makers; but necessity is despotic, and under this hard master we were serving an apprenticeship.

But the rains had already commenced; several times we had been visited by heavy showers, which hastened our remaining occupations. By degrees the horizon became covered with thick clouds, the winds swept fearfully along the coast, the billows rose, and for the space of fifteen days we were witnesses of a scene of whose majesty and terrific grandeur man cannot form an idea. Nature seemed overturned, the trees bent to the terrible blasts, the lightning and the thunder were mingled with the wind and the storm; in one word, it was a concert of nature's many voices, where the deep tones of the thunder served for the bass, and harmoniously blended with the sharp whistlings of the storm. It seemed to us that the storm of last year had been nothing in comparison to it. Nevertheless, the winds began to calm, and the rain, instead of beating down upon us in torrents, began to fall with that despair-inspiring uniformity which we felt would last for twelve long weeks. The first moments of our seclusion were sad enough, but necessity reconciled us to our situation, and we began as cheerfully as possible to arrange the interior of our subterranean habitation.

We had only taken a few of our animals in with us: the cow, on account of her milk, the ass to take care of the little foal, and Lightfoot and Storm, because we would need them in the excursions which might become necessary.

The small department that we had allotted for the stable prevented us from accommodating more, and the sheep and goats were sent to Falcon's Nest, as we were sure they would find abundant pasture and good shelter there; and one of my cavaliers rode over every day to carry them a handful of salt and see that nothing was wanting. Our dogs, the jackal, Master Knips, and the eagle, all found a home with us, and their attention and playfulness enabled us to pass away many of the long hours we were obliged to spend under the roof of our grotto.

We devoted our attention first to a crowd of minor wants, which we only discovered by occupation, but yet were of primary necessity. I have said that our apartments were all on one floor; but the ground had not been carefully levelled, and we set to work to fill up the cavities and cut away the projections, so as to prevent any of us from breaking our necks. The fountain I had made did not answer the purpose, and the one great necessity of a good supply of water was as yet unprovided for. We also made tables and chairs, prepared for all the exigencies of our position, and endeavoured to render our long confinement as supportable as possible. But there was yet an inconvenience. We had not imagined we wanted light. There were but three openings in the grotto, besides the door: one in the kitchen, one in the work-room, and a third in my sleeping-chamber. The boys' room, and all the rest of our habitation, was plunged in the most complete darkness. The light never penetrated into the recesses of the grotto. I discovered that three or four more windows were necessary; but they could not be made before the return of fine weather, and I devised the following remedy for the defect.

Among the bamboos that I had procured as leaders for the water was one of great size, which I had preserved. This bamboo I found by chance was just the height of our grotto. I trimmed it, and planted it in the ground about a foot deep, surrounding it with props to make it steady. I then gave Jack a hammer, a pulley, and a rope, and, appealing to his agility, I asked him to climb the pole. In a moment he was at the top, and, after having driven the pulley into the roof of the grotto, and thrown the cord over it, he descended safely to the ground. I then suspended to one end of the cord a large lantern which we found in the ship. Francis and my wife were officially charged with its supervision, and, thanks to the thousand reflectors which lined the sides of the rock, our grotto was as light as if it had been broad day. The light was an immense benefit to us, and enabled us to carry on our different occupations with zeal and comfort.

Ernest and Francis charged themselves with the task of arranging our library, and disposing, in its different shelves, the works we had saved from the wreck. Jack aided his mother in the kitchen, and Fritz, being stronger than his brothers, assisted me in the work-room.

We arranged there, by the window, a superb English turning-lathe,* with all its equipments. I had often amused myself by turning in my younger days, and I now could put my knowledge of the art to some use. We also constructed a forge: anvils were fixed in large blocks of wood, and all the tools of the wheelwright and the cooper were laid out in long array on the racks I had put up next the wall. Our shop began to assume a businesslike appearance, of which I was proud; and often did I congratulate myself that I had sufficiently acquainted myself, in youth, with mechanics to prevent their being entirely new to me.

The grotto every day grew more agreeable, and we were enabled to wait without ennui for the welcome light of the sun. We had our work-room, our dining-room, and our library, where we could refresh our minds, after the fatigues of the body; for the cases we had saved from the ship contained a quantity of books which had belonged to the captain and officers. Besides Bibles and books of devotion, we found works on history, botany, philosophy, voyages, and travels, some enriched with engravings, which were a real treasure to us. We had also maps, several mathematical and astronomical instruments, a portable globe, an English invention which blew up like a balloon; but the sort of works which prevailed were grammars and dictionaries of different nations: they generally form the chief stock of ship libraries.

We all knew a little of French, for this is as much in use as German throughout Switzerland. Fritz and Ernest had commenced to learn English at Zurich, and I had myself paid some attention to the language, in order to superintend their education. I now urged them to continue their studies, as English was the language of the sea, and there were very few ships that did not contain someone who understood it. Jack, who knew nothing at all, began to pay some attention to Spanish and Italian, the pomp and melody of these two languages according with his character. As for myself, I laboured hard to master the Malay tongue; for the inspection of charts and maps convinced me that we were in the neighbourhood of these people.

It was agreed that we should cultivate the French in common, while I taught English to my wife and Francis, and that the others should learn it for themselves. Our study was not a bad resemblance of Babel,* on a small scale, especially when we recited aloud, in order to break the learned silence that reigned there, passages from our favourite authors. This exercise, strange as it may appear, was productive of great advantage: it brought on questions and answers, and taught the little family many a foreign phrase that otherwise they would not have understood. Ernest reigned chief of literature among us. Memory, intelligence, and perseverance were all united in him; not content with studying English, he continued Latin, which his passion for natural history rendered almost necessary to him; and, so constant was his application, that I was often obliged to tear him from his book, and force him to take some exercise necessary to his health.

I have as yet said nothing of the thousand little comforts we found in the boxes we had saved from the vessel, and which we now looked over. We found all sorts of furniture: mirrors, several very handsome toilet-cases and bureau-tables, in which we found every thing necessary for writing. We even found a splendid clock, with an automaton figure, which, if I could

have put in order, would have sounded the hours: as it was it
made a very handsome show on the marble table in our saloon.
Our grotto grew every day so comfortable that the children
could not think of any name suitable to call it by: some wanted
it called The Fairy Palace, others The Resplendent Grotto; but,
after a long discussion, we came to the conclusion that it
should be called simply, 'Felsenheim', or the dwelling in the
rock. Time rolled away so rapidly in all these occupations, that
two months of the rainy season had elapsed, and I had not yet
found time to make my double yokes, or a new pair of carding-
combs, that my wife had teased me for a long time.

38
End of the Rainy Season; the Whale; the Coral

THE end of the month of August was marked by a renewal of
the bad weather. The rain, the winds, the thunder redoubled
with new fury. How happy we were in the habitation we had
made. What would have become of us in our aerial palace at
Falcon's Nest? And our tent, how could that have withstood the
storm? But at last the weather became more settled; the clouds
dissipated; the rain ceased; and we were able to venture out
from our grotto, to see whether the world yet remained firm.

We promenaded upon the belt of rocks that extended all
along the coast; and, as we had need of liberty and exercise, we
took pleasure in scaling the highest peaks, and looking over the
plain, which was spread out beneath us. Fritz, always daring, and
whose eye almost rivalled that of his eagle, was standing upon
the peak of rocks, when he perceived, upon the little island in
Flamingo Bay, a black spot, the nature and form of which he
could not determine; but he thought it was a shipwrecked vessel.
Ernest, who mounted after him, took it for a sea-lion,* such as
Admiral Anson* speaks of in his voyages. I determined to go
and inspect it myself. We walked down to the sea-shore, emptied
the rain water from the canoe, and all set off, with the exception
of Francis and my wife, who hardly liked our excursion.

The nearer we approached, the more rapidly one conjecture followed another. At last, when we were near enough to
distinguish it, what was our surprise to see an enormous whale,
lying on his side upon the strand.

Being ignorant whether he was dead or sleeping, I did not think it prudent to approach without precaution; consequently we turned around and steered for the other side of the island, which consisted of nothing more than a sand-bank elevated above the waves; but a rank growth of herbs and plants covered it, and it was the resort of numbers of seabirds, whose nests and eggs we found in abundance.

There were two roads to choose, by which to reach the whale: one by climbing over the rocks, which rendered it laborious; the other longer, but far the less fatiguing. I took the first path, and commanded the boys to take the other, as I wished to examine fully this little island, which wanted but trees to render it charming. From this elevation I could see the whole coast, from Tent House to Falcon's Nest, which spectacle made me almost forget the whale; and when I reached the side where my children were, they came running toward me, screaming with joy, and carrying their hats full of shells and coral, which they had picked up on the beach. 'Look papa,' said they, 'what beautiful shells we have found; what can have brought them here?'

'It is the sea, my children,' I answered; 'the sea has thrown them up from its abyss, and it appears to be little cause for astonishment that she should bring such frail, light, things as these shells, when she has thrown upon our shores a monster whose bulk is so immense.'

'Enormous,' replied Fritz. 'It seems strange that we should amuse ourselves with these petty trifles instead of examining our whale. I can hardly estimate its grandeur.'

Ernest did not like to hear the shells called petty trifles, and declared he would rather examine one of them than a thousand such enormous masses as our ugly whale; but I ended the dispute that was arising between himself and his brother by declaring that we would return after dinner, with the necessary articles for attacking the enormous prey that the ocean had thrown for us upon the sand.

I remarked that Ernest followed us to our embarkation with regret. I asked him the cause, and he declared that, if I wished to make him happy, I would leave him alone on this island where he would live like Robinson Crusoe. The idea made me smile, and I instantly replied, 'You foolish boy, do you know that the life of Robinson is but a finely wrought fiction, and that your romantic project has a thousand obstacles attending it? You would not be there long, before you would grow tired of your solitude; sickness would come, and some fine morning we would find the poor hermit dead upon the beach. Thank God he did not separate us at our shipwreck; we are six in all, and we are scarcely able to provide for our well-being. What could you do, alone, upon these rocks?'

The new Robinson was convinced, and quietly followed us to the boat. My little rowers were scarcely able to make any headway against the tide, and they complained bitterly of the hard lot to which they were subjected.

'You ought, dear papa,' said they, 'to devise some plan to render rowing less laborious.'

'You think I have more power than I really possess,' I answered; 'but if you can procure me an iron wheel, about a foot in diameter, I will try what can be done.'

'An iron wheel?' said Fritz; 'there were two among our iron-works: they belonged to a smoke-jack;* and I can easily get them for you, if mamma has not taken them yet.'

I did not want to engage myself too far; so I neither promised nor refused, but urged my rowers to redouble their labours, and we soon glided swiftly over the waves.

Some moments after, the conversation turned upon coral, and Jack asked me what use they made of this production of nature.

'Formerly,' said I, 'coral was in great vogue throughout Europe, as it was used in the head-dresses of ladies; but the coral has now gone out of fashion, and it is only gathered as specimens for museums, which is all we can do with this, among our other curiosities.'

Fritz wished to know to what region coral belonged; 'for', said he, 'I have read that it was the work of a little worm.'

'It is true,' said I; 'all sorts of shells are formed by the viscous deposit of the individual who dwells in them. Coral is formed by a very minute insect, which lives in the water, but is of so frail a nature that it can not subsist but in numberless quantities.'

I then related to my sons the phenomena of the existence of the polypi.* I told them also about the coral fisheries; and, while talking, we arrived at our destination, where my wife and son were ready to receive us. She admired the beauty of our coral, but observed that it was of no use in the household affairs; and when I had told her my resolution to return to the whale that afternoon, she cheerfully declared that she would accompany me. I was enchanted at this resolution, and we hastened to prepare the necessary provisions and articles for a stay of two days; for, perhaps, we might be detained on the island, and I thought it best to make preparations accordingly.

39
Dissection of the Whale

AFTER dinner, which we partook of an hour earlier than usual, I made a search for some barrels in which to put the blubber of the whale. I did not want to take the empty barrels we had at Falcon's Nest and Felsenheim, as I knew that it would be impossible to remove the disagreeable smell of oil. My wife reminded me that we yet had four tubs in our boat, which would answer my purpose very well. I fastened them to the stern of the canoe; and, after having armed my sons with knives, and hatchets, and saws, and all the cutting-instruments I could find, we weighed anchor, and directed our course toward the island where the whale lay. The sea was calm, and we arrived without much trouble, excepting the weight of our cargo, which necessarily delayed our progress.

My first care was to make the canoe and the tubs fast, and protect them against the violence of the waves. My wife and Francis were almost frightened at the first sight of such a monster.

Our whale looked like those of Greenland: the back was greenish-black, the belly yellowish, the fins and tail black. I immediately measured it, and I found that it was between 60 and 70 feet long, and about 40 in diameter, which is about the ordinary size of these monsters of the deep. My children were astonished at the proportions of the head, which formed a third of the whole creature; its mouth was immense, and its jaws, which were full 12 feet in length, were furnished with a sort of beard, formed of those long, flexible appendages called 'dewlaps',* and which in Europe form an article of commerce. One thing which struck Fritz was the smallness of the monster's eye, which was not larger than that of an ox; and the opening by which his immense mouth communicated with his throat, was scarcely the diameter of my arm.

'If the animal is voracious,' said my son, laughing, 'he can not swallow a very large mouthful at a time.'

'You speak truly,' I answered; 'and the whale owes, in effect, its enormous corpulence to a little fish, of which it is very fond, and which it finds in the Polar Seas. It swallows a prodigious quantity at every meal. It also absorbs, at the same time, a large quantity of seawater, which it throws out by the two holes placed below its nostrils; it is to this faculty it owes the name of "blowing fish", which it shares with several other marine monsters, as voracious, and nearly as immense as itself.'

Ernest, in considering the narrow throat of the whale, wondered how Jonah* could have found entrance. I then took occasion to say that it was not right to take the Sacred Books in their literal sense, and that, under the general name of whale, are comprehended other marine monsters of the same force, but whose interior organization may render the miracle of the Sacred Writings probable.

'But', added I, 'let us adjourn to another time all learned dissertations. To work, gentlemen—to work, and let us hasten, if we wish to get anything done before night comes on.'

Fritz and Jack then entered the head of the whale, and, working with the hatchet and the saw, cut out the 'dewlaps', which Francis and his mother carried to the boat; we cut out more than 200 pieces of different sizes. While this was going on, Ernest and I cut several feet deep into the fat which covered the sides of the animal; we literally swam in grease, for walls of solid fat rose on each side of us. But we were not long the only claimants for the whale. A multitude of winged robbers surrounded us, eager to associate in our work. They flew round and round our heads, then, gradually approaching, they were so bold as to snatch pieces of fat from our hands. The birds were very troublesome; but my wife having made the remark that their down would be of use to her, I knocked down some with a club, and threw them into the boat. I cut from the back of the animal a long and large band of skin, out of which I wanted to make a harness for the ass and the two buffaloes. It was a difficult task, the skin was so thick, and so hard to cut.

I would have liked to have carried away some of the intestines,
and the sinews of the tail; but the night was advancing, and it
was time to return. The tubs were placed in the canoe, and we
set out for the coast with the new cargo we had acquired. It
was for us a precious treasure, but far from agreeably obtained;
for the bloody work we were obliged to engage in was anything
but pleasing either to the sight or smell. Arrived at the shore,
we unloaded our cargo, which the ass, the cow, and the buffalo
immediately transported home.

The next morning we again embarked in the canoe; but this
time Francis and his mother were left behind, as they could
have been of no use in the work I intended, which was to
penetrate into the interior of the whale, and, if possible, to
procure some parts of its immense intestines. A fresh wind was
blowing, and we soon arrived at the island, which we found
covered with gulls and other marine birds, who, in spite of the
canvas with which the pieces that had been cut from the whale
were covered, had made a plentiful meal. It was necessary to
have recourse to firearms to drive away this horde of pillagers.

We took care, before commencing our work, to strip off
every article of clothing, excepting our pantaloons; then, like
true butchers, we opened the belly of the animal, selected from
the mass of entrails those which would best suit my purpose: I
cut them in pieces of from 6 to 12 feet long, and, after having
turned them inside out, washed them, and well rubbed them
with sand; they were then placed in the boat.

'Ah,' said Ernest, while we were preparing them, 'what
splendid sausages mamma could make with these entrails!'

'I wish I had one here', said Jack, who heartily approved of
Ernest's idea; and so saying, he commenced to inflate a piece
of entrail which was not less than 1.5 feet in diameter.

We abandoned the rest of our prey to the voracious birds;
and, after having loaded our boat with a new cargo of whale
blubber, we set sail for Felsenheim.

The reason that I had taken so much trouble to obtain the
whale's intestines was, because I wished to use them as vessels
to contain the oil; my sons thought that the idea was wonderful,
and wanted to know how I had thought of it.

'The author of this idea', said I, 'was necessity, that great
impulse of human industry; it was that which taught those

wretched beings who live where no wood or green thing ever grows how to supply their wants. It was necessity that taught the Esquimaux to use the entrails of the whale in place of vessels they had not the means of constructing, and discovered to them in this single animal treasures which the inhabitants of more favoured climes cannot appreciate.'

The conversation then turned upon the many uses of animal entrails, from cords for violins up to the aerostatic globe which elevates man above the earth—the covering of balloons being generally made of the skin of the intestines of animals.

Ernest then tried to explain to his brothers the phenomena of aerostatic ascension.

'Balloons', said he, 'are only kept up in the air because they are lighter than the atmosphere that surrounds them, the same way as bladders filled with air float on the water, the air they contain being lighter than water. This light air is obtained by heat, which expands the particles of air, and causes them to occupy more room; but as that process is long and tedious, balloons are generally filled with hydrogen gas, which is lighter than common air.'

'Papa,' said Jack, 'can't you make me a balloon with this piece of whale's entrail. It would be so nice to ride through the air, over the forests and the rivers.'

I observed to my little aeronaut that his project would be rather difficult, as it would require a balloon to be 80 feet in diameter to lift a little boy 60 pounds' weight, and that even then it would be of no use, as he could not direct it through the air.

When we arrived at home, we found my wife anxiously expecting us; but the sight of our greasy habiliments almost frightened her, and she anticipated the labour of washing them with no very pleasurable sensations; but I consoled her by promising miracles from the rich treasures of whale oil, and the entrails we had brought home. We washed ourselves completely, and, after an entire change of clothing, set out for Felsenheim.

40
*The Rowing-Machine; Excursion to Prospect Hill; the Giant
Turtle*

THE day had scarcely dawned when we were all up and ready
for work. The four tubs of fat were raised from the ground,
and, a strong pressure being applied, we squeezed out as much
of the oil as possible; and as this was the finest and purest, we
filled one or two of the entrails with it.

The rest was emptied into a large iron kettle, and, a slow
fire being applied, it was soon reduced to a liquid state. A large
iron spoon, which we had saved from the wreck, and which had
been originally intended for the sugar factory, served us to
empty the oil into the entrails from the kettle. All these works
were carried on at a distance from Felsenheim, as we did not
wish to perfume the air around our habitation with the fetid
odour of whale oil.

When we had procured a sufficient quantity of oil, we threw
the lumps of tried-out fat into the Jackal's River, where our
geese and ducks found a delicious repast. We also threw our
birds away, after my wife had plucked the feathers: their meat
was too coarse for us. The lobsters, however, were not so
particular, and we profited by the avidity with which they
rushed to the bait to renew our stock.

While we were occupied in our manufacture of oil, my wife
made me a proposition which met my hearty approbation: it
was to establish a new colony on the island of the whale. 'We
will put some fowls there,' said she; 'they will be safe from their
two great plagues, the monkeys and the jackals.'

I liked the project of my wife very much; and the children
were so enchanted that they wanted to start immediately, and
put it into execution. But it was now too late, and I calmed
their ardour by mentioning my promise about the canoe.

'Oh,' cried Jack, 'the canoe will go along without any rowing; how fine it will be!'

'Stop, stop,' said I; 'not so fast. All that I can do will be to
save your arms some labour, and quicken our speed.'

I immediately commenced the work. All my materials consisted of the wheel of a smoke-jack and an iron-toothed axle
upon which it turned. The machine that I constructed was not

a masterpiece of execution; but it answered the purpose very well. A handle attached to the wheel put the machine in motion, and two large, flat pieces of whale-bone, nailed together in the form of a cross, and fixed at each end of the axle, resembled the wheels of a steam-boat.* When the handle was turned, the wings of whale-bone beat against the surface of the water, and drove the canoe forward. Its velocity was in proportion to the power imparted to the wheel.

I will not attempt to describe the transports of joy that my children evinced when they saw the canoe gliding over the surface of the water: they clapped their hands, and jumped with joy as they watched Fritz and myself making a trial of our invention. I was astonished myself at the rapidity of our course. We had scarcely touched the earth, when everyone was in the boat, and begging me to make an excursion to the island of the whale. But the day was too far advanced to admit of such a thing, and I promised them that we would make, on the morrow, a grand trial of our vessel by an excursion, by sea, to the farmhouse at Prospect Hill, for the purpose of inspecting our colony there.

My proposition was well received; and we immediately began to prepare our arms and provisions, so that we could start early on the morrow.

At the first dawning of the day everybody was ready. We did not forget provisions; and my wife put up, in a double envelope of fresh leaves, a piece of the whale's tongue, which, by the recommendation of doctor Ernest, she had cooked and spiced as a delicate viand.

We gaily quitted the shore, and the strong current of the Jackal's River soon brought us into the sea: the breeze was good, and everything promised a favourable sail. We soon perceived Shark Island—the bank of sand where the whale had been stranded—and so well did our machine work, that in a short time we found ourselves in sight of Prospect Hill. I had kept at some distance from the coast, as I was afraid that there might be some hidden rocks inshore, which might destroy our frail bark. The distance was great enough to permit us to take in at one glance the splendid panorama which was unfolded before us. On one side was Falcon's Nest, with its giant trees; in the distance, a range of rocks which seemed to touch the heavens;

while, if your eyes sought a nearer object they fell upon Whale Island, its green verdure contrasting beautifully with the sublime expanse of ocean. Our hearts were overflowing with admiration, and we inwardly thanked the Lord for his goodness.

When we had arrived opposite the 'Wood of Monkeys', I ran the boat into a little creek, and landed, to replenish our stock of coconuts. It was with feelings of the keenest pleasure that we heard the crowing of the cocks through the woods, announcing the neighbourhood of the farmhouse. The sound recalled our dear country to our minds, where often this same noise has made known to the wearied traveller the neighbourhood of some friendly cottage. We re-embarked and rapidly neared Prospect Hill, and could plainly distinguish the bleating of our little herd. We landed, and directed our course toward the farmhouse.

Everything was in order; but what greatly astonished us was the wildness of the sheep and goats, who fled on all sides at our approach. My sons began to run after them; but as the long-bearded ladies were far more agile than they were, they soon grew tired of the chase, and, drawing from their pockets the strings with balls attached, they soon captured three or four of the fugitives. We distributed some potatoes and a handful of salt among them, in return for which they yielded us several bowls of most delicious milk.

My wife wanted to take away with her some of the fine pullets that were so numerous. A handful of rice and oats brought the whole feathered tribe about us; and, selecting those that she wanted, we tied their feet securely and threw them into the boat.

We dined at Prospect Hill. The cold meats we had brought composed our repast; but the whale's tongue was unanimously pronounced most detestable, and only fit for a sailor. We gave it to Jack's jackal, the only one of our domestic animals who had followed us.

He made a delicious meal from it, to judge from his avidity, while we rinsed our mouths with milk, in vain endeavours to remove the oily taste.

I left my wife to make the preparations for our departure, and started out with Fritz to gather some sugar-cane. I also dug up some roots of this precious article to plant on Whale Island.

We weighed anchor, or, at least, we pulled up the stone that secured us, and coasted along in the direction of Cape Disappointment, which I wished to double; but the cape still justified its name, and a long bank of sand stretching out prevented our progress, and we were obliged to turn back. I hoisted the sail, we redoubled our labours at the wheel, and, thanks to a little breeze that sprung up, were soon in sight of Whale Island.

Our passage was signalized by a spectacle entirely new to us. While sailing quietly along, we perceived, in the distance, what appeared to us to be a mass of rocks; on nearing it, however, we heard strange bellowings, and discovered that we had mistaken for shoals a herd of marine monsters, who, apparently, were fighting with one another. We were dumb with fear, and, I need scarcely say, we used all our strength to escape from these monsters before they perceived us.

On landing at Whale Island, my first care was to plant the roots I had brought from Prospect Hill; but my little companions, on whose assistance I had counted, did not think the plantation of sufficient consequence for them, and ran off to the beach to gather shells. My good wife supplied their places, and we two began our labours. We had scarcely commenced when Jack came running up to us, all out of breath. 'Papa, papa,' cried he, 'come here—quick, I have discovered the skeleton of a mammoth!'*

I burst into a laugh, and informed my little boy that his skeleton was nothing more than the carcass of our whale.

'No, no,' replied he; 'they are not fish bones, but those of some immense animal; and, besides, they lie a great deal farther up on the sand than the whale did.'

Jack implored and entreated so earnestly that I consented at last to go; but another voice soon stopped my progress.

'Run, run—this way,' screamed Fritz, from some distance, waving his hand to hasten my arrival. 'Quick—a monstrous turtle, that we are not strong enough to turn.' I caught up two handspikes, and ran, as fast as possible, to the spot, where I found Ernest struggling with a monstrous turtle, which he held by one leg; but which, despite all his efforts, had reached the border of the sea. I arrived just in time; and, throwing one of the spikes to Fritz, we were able to turn the enormous animal on its back.

It really was of prodigious size, about 8.5 feet in length, and could not possibly weigh less than 500 pounds. I did not know how we would be able to carry him away; however, the position in which we had placed him, gave us time for reflection.

But Jack had not forgotten his mammoth, and continued teasing me to go and see it. I soon perceived that it really was our whale; but the birds of prey had not left a morsel of flesh on the bones, which had blanched in the sun, and deceived, by their form, the little boy. I showed him our footprints in the sand, and some morsels of whale-bone which we had forgotten.

'How, in the world,' said I, 'could you imagine that it was the skeleton of a mammoth?'

'Oh, I did not think so; but Ernest told me, and I thought he knew.'

'And so you believe, without scruple, all you hear, without inquiring whether it is true or false.'

'But, papa, I thought that the tide had brought the carcass here.'

'Precisely; that is where your foolishness consists. You did not need much reflection to perceive that it would be very improbable for the sea to take away the carcass of a whale, and in less than a day deposit the skeleton of a mammoth in the same place.'

'True. I never thought of that.'

'And now, for a punishment, tell us all you know about the mammoth.'

He then told us that the petrified bones of these animals were found buried in the earth.

'Why, you are almost a professor,' I replied; 'Ernest will soon have to yield the palm to you.'

I then added some words on the yet doubtful existence of this animal, which, according to all appearances, is but an extinct variety of the elephant species. Jack's credulity procured him some sarcasms from his brothers.

'Oh, the good boy,' cried Ernest; 'he swallowed the bait too greedily, and mistook the carcass of the whale we dissected yesterday for an antediluvian* animal.'

'Plague on it!' exclaimed Jack; 'I know I am not much of a scholar, but surely these are not fish bones.'

'No,' said I; 'and if you would read a little more, you would discover that the whale, like all the fish of its genus, has real bones. Birds, men, all living beings have them; only the structure and composition is different, according to their different destinations. The bones of fishes are formed of an oily matter, lighter than water, and which aids them to preserve their equilibrium. Birds also have their bones, as it were, blown up with air, and appropriate to their course through the superior regions; as for terrestrial animals, their bones are more solid, as they are designed to support more of the weight of the body.'

'Can we not', said Fritz, in considering the skeleton of a whale, 'draw from this mountain of bones some utility?'

'I do not know', said I, 'what use we can make of them; the Hollanders make palings of them, and also rustic chairs, which produce a fine effect; and we will one day, when we have leisure, make a philosophical chair for our museum.' Discussing in this manner, we arrived at our plantation. I perceived that it was too late to finish our work that night. We buried the roots, yet unplanted, in the ground, and deferred till another day this important occupation. The giant turtle was now our grand object: I brought the boat round to where he lay extended on his back, and, forming a circle around him, we debated as to the means of transporting him.

'Zounds! gentlemen,' said I, striking my forehead, 'we need not embarrass ourselves much; instead of carrying this monster, let him conduct us back to Felsenheim. A turtle makes an excellent equipage on the sea: Fritz and I have tried the experiment.'

My idea was a happy one, and everyone was glad. I commenced by emptying out the barrel of water we had brought; then, turning the turtle over on his feet, we fastened the barrel to his back, so that it was impossible for him to sink and draw us with him; a cord, passed through a hole which we broke in the upper shell, served me for reins; and without losing time we all embarked for home. I placed myself in the prow of the canoe, with a hatchet to cut the cord in case of need.

Our course was accomplished rapidly and safely: a handspike that I held in my hand served me for a whip, and a blow well applied would rectify any deviation from the track. Master Ernest, the professor, compared us to Neptune* gliding over the waves, drawn by dolphins.

We arrived safely at Felsenheim, and our first care was to secure our turtle and to replace the empty barrel by strong ropes. But as we could not keep him long in this way, we finished his life the next morning, and his enormous shell was destined to serve as a basin to our fountain in the grotto. The work cost some trouble and time, as it was very difficult to detach the flesh from the shell. It was a superb piece of meat, full 6 feet by 3, and afforded us material for many a delicious pot of soup. We inquired into all our works on natural history, and we came to the conclusion (the professor and I) that our turtle was the giant green turtle, the largest of all.

41

Anniversary of our Deliverance

ONE morning, having arisen earlier than the rest of my young family, I occupied myself by counting up the time that had passed away since our shipwreck. I calculated the dates with the utmost exactness, and I found that the next day would be the anniversary of that event. It was just two years since the hand of God had been extended over us to save us from a watery grave. I felt my mind filled with thanksgiving, and I resolved to celebrate the day with all the pomp our situation would permit.

As I had not yet fixed upon the arrangements for our holiday, I said nothing about it to my family. Breakfast over, we proceeded to our different employments, and it was not until we were seated at supper that I announced, in a pompous manner, the holiday for the morrow.

'Be ready', said I to my sons, 'to celebrate the anniversary of the morrow; let each one prepare himself as is proper for so great a day.'

These last words, joined to the announcement of a holiday, surprised and overjoyed my children. Their mother was not less astonished than they were to find that they had been on the island two years.

'It is the property of labour', said I, 'to shorten the time; the days have leaden wings for the indolent, and they fly with the rapidity of an eagle for the industrious.'

A crowd of questions was now poured in upon me as to the manner in which I had discovered that we had been here two years.

'Very simply,' said I to my sons. 'When we were wrecked, it was the end of August; our calendar had four months yet to run through, and I have followed its course exactly. If my recollection serves me right, and if the dates which I have made are exact, tomorrow finishes the second year. But I think some sort of a calendar had better be adopted.'

'Well, then,' said Ernest, 'let us make an almanac, like Robinson Crusoe,* with a piece of stick, on which he every day made a notch.'

'Precisely; but that would not be sufficient, and your notches would represent absolutely nothing if you did not know the number of days each month contained, and in what order the seasons succeeded to each other.'

My little philosopher then began, in his learned way, to recite to me the rules for the division of time. 'The months have, some 30 days, and others 31: February alone has but 28 or 29. The year has 365 days, the day 24 hours, and the hour 60 minutes, which again are divided into 60 seconds each.'

'Very well,' answered I, 'for common use; but for doctors like you, does the year contain just 365 days?'

'No,' answered he; 'it contains 365 days, 5 hours, 48 minutes, and 45 seconds.'

'Very well; but what do you do with these odd hours, minutes, and seconds?'

'I lay them one side, and every fourth year makes one day more, which, added to the year, gives it the name of Bissextile.'

'Well answered, my dear sir; but it seems to me that, notwithstanding your science, we shall yet have some trouble to compute our time here. Who can tell us when the leap years come? Or who can determine what months have thirty, and what thirty-one days. We must not run the risk of confounding, on our calendar of wood, the times and seasons so strangely.'

Ernest replied, by holding out his clenched fist, and showing to us, by following the bones and cavities where the fingers commence to grow out, the order in which the months alternate. His brothers marvelled at his science, and I congratulated him upon having retained in his memory a thing so puerile in appearance, but which might eventually become useful.

We talked, during some time, of different affairs, until I gave the signal of retirement. My good little men lay extended on their matresses a long time, wondering and talking over the coming holiday.

The wished-for morrow at length arrived, and the day had scarcely dawned before I heard the report of a cannon resounding through the rocks. I was frightened, and instantly rose up to discover the cause. I found my sons stretched on their beds, apparently fast asleep, and Jack was snoring with all his might; but it was impossible for him to play the part of 'sleeper' long, and he no sooner perceived me than he cried out, 'Oh, how well she went off!'

I understood him; but, far from partaking in the enthusiasm of the thoughtless boys, I put on a frown, and reprimanded them severely for wasting a thing so precious as powder. They begged my forgiveness; and, as I did not wish any cloud to dim the brightness of our holiday, I let the matter drop.

We rose and dressed as decently as our scanty means afforded, and proceeded to breakfast. After our daily prayer, I turned to my children, and said, 'Two years have elapsed since we arrived here, and now is the moment to cast a glance upon the past.' I then took the journal of each day, which I had always kept, and read it aloud, dwelling on, and explaining the principal circumstances of our life. When I had finished, we all again thanked our Almighty Father for the many blessings he had granted us. It was an interesting spectacle to see those four children kneeling upon the sea-shore, and, in their childish form of words, thanking the God who had preserved them.

After our devotions were over, I announced to my family that the amusements of the day would conclude with the exercises which always terminated our holidays.

'You have practised for some time now', said I, 'in wrestling, running, slinging, and horsemanship; the time has come when you shall receive the reward for your labours. You shall this day contend, before your mother and me, and the crown shall be given to the victor. Come, champions,' I added, elevating my voice, 'the barrier is open, enter the lists; and you, trumpets, sound the horn of combat.' As I said these last words, I turned to the little inlet where our geese and ducks were feeding, and the whole troop, frightened by my gestures, and the tone of my voice, commenced a most deafening clamour, and furnished my sons with a good joke to excite their risible faculties.

I then organized the different combats which were to take place. First came firing at a mark; the materials for this were soon arranged by fixing in the ground a rudely shaped piece of wood, with two bits of leather at each side of the top, which we called a kangaroo. Jack did wonders, either by chance or skill: he shot away one of the ears of our pretended kangaroo! Fritz just grazed the head, and Ernest lodged his ball in the middle of the body. The three shots were all worthy of praise. Another proof of skill was then made: it consisted in firing at a ball of cork which I threw up in the air; Ernest had the advantage here: he cut the ball all to pieces. Fritz also shot well, but Jack could not hit it. We then tried the same thing with pistols, shortening the distance, and again I complimented my little boys upon the progress they had made in a year.

Slinging succeeded to the pistol exercise: Fritz carried off the prize. After that came archery; and here all—even little Francis—distinguished themselves. Next came the races; and I gave them for a course the distance between 'Family Bridge' and 'Falcon's Nest'.

'The one that arrives first', said I to the runners, who were gathered about me, 'will bring me, as proof of his victory, my knife, which I left on the table, under the tree.' I then gave the signal, by clapping my hands three times. My three sons set out, Jack and Fritz with all the impetuosity that marked their character; on the contrary, Ernest, who never did anything without reflecting, set off slowly at first, but gradually augmented his pace. I perceived that he had his elbows pressed firmly against his body, and I augured well from this little mark of prudence.

The runners were absent about three-quarters of an hour. Jack returned first; but he was mounted on his buffalo, and the onagra and the ass followed him.

'How now,' said I, 'is this what you call racing? It was your legs, and not those of the buffalo, that I wished to exercise.'

'Bah!' cried he, jumping from the back of his courser; 'I knew I would never get there, so I left the course; and, as the trial of horsemanship comes next, I thought that, as I was near Falcon's Nest, I would bring our coursers back with me.'

Fritz came next, all out of breath and covered with sweat; but he had not the knife, and it was Ernest who brought it me.

'How came you to have the knife?' said I, 'when Fritz got here before you?'

'The thing is simple,' answered Ernest; 'in going, he could not long keep up the pace he started with, and soon stopped to breathe, while I ran on and got the knife; but in coming back, Fritz had learned a lesson; he pressed his arms against his sides, and held his mouth shut, as he had seen me do, and then the victory depended upon our relative strength: Fritz is 16 while I am but 13, and of course he arrived first.'

I praised the two boys for their skill, and declared Ernest conqueror.

But now Jack, mounted on his buffalo, demanded that the equestrian exercises should commence, and he be allowed to repair the injury his reputation had sustained.

'To the saddle, to the saddle, my lads,' he bellowed with all his force, 'and you shall then see who can best manage a courser; we shall then know whether you can sit your horse as well as you can exercise your legs.'

I hastened to comply with the request of the little braggadocio: Fritz mounted his onagra, and Ernest took the ass; but although they tried all their skill, Jack distanced them both. I was frightened myself to see with what boldness the boy abandoned himself to the powerful animal that bore him. To stop, charge, and turn, was but a trifle to him; a practised groom could not have managed a thoroughbred horse with more ease and grace than he did his bull. Often, when the animal was in full gallop, he would rise, and, throwing himself on his back, extend his arms on each side, as the circus-riders do. But I expressly forbade this useless and dangerous experiment. Just as I had declared the contest over, and was about to proclaim Jack victor, the little Francis rode into the arena, mounted on his young bull 'Broumm', who was not more than three or four months old; my wife had made him a saddle of kangaroo-skin, with stirrups adjusted to his little legs, and there he sat, a whip in his right hand, and the bridle of his animal in the left.

'Gentlemen,' said the little cavalier, saluting us with a gracious bend, 'I have not contended with you thus far in any of the exercises of the day; will you now permit Milo* of Crotona to make a trial of his horsemanship before you?'

The assembly loudly applauded this little harangue, and the cavalier commenced to manœuvre his courser. The boy was more cool and calm than those of his age are apt to be. But what I admired most was the docility of the animal. My wife looked on with maternal pride to see the success of her dear pupil, and Francis was unanimously proclaimed an excellent horseman.

After the horsemanship, the swimming occupied some time; they also climbed the trees; and, after we had finished our gymnastics, I announced that the rewards would now be distributed, and that the crowns would shade the brows of the victors.

Everyone hastened to the grotto, which had been lighted up with all the torches we possessed: my wife, as queen of the day, was pompously installed in an elevated seat, decorated with flowers, and I called up the laureates to receive the rewards, which their mother distributed to each one as she impressed a tender kiss upon his forehead.

Fritz—conqueror at shooting and swimming—received a superb English rifle, and a hunting-knife, which he had long wished for. Ernest had for the reward of the race, a splendid gold watch. Jack—the cavalier—obtained a magnificent pair of steel spurs, and a whip of whalebone. Little Francis received a pair of stirrups and a box of colours, as a reward for the industry he had displayed in educating his bull.

When this distribution was finished, I rose, and, turning to my wife, presented her a beautiful English work-box, in which was contained all those little things that add so much to the comfort of an industrious woman, such as pins, needles, scissors, etc.

'Receive,' said I, 'my excellent companion, also a reward; for your services and endurance during these past two years well deserve one, even though the tender love of myself and children may be in itself a sufficient reward.'

The day was finished as it had begun—with songs and expressions of joy; we were all happy, all contented: we all

enjoyed that pure felicity which a life free from reproach had
given us; and we all thanked in our hearts the Lord who had
been so merciful toward us.

42
Domestic Affairs; the Palanquin; the Boa

WE had had so much trouble in harvesting our crops the last
season, that we had resolved, instead of trusting them to the
ground, without any order or regularity, to prepare a field
which could receive them all at the same time, and where they
could ripen together. But as our animals were not yet sufficiently accustomed to the yoke to warrant our undertaking
the task, I was obliged to defer it till some future period.

In the meantime, I employed myself in constructing a
weaving-machine for my wife: our garments had become so
tattered and torn, that the machine was of incalculable benefit
to us. It was neither perfect nor handsome, and nothing more
could be expected. As we had none of the wheat flour that the
weavers use to make paste, which they employ in hardening
the warp, and preventing the threads from tangling, I substituted the glue of fish; and I may confess, without self-praise,
that my composition was better than that of the weavers, for
the fish-glue preserves a humidity that the ordinary glue does
not, and by employing it one can weave in a dry situation,
instead of descending into cellars, where the weavers from time
immemorial, have been obliged to confine themselves. From
this fish-glue I also made window-panes, not well calculated, it
is true, for windows exposed to the rain; but they answered for
ours, which, on account of their deep embrasures, were protected from storm.

These two successes encouraged me, and I resolved to try
my hand at another thing, or, to speak poetically, 'add another
flower to my wreath'. My little cavaliers had long tormented
me to make them saddles and bridles, and our beasts had need
of yokes and other harness. I commenced my work, and established myself as saddler; kangaroos and sea-dogs furnished me
with the necessary leather, and I used for wadding the moss
that the Molucca pigeons had discovered to us. But as this moss
would have matted together, and grown hard under the rider,

I employed my sons in twisting it into cords, in which state it was left some time, and then untwisted; by that means we obtained frizzed hair, as elastic as that of horses. In a short time we had saddles and stirrups, bits and bridles, yokes and collars, each adapted to the strength of the animal for which it was intended.

Instead of attaching the plough, or any other implement, by cords, to the horns of the animal, as they do in France and Germany, I resolved to adopt the Italian method, and put the yoke on their necks. Besides, it appeared to me that the pressure was less on the forehead than on the shoulders of the animal, and future experience confirmed my opinion.

These labours were not yet terminated when we received, as we had the preceding year, the visit of a bank of herrings. We had found them so agreeable during the rainy season, that we did not let these pass unheeded.

The herrings were followed by the sea-dogs, which we also received thankfully; their bladders and skins had become too precious to allow us to neglect them. We killed about twenty or twenty-four, of different sizes; the skin and fat were all put to use; nothing was left but the tough flesh, which we abandoned to the crayfishes in the Jackal's River.

But this sedentary course of life did not suit the restless minds of my young people, and they earnestly begged me to take them hunting in the country. I put the matter off, and took in hand another sort of work, the want of which we felt sensibly. I speak of the making of baskets, a number of which articles we needed to carry our rice, roots, grain, etc. Our first attempts were clumsy enough, and we reserved them for our potato baskets. We gradually improved, and, when I thought we were skilful enough, I ventured to use those Spanish rushes that had cost Jack so dear, and we made a number of fine baskets; they were not as finished workmanship as more skilful hands would have effected, but they were light and strong, and that was all we cared for.

My sons had made a large basket to put manioc roots in, and, in a fit of mischief, Jack and Ernest had passed a bamboo cane through the handles, and, putting little Francis in the basket, set off on a full run, while the poor fellow endeavoured in vain to stop them.

Fritz, who had been looking at them, turned to me, saying, 'An idea has struck me, papa; why cannot we make a litter of rushes for mamma, and then she will be able to accompany us in our distant excursions?'

'Really,' I replied, 'a litter would be much more convenient than the back of the ass, and much easier than the cart: we will try what can be done.'

My children were delighted at the plan; but my wife laughingly observed 'that she would make but a poor figure, seated in a wicker basket.'

'Never fear,' said I; 'we will make you a fine palanquin,* such as are used in Persia and Hindustan.'

'Heyday!' cried Ernest; 'a palanquin supposes slaves to carry it: are we to act in that capacity?'

'Be quiet, my dear children,' answered their good mother; 'I will never take you for my slaves; and if ever I consent to ride in such a machine it will be on condition that it shall be carried on stronger shoulders than yours.'

'Really,' said Jack, 'we need not trouble ourselves about being porters; have we not the buffalo and the bull? Mr Storm, my courser, will do all that is asked of him.'

I approved of his idea, and I complimented the boy on his bright thought, a rare occurrence for him.

We immediately put the design of the palanquin into execution: the two animals were brought out; two poles, which supported a large basket, were suspended by cords on each side, and Ernest jumped in to make the first trial. Jack mounted Storm, who was placed at the head, and Francis Broumm, who supported the hinder part, and they set off. The first steps answered admirably; the basket, balanced between the poles, resembled a luxurious carriage on its springs of steel. But it was not exactly a pleasant carriage-drive that Ernest enjoyed; for, at a given signal agreed on by his coachmen, they whipped up their beasts, and set off at full gallop, subjecting Ernest to a punishment as novel as it was ridiculous, and consisted in forcing him to perform a sort of *basket-dance* at each jump of his conduc. rs. The fun was violent, but it was harmless, and we could not help laughing to see the phlegmatic Ernest so tossed about.

'Hold on! hold on! Stop! stop!' he screamed, at the top of his voice.

But they turned a deaf ear to his entreaties, and the poor patient was obliged to undergo this agreeable amusement during the time it took them to gallop from our sides to Jackal's River. One can easily imagine his anger when they dumped him over on the sand; and the quarrel went so far, that I was obliged to interfere. I reprimanded Jack, and that was satisfaction enough for the pacific Ernest, whom I saw, a moment after, helping his brother to unharness the animals, and put them in the stable. He even went to find some salt for these innocent instruments of his memorable ride. We then all returned to our basket-making; but we had scarcely recommenced, when Fritz, whose eagle eye was always making discoveries, suddenly started up, as if frightened at a cloud of dust which had arisen on the other side of the river, in the direction of Falcon's Nest.

'There is some large animal there,' said he, 'to judge from the dust it has raised; besides, it is plainly coming in this direction.'

'I cannot imagine what it is,' I answered; 'our large animals are in the stable, resting themselves after the experiment of the palanquin.'

'Probably two or three sheep, or, perhaps, our sow, frolicking in the sand', observed my wife.

'No, no,' replied Fritz, quickly; 'it is some singular animal:* I can perceive its movements: it rolls and unrolls itself alternately; I can see the rings of which it is formed. See, it is raising itself up, and looks like a huge mast in the dust; it advances — stops—marches on; but I cannot distinguish either feet or legs.'

I ran for the spy-glass we had saved from the wreck, and directed it toward the dust.

'I can see it plainly,' said Fritz; 'it has a greenish-coloured body. What do you think of it, papa?'

'That we must fly as fast as possible, and entrench ourselves in the grotto.'

'What do you think it is?'

'A serpent—a huge serpent, advancing directly for us.'

'Shall I run for the guns, to be ready to receive him?'

'Not here. The serpent is too powerful to permit of our attacking him, unless we are ourselves in a place of safety.'

We hastened to gain the interior of the grotto, and prepared to receive our enemy. It was a boa-constrictor; and he advanced so quickly that it was too late to take up the boards on Family Bridge.

We watched all his movements, and saw him stretching out his enormous length along the bank of the river. From time to time the reptile would raise up the forepart of his body twenty feet from the ground, and turn his head gently from right to left, as if seeking for his prey, while he darted a triple-barbed tongue from his half-open jaws. He crossed the bridge, and directed his course straight for the grotto: we had barricaded the door and the windows as well as we were able, and ascended into the dove-cot, to which we had made an interior entrance; we passed our muskets through the holes in the door, and waited silently for the enemy—it was the silence of terror.

But the boa, in advancing, had perceived the traces of man's handiwork, and he came on hesitatingly, until at last he stopped, about thirty paces directly in front of our position. He had scarcely advanced thus far when Ernest, more through fear than through any warlike ardour, discharged his gun, and thus gave a false signal. Jack and Francis followed his example; and my wife, whom the danger had rendered bold, also discharged her gun.

The monster raised his head; but either because none of the shots had touched him, or because the scales of his skin were impenetrable to balls, he appeared to have received no wound. Fritz and I then fired, but without any effect, and the serpent glided away with inconceivable rapidity toward the marsh which our ducks and geese inhabited, and disappeared in the rushes.

A general exclamation accompanied his disappearance. We were inexpressibly relieved. We commenced to speak. Everyone was sure that they had hit him; but all agreed he was as yet unwounded. We also concurred as to his immense proportions; but as for the colour of his skin, everyone embroidered it according to his own taste.

The neighbourhood of the boa threw me into the most unenviable state of mind; for I could think of no way to rid ourselves of him, and our united forces were as nothing against such an enemy. I expressly commanded my whole family to remain in the grotto, and forbade them opening the door without my permission.

43
The Ass and the Serpent

THE fear of our terrible neighbour kept us shut up three days in our retreat—three long days of anguish and alarm, during which time I suffered no one to break the rule I had established; the interior service of the grotto was the only consideration that could induce me to break it, and even then I allowed no one to go beyond the reservoir of the fountain.

The monster had given us no signs of his presence, and we would have supposed him departed, either by traversing the marsh, or by some unknown passage in the rock, if the agitation which reigned among our aquatic animals had not assured us of his presence. Every evening the whole colony of ducks and geese would direct their course to the bay, making a terrible noise, and sail away for Whale Island, where they found a safe asylum.

My embarrassment daily increased; and the immovability of the enemy rendered our position very painful. I was afraid that a direct attack might cost us the lives of one or more of our little family. Our dogs could do nothing against such a foe; and to have exposed any one of our beasts of burden, would have been certain destruction to it. On the other hand, our provisions daily diminished, as the season was not yet far enough advanced to have laid in any winter stores. In a word, we were in a most deplorable situation when Heaven came to our aid. The instrument that effected our deliverance was our poor old jackass, the companion of our wanderings, and faithful servant.

The fodder that we happened to have in the grotto had diminished frightfully: it was necessary to nourish the cow, as she contributed in great part to our subsistence, and some must be taken from the other animals. In this dilemma I resolved to set them at liberty, and let them provide for their own nourishment. As inconvenient as this measure was, it was better than to see us all dying of hunger, shut up in the grotto. I thought that if we could get them on the other side of the river they would find a plentiful supply of food, and be in safety as long as the boa remained buried in the rushes. I was afraid to cross Family Bridge, lest I should arouse the monster, and I decided to ford at the spot where our first crossing was made. My plan was, to attach the animals together. Fritz, mounted on his onagra, would direct the front of the procession, while I would take care that the march was effected in good order. I recommended to my son, at the first sign of the serpent's presence, to fly, as fast as his beast would carry him, to Falcon's Nest. As to our animals, I left to Providence the care of watching over and saving them. For my part, I proposed to post myself on a rock that overlooked the marsh, and, in case of an attack on the part of the serpent, retreat to the grotto, where a well-directed discharge of firearms would rid us of him.

I then loaded all our arms; my sons were placed as videttes,* in the dove-cot, with orders to observe the movements of the enemy, while Fritz and I arranged our beasts as aforesaid. But a little misunderstanding put an end to all my plans. My wife, who had charge of the door, did not wait for the signal, and opened it before the animals were attached together. The ass, who had grown very lively, considering his age, by his three days rest and good feed, no sooner saw a ray of light than he shot out of the door like an arrow, and was away in the open plain before we could stop him. It was a comical sight to see him kicking his heels in the air; and Fritz would have mounted his onagra, and rode out after him, but I restrained him, and contented myself by trying every manner of persuasion to induce the poor animal to come back. We called him by his name; we made use of our cow-horn; but all was useless—the unruly fellow exulted in his liberty, and, as if urged on by some fatality, he advanced direct to the marsh. But what horror froze our veins when, suddenly, we saw the horrid serpent emerging from the rushes; he elevated his head about ten feet from the ground, darted out his forked tongue, and crawled swiftly on toward the ass. The poor fellow soon saw his danger, and began to run, braying with all his might; but neither his cries nor his legs could save him from his terrible enemy, and in a moment he was seized, enveloped, and crushed in the monstrous rings that the serpent threw around him.

My wife and sons uttered a cry of terror, and we fled in haste to the grotto, from whence we could view the horrible combat between the boa and the ass. My children wanted to fire, and deliver, said they, the poor jackass; but I forbade them to do it.

'What can you do', said I, 'with firearms? The boa is too much occupied with his prey to abandon it, and besides, if you wound him, perhaps we may become the victims of his fury'. The loss of our ass was great, it was true, but I hoped that it would save us from a greater. 'Let us remain here, and the enemy will fall an easy prey to us; only wait until he has swallowed the victim he is now strangling.'

'But', said Jack, 'we will have to wait a long time; for it will be a great while before the snake can tear in pieces, and swallow our poor ass.'

'No; the serpent never tears his prey in pieces, and the teeth with which he is armed serve but to seize it; and when he has prepared it, he makes but one mouthful of it.'

'What!' asked Francis, in a tone stifled with terror; 'a single mouthful! Is it venomous?'

'No,' replied I; 'the boa is not venomous, but it is not the less terrible: he is endowed with extraordinary strength, and when he has become master of an animal, he crushes it, and, mixing the bones and flesh together, buries the whole in his body.'

'Impossible!' answered Jack; 'the boa can never break the bones of our ass; and as for swallowing him whole—why, the ass is larger than the serpent.'

'Impossible!' interrupted Fritz. 'Look, the monster is already at his work; do you not see how he is torturing our poor animal?—look how he fashions it to the dimensions of his throat!'

In fact, the boa proceeded with horrible avidity to his repast. My wife would not behold the mournful spectacle, and she retired to the interior of the grotto, taking little Francis with her. I was glad of this, as the sight became so horrible I could scarcely bear it myself. The ass was dead; we had heard his last bray, half stifled by the pressure of the boa, and we could now distinctly hear the cracking of his bones. The monster, to give himself more power, had wound his tail about a piece of rock, which gave it the force of a lever, and we saw him kneading like dough, the deformed mass of flesh, among which we could distinguish nothing but the head, dripping with blood and covered with wounds. When the monster judged his preparation sufficient, he commenced to swallow the prey he had secured. He placed before him the mass of flesh, and, extending his immense length along the ground, by a sudden effort distended his body frightfully; then, squirting a stream of saliva over the carcass, he began. Seizing the ass by the hind feet, by little and little, we saw the whole body buried in the insatiate maw of the monster. Every few moments he would eject a flood of saliva over his prey, as if to render the operation of swallowing it more easy. We observed that, as he advanced, the animal lost his strength; and when all had been swallowed he remained perfectly torpid and insensible.

The operation had been long: at 7 o'clock it had commenced, and at noon had just finished. I saw that the time had now arrived, and I exclaimed, 'Now my children, now the serpent is in our power!'

I then set out from the grotto, carrying my loaded gun in my hand; Fritz followed close by my side; Jack came next, but the more timid Ernest lingered behind. I thought it best to pay no attention to him until all was over. Francis and his mother remained at home.

On approaching the reptile, I found that my suppositions were right, and that it was the giant boa of the naturalists. The serpent raised his head, and, darting on me a look of powerless anger, again let it fall.

Fritz and I fired together, and both our shots entered the skull of the animal; but they did not produce death, and the eyes of the serpent sparkled with rage. We advanced nearer, and, firing our pistols directly through the eye, we saw his rings contract, a slight quiver ran through his body, and he lay dead upon the sand before us, stretched out like the mast of a ship.

We set up a shout of victory, and we huzzaed so long and loud, that Ernest, Francis, and my wife came running down toward us, trembling with fright, for fear a band of howling savages had landed on the coast.

'Let us rejoice', said I, 'after such a victory. Once more we owe our lives to the providence of God.'

'For my part,' said Fritz, 'I must say that I felt strange sensations of fear while the serpent was devouring his prey. Poor jackass! He devoted himself for our deliverance, as Curtius did, a long time ago, for the Roman people.'

'And', replied Ernest, 'how often the things we prize least become the most useful.'

'Poor dear ass,' added little Francis, 'we shall never ride him again.'

''Tis true, my child,' answered his mother; 'we should regret the loss of our good and faithful servant; but it was necessary that one of our animals should die for us. Let us thank God that our poor ass was chosen; for we could spare him the best, and he was so old that he soon must have died a natural death. The dragon has but anticipated it by a few months.'

Francis took notice of the name by which his mother had called the serpent, and asked if it was one of those dragons that used to infest Switzerland.

I laughed at his simplicity, and replied that the dragons of the story-books and mountain-ballads never existed but in the minds of those who wrote them.

'But', continued the little boy, 'do they eat serpents; if so, we will have enough for all winter.'

'Oh, fie!' exclaimed everybody, with an expression of disgust.

'I think that it will look well stuffed, and put up in front of our grotto, to frighten away all intruders', said Fritz.

'But it will also frighten our domestic animals,' interrupted Ernest. 'Its place is in our library, where it will figure well by the side of our coral and other curiosities.'

My wife asked me if the flesh of serpents ever was eaten.

'The Indians', said I, 'eat the flesh of the rattlesnake, the most venomous of the whole tribe, and they also never receive any injury from the flesh of those animals that they shoot with poisoned arrows.'

'I would not eat a morsel for the world', replied my wife.

'Nothing but prejudice; and I can assure you that if I had nothing else I would gladly put up with a steak from a nice fat boa.'

This was an appropriate occasion to give my sons a lesson in the natural history of the serpent, and I answered their numberless questions with pleasure. I recounted to them how some pigs* that had been left on an island in America, so infested by rattlesnakes that it was impossible to land there, completely cleared it of every serpent.

Ernest wished to know whether the rattlesnake really possessed the power of charming birds, and killing them by his breath.

'Very learned men', I replied, 'have thought so; but it is probable that the charm consists in the terror of the birds, which renders them unable to fly. Besides,' said I, 'there is a bird found in Africa called the secretary bird, on account of a feather projecting from the ear, whose sole food is serpents.' I then explained to my children how the poison of the serpent is infused into any object.

'There are two little bags fixed in the upper jaw, corresponding to two of the lower teeth, which are very long and pointed, and which the serpent has the power of drawing down in the gum. When the animal only wishes to take hold of anything, he does not use them; but when he wishes to inflict a death-wound, then he strikes their sharp, hollow points into the two little bags of poison: the venom runs down into the teeth, and from thence into the wounds inflicted by them.'

I then spoke of the 'Hooded Snake',* that the Indian jugglers teach to dance for the amusement of an ignorant populace. In short, all my knowledge of serpents was unfolded to the little boys, and the presence of the boa added much to the interest of the lesson.

44
The Stuffed Boa; the Soap; the Crystal Grotto

AFTER the three days of anguish that we had spent in the grotto, we felt the extreme pleasure of regained liberty; it was a second deliverance, almost as great as that from our shipwreck. Men never feel the magnitude of the gift of life until after some danger has threatened its extinction.

As I thought it best to finish immediately with the boa, I sent Fritz and Jack to the grotto with injunctions to bring back the buffaloes. I remained with Ernest and Francis, to keep off the birds of prey which already hovered around it; for I wished to preserve the brilliant-coloured skin with which it was covered.

When we were alone, I gently reproached Ernest with the timidity he had shown in attacking the serpent, and, as a punishment, I laughingly condemned him to compose an epitaph on our poor departed jackass. The punishment was as good as a reward to the doctor, for it was he who composed all the odes for our holidays.

He set about the work. Resting his head on his hand for about ten minutes, he arose and recited, with a sort of half-timid, half-satisfied air, the following stanzas:

'Here rests a faithful ass
    Who his master once disobeyed,
And was devoured by a snake at last,
    Who of him a breakfast made.
A family, shipwrecked on the isle,
    Mother, father, and children four,
Tried to save him, but in vain;
    And now our faithful slave's no more.'

'Wonderful! wonderful!' cried I. 'Here are eight lines, of which some have as many feet as the centipede; but as they are the best verses ever composed on the island, they will figure admirably on this monument to our jackass.'

So saying, I drew a piece of red lead from my pocket and scrawled the verses, after his dictation, on the surface of the rock.

I had scarcely finished when Fritz and his brother returned with the buffaloes. The new epitaph was the subject of conversation; but my young critics were so severe upon the poetry, and showered so many sarcasms upon the author, that he was obliged to give up the defence, and join in the general laugh.

We commenced our work by attaching the two buffaloes to the head of the ass, which yet projected from the mouth of the boa. While we held the serpent by the tail, they pulled from its stomach the disfigured remains of our unfortunate jackass. We buried them in the earth near by, and piled some pieces of rock over them for a monument.

The buffalo and his companion were then attached to the tail of the monster, and we set out for the grotto, supporting the head to prevent it from trailing on the ground.

'How shall we go to work to get the skin off?' asked my sons, as we deposited our heavy burden before the grotto.

'See if you can not find a way yourselves', said I, good-humouredly.

'I remember', said Fritz, 'to have read, in the travels of Captain Stedman,* that a negro having killed a boa, the skin of which the captain was very desirous to preserve, took a very ingenious method to deprive the serpent of it. He passed a rope round its neck, the end of which he threw over a strong branch, and then, having hoisted the serpent, he climbed the tree, and, throwing one arm around the animal, with the other he made a deep incision in the throat; then, holding his knife steady, he slid gently down the body of the animal, cutting the skin all the way down, and thereby rendering it much more easy for removal.'

'Admirable!' cried all the boys, simultaneously; 'but there is one difficulty: we are none of us as heavy as the negro; we could not make an incision.'

'I have thought of a simpler method than that,' cried Ernest; 'one I have often seen employed to skin eels, and which will serve admirably for our present purpose. It is this: to cut the skin around the neck, and, loosening the first part, attach strong cords to it, fasten the cord to the buffaloes, and, taking care to secure the head of the serpent strongly, drive the animals in the opposite direction, and, by that means, draw off the whole skin.'

'Oh,' said Jack, 'that will not be half as amusing as the negro's method: I would like to take a slide down the body of the snake.'

'Amusement must give place to utility; and I think that the idea of Ernest is more simple and easier than the other. Come, gentlemen, to the work. I leave the whole *labour* and the honour of the invention to you alone. As for the preparation of the skin, nothing can be easier: after you have cleaned the head as well as possible, you can wash the skin with salt water, sand, and ashes; then you must expose it to the sun's rays to dry, and, finally, fill it with hay, cotton, and all sorts of light materials.'

Fritz assured me that he understood all that I wished done, but that he was afraid it would not succeed. I encouraged him by saying that it would never do for men in our situation to give way to difficulties—we should never accomplish anything.

They at last undertook the work. The skin was washed, dried, and prepared as I had directed; and I could not see without laughing the strange method they adopted to stuff it. They hoisted the snake up to the limb of a tree, and Jack, in his swimming-costume, jumped into the long, hollow skin, and trampled down the hay, moss, and cotton that his brothers threw in to him with their pitchforks. When the skin was full, we saw him sticking his head out of the hole, and hurrahing with all his might.

When this work, which had occupied a whole day, was finished, we mended the holes that our balls had made in the skin; and, with a piece of cochineal, gave to the tongue that blood-red colour of which death had deprived it; then we elevated it on a wooden cross, arranging its body as gracefully as possible around the pole, and fixing the jaws half-open. Our dogs began to bark as soon as they saw it: and our animals recoiled from it as if it were a living boa. So arranged, it was solemnly installed in our library, where it took the first rank among our curiosities; at the same time Ernest wrote over the door the following legend: 'Asses cannot enter here.' We took the legend in two different ways, and it was agreed that the inscription should signify that the library was the sanctuary of science and study, and forever interdicted the entrance of idleness or ignorance.

We had nothing more to fear from the neighbourhood of the boa; but I was afraid it might have either left its mate (it was a female) behind it, or else a nest of little ones, which in time would spread terror through the land. I resolved, in consequence, to undertake two expeditions—the one through the marsh, the other toward Falcon's Nest, through the passage in the rock, where I supposed the boa had got through. We commenced by the marsh; but, at the moment of starting, Ernest and Jack begged me to allow them to remain at home.

'I shiver with fear', said Jack, 'to think of meeting one of those horrible serpents in the rushes.'

I endeavoured to overcome this childish fear, and reminded my sons that pusillanimity was a sentiment unworthy of the mind of man.

'When we have triumphed over a real danger,' said I, 'we ought not to recoil from one that exists but in our imagination. It would have been of very little advantage to us to have killed a snake if we should be surprised tomorrow by one of the same size, or, in a few weeks, should behold a whole army of small ones issuing from the rushes. He does nothing,' I continued, 'who stops in the middle of his work.'

We then set out, loaded with our hunting equipage. We carried, besides our arms, some boards, and the bladders of sea-dogs to sustain us on the water, if necessary. The boards we wanted to assist us in our march; for, by placing one before the other, and taking them up, we made a solid walk of wood. This was a great convenience, and enabled us to search the marsh thoroughly. We easily recognized the traces of the boa; the rushes were bent down where she had passed through, and there were deep spiral impressions in the wet ground where she had rested her enormous rings. But we discovered nothing that induced us to believe that the boa had a companion; we found neither eggs, nor little ones—nothing but a nest of dried rushes, and I did not think that the boa had constructed even that. Arrived at the end of the marsh, we made an interesting discovery; it was that of a new grotto, which opened out of the rock, and out of which flowed a little stream that passed on among the rushes of the marsh.

The grotto was hung with stalactites, which rose in immense columns on each side, as if to sustain the vault, and formed themselves into singular and beautiful designs. We remained some time in admiration of this miracle of nature, and, as we walked on, I remarked that the ground upon which we trod was composed of an extremely fine and white sort of earth, which, after examining it, I recognized as being 'fuller's clay'. I immediately gathered some handfuls, and carefully placed them in my pocket-handkerchief.

'Here,' said I to my sons, who were regarding me with astonishment, 'here is a discovery that will be very welcome to your mother, and henceforth if we bring her dirty clothes, we will bring her something to wash them, for here is soap.'

'I thought,' said Ernest, 'that soap was the result of human industry, and not a production of the earth.'

'You thought rightly; the soap that is ordinarily used is composed of a certain salt, the acidity of which is corrected by the addition of grease, and so forth, which weakens its power greatly. But this fabrication is tedious and costly, and men have been fortunate to find a sort of earth in which is united certain qualities of the soap; it is this we have found, and it is called "fuller's clay" because it is used to clean woollen goods.'

We had approached the source of the spring while conversing; and Fritz, who was a little in advance, cried out that the rock had a large opening on one side. We ran forward and soon found ourselves in a new cavern. We fired off a pistol, and we were able to judge by the echo that the grotto extended to a great distance. We then lighted two candles, with which our knapsacks were provided; they burned without obstacle, and the pure light assured us of the salubrity of the air. Having left the others behind, Fritz and I continued to advance, when suddenly we saw our torches reflected from every side of the rock.

'Ah, papa,' cried Fritz, in a transport of joy, 'see! see! a salt grotto! Look at the enormous blocks of salt laying at our feet.'

'You are very much mistaken, indeed,' I answered; 'these masses are not salt: if they were, the water which drips from the rock would have melted them long ago; instead of salt it is crystal: we are really in a palace of rock-crystal.'

'Better yet—a palace of crystal! What an immense treasure for us!'

'Yes; such a treasure as the gold mine* was to Robinson Crusoe.'

'Look, papa, here is a piece I have broken off: it is not salt, as you said; but it is not transparent like crystal.'

'That is your fault; you *troubled* it by breaking it off.'

This expression appeared new to my son, he could not comprehend how you could *trouble* a piece of crystal. I then explained the whole theory of the formation of crystals.

'The masses which we see before us', said I, 'all form columns and pyramids of exactly six sides; the greatest care is necessary in transporting them, as rough usage loosens long, needle-like particles in the interior, which cross each other and produce obscurity. The crystal then is called "troubled". Considerable blocks are to be seen in many museums of Europe.'

'I begin to think,' said my son, with an air of chagrin, 'that our discovery will not be of much use to us, except to add to the curiosities in our museum.'

The curiosity of my son had been excited by what I had said upon the subject of crystals, and I saw with satisfaction that his youthful mind eagerly grasped at the knowledge of the wonderful works of nature. I informed him, that the crystals were formed from the sediment, deposited by the emanations of water, which attaches itself to the rock, coagulates, and in time becomes as hard as any of the metals. 'The ancients thought it was petrified ice, but modern science has investigated the matter more fully. The art of fashioning and moulding rock-crystal has been discovered, and many medical and chemical instruments are made from it.' But the light of our candles commenced to grow dim, and I thought it prudent to retreat.

When we reappeared at the entrance of the grotto, we found Jack there weeping bitterly. When he saw me, he uttered a cry of joy and flew into my arms, loading me with caresses.

'Well,' said I, 'what is the matter, that you mingle together your tears and smiles so freely?'

'Oh, I am so glad to see you, papa; I heard two deafening reports, and I was afraid you were both buried under the rock and that I would never see you again.'

I pressed the poor little boy tenderly to my heart. 'Thank God,' said I, 'my poor Jack, that nothing happened to us; the reports you heard were the noise of two pistols that your brother fired to test the solidity of the rock. We have discovered a new palace, as brilliant as that of Felsenheim. But what have you done with Ernest—where is he?'

Jack conducted us to the border of the swamp, where we found our phlegmatic philosopher, who had not heard the two explosions, tranquilly employed in making a rush basket such as fishermen use, consisting of a frame of long stalks, terminated at the end by a funnel, through which the fish passed, but could not return.

'Quick, quick!' he cried, when he saw us approaching. 'I have killed a young boa.'

We had been talking so much of serpents, etc., that the poor boy had mistaken a superb eel, 4 feet in length, for a snake; he had walked straight up to it, hit it two or three blows on the head with his gun with as much courage as would have sufficed to kill a dozen boas.

The examination that I made of the snake humbled the pride of the victor; but the eel was a great treat for us, and we returned home to Felsenheim, where my wife and little Francis were eagerly expecting us. I presented to my wife the 'fuller's clay', and we commenced to relate, in their minutest details, the adventures and discoveries of the day.

45
Excursion to the Farmhouse; the Cabiai; the Muskrat

I HAD as yet only half accomplished my design, and there remained all the country about the farmhouse yet unexplored; and besides, I wished, if it was possible, by fortifying the passages in the rock, to keep out all such visitors as the one we had lately received. I made sure, before we set out, against any accident that might happen: we took plenty of provisions, arms, vessels of all sorts, torches to scare away all intruders on our night encampments, in short, everything that would render our excursion safer and less disagreeable; and anyone that could have seen our departure from Falcon's Nest, would have thought that we were moving into the country, so loaded was the wagon with all sorts of articles.

The good mother took her place in the cart. Storm and Valiant were attached to it, carrying also their young masters; the cow drew our baggage, and Fritz, mounted on the onagra, rode in front of the caravan, as a sort of advanced guard, while Ernest and I brought up the rear on foot. My little philosopher liked much better to wander along by my side than to take a place in the cart by his mother; it gave him an opportunity for conversations and scientific discussions upon whatever object we encountered. The dogs sustained the wings of our army, and Rapid, the name of our little onagra, gambolled gaily about us.

We advanced in good order along the avenue of Falcon's Nest, and discovered the marks of the boa's progress, half effaced by the wind. We found everything in good order at the Nest; the harvests and the fruit-trees gave promise of an abundant crop. The goats and sheep received us joyfully, and came up of their own accord to receive some salt we threw them in passing. We did not stop, as the Lake farmhouse was the object of our expedition, and we wished to arrive as soon as possible, in order to gather, before night, cotton enough to make some pillows and matresses that might render our slumbers more agreeable.

The farther we advanced, the fewer traces we found of the serpent. We could not see a single monkey in the coconut wood; and the crowing of our cocks, mingled with the bleating of our herds, gave promise of good order at the farmhouse; and we were not disappointed. As soon as we arrived, our good housekeeper set about procuring us some dinner, while we went to gather the cotton.

After dinner I announced that we would immediately commence our search, and we divided into three parties, each one charged to explore a part of the country. Ernest and his mother had, for their division, the guard of the provisions and the collection of all the ripe blades in the rice-field; to defend them we left our brave dog Billy. Fritz and Jack, accompanied by Turk and the jackal, took the right bank of the lake, while I followed the left, with Francis and his two young dogs. It was the first time that the little fellow had shared in any of our expeditions, or had had a gun entrusted to him; he marched along with his head up, and as proud as a new-made officer; and he burned with ardour to make trial of his new weapon. But the noise of our steps among the dried rushes frightened only some herons, and they flew so suddenly and quickly, that it was impossible to shoot them. Francis began to grow despairing at his ill success, when suddenly we found ourselves in presence of an innumerable quantity of wild geese and black swans, which covered the waters in all directions. Frank was just about to fire into the mass, when suddenly a sort of deep, prolonged cry, like a bellow, issued from the middle of the rushes. We stopped, astonished, and a second after the cry was repeated.

'I am sure,' said Francis, 'that it is the little onagra.'

'Impossible,' said I; 'he would not leave his mother; and, besides, we must have heard him as he passed along. It is more likely to be a swamp-bird, called a "bittern".'*

'How could a bird make such a noise as that? Why, it sounds like the bellowing of a bull; it must be of prodigious size.'

'Not at all; it is not as large nor as strong as other herons of the same family. But your supposition proves that you are ignorant that the voice of an animal has nothing to do with its size, but only with the conformation of its throat and the muscles of the breast. Thus, the canary and the nightingale, two very small birds, fill the air with their song, and modulate their voice to such prolonged notes as we would not think so feeble an animal capable of. As to the bittern, it is related that, when he sings, he buries the extremity of his beak in the mud of the swamp; and this gives it that deep, sonorous tone which resembles more the voice of a bull than a bird.'

'Oh, how I would like to kill it!' said my little hunter. 'How proud I would feel to bring down such a prey with my first shot!'

'Attention, then, and in a few moments you can have that pleasure.'

I called the dogs to my side, and, setting them in the rushes, presently heard the report of Francis's gun. But, instead of firing in the air, he had discharged his gun right into the thickest part of the rushes, and I saw the birds that the dogs had disturbed flying away safe and sound.

'You awkward fellow,' said I; 'you have let your game escape you.'

'On the contrary, papa, I have him! I have him!' repeated he, with passionate emotion. 'Look!'

So saying, he pulled out of the rushes an animal resembling an agouti, and which the little hunter had already christened by that name.

I examined it with attention, and discovered that there was much difference between it and the agouti.

This one was about two feet in length, had incisor teeth like the rabbit, webbed feet, long snout, but no tail at all.

'You have killed a rare and curious beast,' said I to my little boy. 'It is an inhabitant of South America, of the same family as the agouti and peccaries, but much rarer. It is a cabiai,* and, what is more, a cabiai of the largest size.'

'And what sort of an animal is this cabiai? I never heard of him before.'

'Oh, yes; you heard him bray just now; for it was his cry that I attributed to the bittern. This animal profits by the darkness of night to provide his food: he runs fast; can swim well, and has the power of remaining a long time under water; he eats seated on his hind legs; and as to his cry, it sounds exactly like the braying of an ass.'

But it was now time to return home, and Francis rejoiced at the prospect of his triumph over his brothers. He took up his cabiai, threw it over his shoulder, and although I saw that it was much too heavy for him, I thought I would let him have the merit of the whole affair. We trudged on a little way in silence, when Francis exclaimed, 'How foolish I am to carry such a load. If I clean my game 'twill be twice as light.'

'A good idea,' said I. 'We do not eat the entrails; and the dogs, to whom they belong, can just as well eat them here as anywhere.'

The little fellow then set about opening the cabiai. During this operation I tried to impress upon his mind that pain always accompanies pleasure in every glory. But my instruction was all wasted: my little man was under the influence of the charm of victory, and he scarcely heard me.

When he had finished, we resumed our march; but the cabiai was yet too heavy for his feeble shoulders. Another idea then occurred to him to put it on the back of one of the dogs.

We then fastened it strongly on the back of Braun; and the dog, proud of his charge, set off in advance.

We arrived at the pine wood, and set to work to gather a quantity of canes, which we had discovered to be eatable. We perceived some monkeys in the distance, who disappeared as we approached, apprising us that the hated race was not yet exterminated. But as to the boa, nothing indicated that it had passed through, or that it had left any young ones.

We found, on returning, Master Ernest tranquilly seated on the bank of the river, and surrounded by a prodigious number of enormous rats, which he had killed. The phlegmatic philosopher then recounted to us the history of this massacre.

'We were occupied,' said he, 'my mother and I, in collecting the ripe rice-blades, when I discovered, at a little distance, a sort of high, solid causeway, which looked like a road constructed in the middle of the swamp. I immediately set off to discover what it was, and Master Knips with me. But we had scarcely advanced one step when he darted from my side, in pursuit of an animal that quickly disappeared in a sort of hole bored in the causeway. I remarked, on advancing, that the two sides of the bank were pierced all along with these holes, all of the same form and size. I was curious to know what they contained, and I introduced into the opening a long bamboo cane that I had in my hand. I had scarcely drawn it out when there issued forth a legion of animals similar to the first. Knips ran after them; but the rice grew so thick that he could not get along. An idea then occurred to me to place my rice-sack over the hole. I did so; and, beating the top of the causeway with a stick, a great number ran into the sack. I then began to beat the bag with my stick, so as to kill the prisoners. But imagine my surprise when I found myself assailed by a whole army of rats, who emerged from every side, and began to run up my pantaloons. Knips made most desperate attempts. I could do nothing with my stick, and I cannot tell what might have happened if Billy had not heard my voice, and come to my assistance. He rushed bravely upon the army of rats, and made so terrible a slaughter that the enemy fled in terror. Those that you behold fell victims to my stick and the formidable teeth of Billy; the rest of the army took refuge in their holes.'

The narration of Ernest excited my curiosity, and I wished to see, for myself, the causeway, with its inhabitants, and I recognized a series of works similar to those of the beaver, with the single difference, that they were not so extensive. I made my sons observe the conformity that existed between these animals and the beaver of the north; both had the same membrane at the feet to facilitate swimming; both had the flat tail, and both were provided with two little bags of musk.

Fritz and Jack returned during these conversations; they brought back a ruffled moorhen and a nest of eggs: we placed them under one of our hens that happened to be sitting at the time. We then all united around a savory mess of rice that my good wife had prepared. She had cooked a small piece of the cabiai; but it was detestable, and we abandoned it to our dogs, who would not taste the flesh of the rats on account of the smell of musk.

The repast was a merry one. We were all delighted to have found no traces of the boa; and my mischievous little boys showered a flood of epigrams upon the 'Conqueror of Rats', as they called poor Ernest.

The conversation naturally turned upon what we should do with our rat-skins; and it was determined upon to make a carpet of them, to preserve the floor of our house dry. Our first care was to clean them with sand and ashes, as we were accustomed to do.

The two little bags of musk, with which these animals were provided, excited the attention of the children, and question followed question, in quick succession, as to the manner in which the Europeans gather this precious substance.

I informed them that many animals were provided with musk: the gazelle, the beaver, the ondatra* (which was the true name of the rats which Ernest had killed), the polecat, the civet, and the muskrat. I explained to them, at the same time, the different processes in use to procure their musk, and how the Dutch, who understand taming these animals, shut up, at certain periods, civets and muskrats, and after they have deposited the contents of their bags they are let loose until they are replenished.

But this dissertation on the civet and ondatra had not made us forget the detestable taste of the cabiai.

'Ah,' said Ernest, who loved good eating—'ah, if we only had a little dessert now, to take away the fishy taste of that abominable beast!'

At this exclamation, Jack and Francis ran to their knapsacks.

'Look here, sir!' said the youngest, as he threw some pine-cones before the philosopher.

'Look here, sir!' said Jack, placing on the table some little shining apples, of a pale green, and which exhaled a strong odour of cinnamon.

A general cry of admiration greeted them.

'Stop!' cried I. 'Before tasting this fruit, Master Knips must undergo the customary trial, for I am afraid these are the fruit of the manchineel-tree;* and the manchineel apples produce most terrible colics.'

I then opened one of the fruits, and discovered that I had been deceived by the appearance. The manchineel apple has a nut, and these had very small seeds, like the common apple. While I was showing this difference to my children, Master Knips snatched one of the apples from the table and commenced to eat it, smacking his lips as if it were something excellent. This determined the matter. I distributed the fruit, and on tasting them we declared them most excellent. Fritz wished to know their name.

'They are', said I, 'cinnamon apples; I think you gathered them from a low shrub; did you not, Jack?'

'Oh, yes, yes—shrub—cinnamon. I am falling asleep,' stammered out the stupid fellow.

I then gave the signal for retiring. We took all necessary precautions for safety during the night, and we sought, on our matresses of cotton, the repose that the fatigues of the day had rendered necessary.

46
The Otaheitan Roast

THE next morning, at break of day, we renewed our search. We directed our course to the sugar-cane plantation, where I had built a hut of branches; but we found it all blown down; and, setting up our tent, we resolved to pass the forenoon there.

While my wife was making preparations for dinner, we explored the sugar-canes, as I thought it would be a natural retreat for the serpents, if there yet remained any in this part of the country. Happily, our investigation was without any result; and we were turning to quit the sugar-canes, when suddenly our dogs began to bark, as if they had taken some dangerous animal. We could perceive nothing; but as it was not prudent to venture among the canes, I ordered my sons to direct their course toward the plain, and we soon found ourselves clear of the canes. At the same time there emerged from them a troop of pigs of quite respectable size and strength. I at first thought it was the young family of our old sow; but their number, the grey colour of their skin, and the singular manner in which they walked soon banished that idea. They trotted one after the other with a precision and regularity that would have done honour to a troop on parade. I took good aim, fired both barrels of my gun, and two of the animals fell. The loss did not seem to make much impression on the rest of the troop, who trotted on as before. It was a singular spectacle to see the whole family marching along the borders of the sugar-cane, with an imperturbable tranquillity; everyone followed exactly in his place, without any pushing for precedence; and, on examining them more closely, we found that there was but one footstep in the sand, so regularly did they march.

But Jack and Fritz, who were a little in advance, could not let them pass unheeded. Bang, bang, went their pistols, and two more animals bit the dust; the dogs also had their part in the victory, and each one strangled a victim.

I immediately recognized them as being a sort of hog called *Tujacus*,* and as I knew they had two small glands under the belly, which, if not immediately removed, would render the meat uneatable, I immediately set about the operation, and my sons gladly aided me, so rejoiced were they at such a splendid booty; for we had six pigs, each, on an average, three feet in length.

While we were thus occupied, I heard the reports of two other guns in the distance; I conjectured them to have come from Ernest and Francis, who had overtaken the pigs. I was right; and Ernest, who soon returned with the wagon for which I had sent Fritz, confirmed me in my conjecture. We thus had in possession three more pigs, for Billy had also done his duty.

The arrival of the doctor naturally provoked a discussion on the name of our game. Fritz thought that these were the Otaheitan pigs* of which Captain Cook speaks. Ernest was of another opinion, and maintained that they were peccaries.* This animal is very common in Guiana and South America. Before loading the wagon we resolved to clean our game to diminish the weight. Although we worked as diligently as possible, the work was not yet finished at dinner-time, and we were glad to find in the sugar-cane a cordial that refreshed and nourished us. We abandoned to the dogs an enormous quantity of entrails, which they soon devoured, and we set out for the tent; but my boys were so proud of our chase that they determined to convert our convoy into a triumphal march: they cut some green boughs and decorated our equipage; they adorned their caps and guns with flowers, and we made our entrée chanting a song of victory.

'You have kept me waiting long enough, gentlemen,' said my wife; 'your dinner is all spoiled; but, bless me! what a quantity of meat. Why should you abuse the provision nature has so liberally provided by killing more than we require?'

We justified ourselves as well as we were able, and my children offered their mother the sugar-canes they had brought home.

Fritz proposed to regale the family with an Otaheitan roast. We received his proposition, but it was put off till the morrow, as the preparation of our pigs precluded all thoughts of anything else.

I sent the two smaller boys to gather a quantity of green branches and leaves, with which to smoke our pork, and we then set to work. Ernest skinned the pigs. Fritz and I cut them up, and my wife salted the pieces. I piled the hams and all together, so that the salt would penetrate every part, and we also poured salted water over them, and allowed them to remain until the hut for smoking was constructed. As for the heads and bones, they were abandoned to the dogs.

The next morning, Fritz recalled to my memory the promise I had made to allow him to serve us for dinner an Otaheitan roast. He began by digging a deep ditch; he then took the pig he had reserved for the purpose, washed it with care, rubbed the interior with salt, and filled it with a sort of stuffing made of meat, potatoes, and different roots.

When the ditch was full of combustibles, Fritz set it on fire, and from time to time the boys threw in, by the direction of their elder brother, a quantity of pebbles, which soon became red hot.

My wife observed all these particulars with a look of incredulity.

'Beautiful cookery you will make,' said she, 'with an entire pig, some ashes, and a hole in the ground—delicious eating I do not doubt!'

Nevertheless, notwithstanding her sarcasms, she could not help but give the boys some advice, and she aided Fritz in giving his pig the graceful turn that roast pigs should always have to taste good.

When these preparations were finished, our cook-in-chief enveloped his 'roast' in leaves and pieces of bark; a hole was made in the burning cinders large enough to receive it, and it was then covered with red-hot stones, and the hole filled up with earth to prevent the air from penetrating.

At the sight of this last ceremony my wife gave up, and exclaimed, in a tone of discouragement.

'Oh, what a mess! It may be very good for savages; but do

not think, I beg of you, that a Swiss woman, who piques herself on her knowledge of cookery, is going to touch such a dish of burned meat as will come out of that hole.'

But Fritz did not despair, and he made a learned appeal to the testimony of navigators in favour of the delicious taste of the Otaheitan roasts. I interrupted his erudition by pointing to the hut ready to receive our meat. We had about twenty superb hams—a nice treasure for us during the rainy season. We filled the hut we had constructed with green leaves and wet branches, set them on fire, and made preparations to remain until the meat was all smoked.

Fritz let his 'roast' cook for about two hours; and it was not without astonishment that, after having taken off the triple layer of earth, cinders, and stones, the most delicious odour saluted our olfactories. I scarcely expected anything eatable, and we had before us meat cooked to a nicety, combined with a spicy perfume that would have done honour to a Parisian cook. Fritz triumphed: his good mother avowed that she was conquered, and everyone proceeded, without delay, to prove the pig. Some ashes which had fallen on it were carefully removed, and the meat was pronounced delicious. That which astonished me most was the spicy odour with which it was impregnated, and I attributed it to the leaves with which it had been cooked. I made an examination, and came to the conclusion that it was the ravensara* of Madagascar, the root called by naturalists 'agathophyllum', signifying 'good leaf'. I threw a certain quantity into the smoking-hut, in hopes that they might impart an aromatic odour to the hams.

During the three days of fumigation, I had every day, with my sons, explored the country. These excursions discovered to us no traces of the boa; but they very seldom ended without their bringing home some little addition to our comforts and luxuries.

One day we directed our course toward the wood of bamboo, and returned home loaded with cups of all dimensions, formed from rushes, which we sawed apart at every knot; some of them were very large, twenty inches in diameter. We also made another discovery the same day: it was that each knot of the rushes distilled a sugary matter which crystallized in the sun, and resembled candied sugar. These rushes also furnished us a quantity of long, strong thorns, which filled the place of nails admirably.

We also made an excursion to Prospect Hill; but we found everything there in the greatest disorder: the walls of the farmhouse were pulled down, and the cattle gone. The monkeys had passed that way, and left unequivocal traces of their progress. I resolved to undertake a war of complete extermination against this hated race, who appeared to dispute the possession of the country with us; but this important question was left until another time.

We then surrounded the hut, where our hams were suspended, with a rampart of earth, and fortified it with branches and stones, so as effectually to keep out all intruders; and we arranged everything so that we would be ready to set off on the morning of the fourth day, and commence our explorations beyond the defile that had been the barrier between the district we had inhabited for two years and an unknown land, which we had but once entered, and then were nearly destroyed by a troop of buffaloes.

47
Excursion into the Savannah; the Ostriches and their Nest; the Little Land Turtle

WE began our march at daylight, and, after having journeyed on for about two hours, I gave the signal for a halt, as about gunshot from the defile which separated the two countries appeared to me to be a favourable spot for our encampment. It was situated on an elevated point that commanded a far-extended prospect, and was defended on one side by a thick pine forest.

'Here', said Fritz, 'is a spot where we can defend ourselves against all enemies; and, if you take my advice, father, you will establish a post here.'

Jack, in accordance with his praiseworthy custom of never attending in the least to the conversation of those around him, caught at the last words his brother had spoken, and, confounding a military post with a letter post, bellowed out, 'A post-office! Why, where can we send the letters to?'

'Sydney, Port Jackson, and New Holland',* replied I, as gravely as possible.

This answer attracted the attention of Francis, who asked me why I had named these places—whether by chance, or because I really thought we were near them.

'The more I consult the charts of the captain,' said I, 'the more I think we are in the latitude of New Holland. The circumstance of our shipwreck, the route which the vessel had followed, the tropical rains, the productions of the coast, the sugar-canes, the spices, the palm-trees, all confirm me in this opinion. But in whatever land we may be, we belong to the great family of God, and we ought to thank him for the treasures which have been so lavishly bestowed on us.'

Fritz wanted me, before quitting the place, to leave, as a mark of our passage, a fortress, after the manner of the Kamtschatkans,* which is simply composed of some boards, elevated by stones at each corner, to a height sufficient to keep out all savage animals. Before commencing this work, we made an investigation of the forest round about; but we discovered nothing but two 'margays', or wild cats, who fled into the forest before we could level a gun at them.

The rest of the morning was devoted to the fortification of our encampment. We then dined; but the heat was so powerful that we were obliged to postpone our excursion into the savannah until the morrow.

Nothing troubled the repose of the night. We were up at daylight, and in a few moments our preparations were complete. I took with me my three eldest sons, as I wished to be in force on entering into a country as yet unknown. My readers may laugh at this expression, applied to an army of four persons, and three of those boys; but, such as it was, this army comprised all our resources. Francis remained with his mother to take care of the baggage; and, after breakfast, we packed some provisions, and took leave of our good mother, who saw us depart without uneasiness.

We passed through the defile, at the extremity of which we had erected a palisade of bamboo and thorny palm; but it had all been torn down, and we could easily trace on the sand the spiral imprints of the boa, clearly demonstrating that she had come from the savannah through this passage. I intended to erect a solid rampart here, that should be proof against the attack of any animal; but I was obliged to defer the execution of the plan until some other time.

We had now ventured into a country we had entered but once before. Jack recognized the place where we had taken the buffalo; the river, which divided the plain, was bordered by a rich line of vegetation. We followed its course for some time, and arrived at the grotto where my son had taken the young jackal; but the farther we advanced vegetation disappeared, and we soon found ourselves in the middle of an immense plain, only bounded by the horizon. The sun beat right down on our heads, the sand burned our feet—in one word, it was a desert—a desert without a single tree—a desert of sand, the only green things being a few withered geraniums and some sort of grass that contrasted strangely with the aridity of the soil. On crossing the river, we had filled our gourds with fresh water, but the sun had heated it so that we could not drink it, and we were obliged to throw it away.

'What a difference between the country we have just left and this!' said Jack, sighing.

'It is Arabia Petrea,'* replied Ernest—'a volcano. My feet burn me as if I was walking on hot irons.'

I endeavoured to sustain the sinking courage of my poor children. 'Patience,' said I, 'patience; nothing is obtained without work: remember the Latin proverb, "Ad angusta, per angusta."* Look! the farther we march the less uniform the ground appears; I can distinguish a hill in front of us. Who knows, perhaps there is another Eden behind it.'

After two hours of painful journeying we arrived at the foot of the hill that we had perceived afar off: it was a rock that elevated itself in the middle of the desert, and afforded us a refuge against the rays of the sun. We were too fatigued to climb the rock and reconnoitre the country: we could scarcely stand against the overpowering rays of the sun, and our dogs were as tired as ourselves; we were isolated in the middle of the desert, and could see the river in the distance, like a silver thread, winding through its green banks. It was the Nile, beheld from a mountain under the burning sun of Nubia.*

We had scarcely been seated five minutes when Master Knips, who had accompanied us, suddenly disappeared over the rock, having probably scented some brother monkeys in the neighbourhood; our dogs also, and the jackal, deserted us; but we were too tired to call them back.

I brought out some morsels of sugar-cane, and distributed them among the boys, for our thirst was terrible. This refreshment restored our appetites, and some rounds of roast peccary furnished us with an excellent repast.

'Confess,' said Fritz, laughingly, 'that a piece of ham, roasted à l'Otaheitan, does not taste bad in a desert like this.'

'It is a little better', said Ernest, 'than the mortified flesh of the Tartars,* who, it is said, put the meat which they eat under the saddles of the horses they ride, and thus carry their provisions with them.'

This trait of erudition on the part of Ernest brought on a discussion; and I was endeavouring to explain my reasons for discrediting this story, when suddenly Fritz, whose excellent sight was always making discoveries, cried out, 'What do I see!' said he. 'There are two horsemen galloping up to us. There, a third has joined them—doubtless they are Arabs of the desert.'

'Arabs!' said Ernest; 'Bedouins* you mean.'

'Bedouins are but one division of the great family of Arabs, and your brother was right,' said I; 'but take my spy-glass, Fritz; your news astonishes me.'

'Oh, I see now a number of wagons loaded with hay; but they are so distant I can scarcely distinguish anything; something extraordinary is certainly going on.'

'Let me have the glass,' cried Jack, impatiently; and he declared he saw a crowd of cavaliers who carried little lances, with banners at the point.

'Come, give me the glass now,' said I; 'your imaginations are too poetic to be relied upon.'

I applied the glass to my eye, and, after having looked some time attentively, 'Well,' said I to Jack, 'your Arabs, your cavaliers with lances, your hay-carts, what do you think they have been transformed into?'

'Camelopards,* perhaps.'

'No; although not a bad idea, yet they are ostriches, and chance has thrown a splendid chase into our hands; and if you will take my advice, we will not let these beautiful inhabitants of the desert pass us by without measuring our strength with theirs.'

'Ostriches!' cried Jack and Fritz: 'How grand! We will capture one: his feathers will figure beautifully in our caps.'

'Yes,' replied Ernest; 'they would look very nice; but the thing is, to catch the bird.'

But the ostriches were rapidly approaching, and it was time to think of some mode of capturing them. It seemed to me that the best way would be to wait until they came up, and then attack them by surprise. I ordered Fritz and Jack to go in search of the dogs, while Ernest and I sought some shelter to conceal us from the ostriches. We threw ourselves down behind some large tufts of a plant that grew among the rocks, and which I recognized as the euphorbia,* commonly called wolf's milk, and the juice of which is one of the most active poisons in the world.

Jack and Fritz now returned with our faithful companions, whom, from their wet skins, we easily judged to have been taking a bath somewhere.

The ostriches were now within eyesight, and I could distinguish that the family was composed of three females and a male, who was easily recognized by the long, white feathers of his tail. We crouched closer to the ground, and held our dogs close to our sides, for fear lest their impatience should defeat our stratagem.

'Make ready your eagle,' said I to Fritz; 'for if our legs are not sufficient we will have recourse to his wings.'

'Do ostriches run so very fast?' asked Jack; 'if they do, however, neither I nor Fritz are *quite* snails, and Master Ernest was long ago crowned victor in our races.'

'Oh,' answered I, 'Ernest's legs, good as they may be, will not make much difference to the ostrich: he does not even fear a horseman at full speed.'

'How are we going to capture them, then? We have no horses.'

'That is true; but it is more by the sagacity than by the mere swiftness of the horse that they succeed. The ostrich is never attacked in front, or behind, only at the side. It is known that, when this bird is pursued, he describes a circle, more or less vast, but yet always returning to the point from whence he set out. The whole science of the chase consists in constraining the bird, if possible, to narrow this circle. The hunter keeps at his side, follows him, presses him, and torments him until, wearied out, the poor ostrich falls into the hands of his enemies. But as the circle it describes is sometimes very extensive, and as one horse gets exhausted before the bird is captured, hunters always take care to have a relay of horses all along the course, and sometimes one ostrich will tire out four or five horses.'

'Is it true,' said Ernest, 'that on the approach of danger it is usual for the ostrich to hide its head behind a stone, in order, as it thinks, to render itself invisible?'

'In order to answer you,' said I, 'one ought to have the key to an ostrich's thoughts; but I have no doubt it is some exaggerated traveller's tale. It is much more probable that, if it really does conceal its head, it is in obedience to the great law of instinct, which teaches all animals to take the greatest care of the most vulnerable parts of their body; or, perhaps, the ostrich buries his head, to give the blows he makes with his feet more force. For my part, I think that the ostrich has always been a much calumniated bird.'

I now perceived that the ostriches were aware of our presence—they appeared to hesitate in their march; but, as we remained immovable, they at last seemed reassured, and were advancing directly to us, when our dogs, whom we could not keep quiet, suddenly sprung out upon them. Away went the timid birds, with a rapidity that can be compared to nothing else but the wind driving before it a bundle of feathers. Their feet did not appear to touch the ground, their half-extended wings had the appearance of sails, and the wind greatly accelerated their velocity. I then ordered Fritz to unhood his eagle; he did so, and the noble bird soon lit upon the head of the male ostrich, and, attacking his eyes, soon brought him to the ground. The dogs and the jackal ran up, and when we arrived the gigantic bird was just expiring under the numerous wounds that the ferocious animals had inflicted.

We were greatly disappointed at this issue of our chase; but, as the evil was without remedy, we contented ourselves with preserving the lifeless corpse. The eagle and the jackal were immediately taken away, as being the most ferocious. We then deprived the unfortunate animal of the white plumes in the tail, and we placed them proudly in our hats. The rich and sumptuous feathers contrasted strangely with our old worn-out beavers; but they were an excellent protection against the rays of the sun.

'What a pity,' said Fritz, as we examined the gigantic proportions of the bird, 'what a pity to have put such a magnificent bird to death!—How beautiful it would have looked stalking among our domestic animals.'

'How can such an immense bird find sufficient nourishment in the desert?' demanded Ernest.

'You are reasoning', said I, 'under a European prejudice. That which is called a desert by us, is not a desert to all the animals of the creation; and the most arid plains always produce some scattered herbs, or grass, that suffice for the subsistence of those animals that inhabit them. Besides, the ostrich resembles all the animals of unproductive countries: it is extremely frugal, and capable of supporting great degrees of hunger. Remember one thing ever, my child, that the Divine Author of all things will not take more care of those beings he has placed in the midst of plenty than of those who dwell in a dry and sandy desert.'

The conversation concerning the ostrich was continued for some length of time. We remarked the sharp points on the ends of their wings, like spurs, and which serve to accelerate their speed when pursued. I also showed my sons the falsity of the idea that ostriches throw stones at the hunters with their feet.

'The same can be said of the horse,' said I, 'for that also in galloping throws up stones and mud as it passes along.'

Fritz wished to know whether the ostrich had a peculiar cry. I informed him, that during the night it sends forth a sort of plaintive groan, and, at other times, a loud growling sound, like that of the lion.

While we were thus talking, Jack and Ernest, who had followed the jackal, made some great discovery, and we soon saw them waving their plumed hats in the air, and shouting to us to hurry on.

'A nest,' they cried, 'an ostrich's nest! Quick—quick.'

We hurried on, and found the two boys standing over a large ostrich-nest—if we can dignify a hole dug in the ground by the name of nest—in which were symmetrically arranged from 25 to 30 eggs,* each as large as a child's head.

'Take care,' said I to my young companions, who were going to meddle with the nest; 'You will disarrange the order, and then the female will desert her nest.' The jackal had broken one of the eggs when the nest was discovered, and I observed to Jack that one thing was yet wanting in the education of his pupil, and that he needed a good whipping to cure him of his destroying propensity.

My sons wanted to carry away the ostrich eggs; they would hatch them, they said, by exposing them in the daytime to the rays of the sun, and wrapping them up as warm as possible at night.

I observed to Fritz, who made the proposal, that each one of these eggs weighed about 3 pounds, and the whole number about 100 pounds, and that, having neither equipage nor beast, it would be impossible to transport them across a desert, through which we could hardly drag our arms and knapsacks; besides, I doubted whether artificial heat could replace the natural influence. But the children had got the idea into their heads, and they agreed that each one should take one egg, which he should carry in his pocket-handkerchief. The little boys soon repented of their agreement, and they changed their burden from hand to hand, with all the signs of ennui and fatigue. I came to their assistance, and advised them to cut some branches from a low sort of pine that grew about the rocks, and make a basket to carry their eggs, as the Dutch milkwomen carry their milk-pots. My plan succeeded admirably, and my little boys began their march without the slightest complaint.

We then arrived at the borders of a swamp that seemed to be formed by the confluence of several springs that flowed from the rocks; we could trace the marks of the dogs and the monkey, and recognized this as the place where they had wet themselves. We could perceive, in the distance, troops of buffaloes, monkeys, and antelopes, but so far from us that we took no further notice of them; nothing, however, indicated to us the presence of a boa, or that such animals resided here. We halted at this marsh, and refreshed ourselves with some provisions; and then, filling our empty gourds with water, prepared to depart, when we perceived the jackal had made a discovery. It was a round object which he had dug out of the sand with his paws; it resembled a mass of moist earth, and I threw it into the water to clean it, when, what was my astonishment to see it move! I took it out, and, on examining it, discovered it to be a turtle of the smallest kind, scarcely as large as an apple.

'How is this?' said Fritz. 'I thought that turtles inhabited the sea only.'

'Who knows?' said Ernest; 'perhaps there has been a shower of turtles here, as the Romans formerly had a shower of frogs.'*

'Stop there,' said I to the philosopher; 'your irony does not show your learning. Perhaps you do not know that there are land as well as sea turtles. They are not only found in swamps, but even in gardens, where they subsist on snails, caterpillars, and all sorts of insects.'

'Well, then,' replied Ernest, 'let us carry some home to mamma. She would like them to put in her garden; we will also put one in our cabinet of natural history.'

The jackal still continued his investigations in the sand, and we soon had a dozen of the little turtles crawling around us, some of whom I picked up, and put in my knapsack. Fritz reiterated his question concerning the different kinds of turtle.

'This species', said I, 'is ordinarily found in plains alternately dry and moist. At the Cape of Good Hope,* during the summer, when the sun parches up the plains, giving them the appearance of vast arenas of burning sand, the turtles bury themselves in it to the depth of several feet; and when the rainy weather returns, they come from their holes and enjoy the freshness of the air. The turtle is one of those animals which pass a portion of the year dormant in the ground. The frogs bury themselves in the mud of the swamp, and remain there during the winter months. In our mountains, the marmots never leave their holes during the whole of our long, cold winters.

48
Combat with Bears; the Porcelain; the Condor

WE quitted the borders of the swamp; but instead of directing our steps through the desert, we followed a little stream of water that led us to the rock where we had reposed on our first excursion into the savannah. It was a delicious route in comparison with our painful journey of the morning. We found trees, grass—in short, it was a little oasis in the desert, and we named it 'Green Valley'. We soon, however, left its verdure far behind us, and again we were in the desert; but the heat was not as violent, and our rest had recruited our weary strength, so that we found the route less painful, and journeyed tranquilly on, carrying our only conquest, the ostrich eggs. But it was not our fault: it was because we had seen no game. As we had remarked that the animals of the country were far more afraid of our dogs than of us, we took the precaution to hold our faithful, yet too quarrelsome, companions in leash. I took charge of Turk, Fritz of Braun, Ernest of Folb, and Jack of his jackal; as for Billy, he always carried Master Knips on his back, so we were obliged to let him go free.

We were yet distant about half an hour's journey from the jackal's grotto; Jack and Fritz had stopped a moment to adjust their burdens and I also stopped with them, while Ernest marched forward, followed by Folb.

'The philosopher is in a hurry to get home,' said Jack, laughing; 'he runs that he may be rested first.'

But scarcely had the fellow finished his sentence, when we heard a cry of distress; it was the voice of Ernest, followed by two terrible howls, mingled with the barking of the dog. A moment after Ernest reappeared; he was running at full speed, his face deadly pale, and he cried out, in a voice stifled with fear, 'Bears, bears! They are following me', and the poor boy fell into my arms more dead than alive. I had not time to reassure him, and I felt myself seized with a sudden shiver, as an enormous bear appeared, immediately followed by a second.

'Courage, children,' was all I could say. I seized my gun, and prepared to receive the enemy. Fritz did the same; and, with a courage and coolness far above his years, he took his place by my side. Jack also took his gun, but remained in the rear; while Ernest, who had no arms—for in his fright he had let his gun fall—took to his heels and ran away.

But our dogs were already at the attack, and they had commenced to measure themselves with their terrible adversaries. We fired together; and, although our shots did not bring down the enemy, they nevertheless told well: one of the bears had a jaw broken, the other a shoulder fractured. But the combat was not yet finished: they were only partially disabled. Our faithful servants did prodigies of valour; they fought most desperately, rolling in the dust with their enemies, while their blood poured in streams on the sand. We would have fired again, but we were afraid that we should kill the dogs, it being impossible, during the changing contest, to take any aim. We resolved to advance nearer, and, at about four paces from the bears, we discharged our pistols direct at their heads. The huge animals gave a groan that caused us to shudder, and then fell back motionless on the sand.

'Oh,' cried I, 'we are saved; thank God that once more he has preserved us.'

We remained some time dumb with astonishment and terror before our two horrible adversaries. Our dogs, covered with bleeding wounds, were still tearing the bears as if they were alive; and, fearing a feint, I again discharged my pistols at the heads of the two beasts. Jack was the first to sing our victory, and he brought back poor Ernest, who yet trembled all over. I asked him how he had happened to discover these terrible enemies. He answered, with tears in his eyes, that he had run on before us in order to frighten Jack, by imitating the growling of bears. 'I thought I should perish with fear,' said he, 'when I found my imitation transformed into a reality, and so agitated was I that I cannot tell how I ever got back.'

I then reprimanded him severely for his conduct, and tried to impress on his memory that all such attempts to frighten people were generally attended with evil consequences, and the safer plan was never to indulge in them.

Jack was the first to remark that the presence of bears in a country as warm as the one we inhabited was rather extraordinary.

'I cannot explain it to you,' said I, 'not having knowledge enough of zoology to judge whether these are European bears, or whether they are of the American family, or of the Tibetan.'*

During this discussion my little boys had approached the two animals, and stood contemplating them with a mixture of fear and admiration. They passed their hands over the long line of sharp teeth with which their jaws were furnished, raised their huge paws armed with terrible claws, and admired their fawn-coloured skins. The conclusion of this examination resulted in a general question of what was to be done with them.

Jack wished to make two helmets out of their skins, 'which', said he, 'will make us appear very terrible to our enemies'. Ernest, less warlike, proposed that we should wear them as cloaks during the rainy season, or else make matresses from them.

It was too late to meddle with the animals, and we took the precaution, before leaving, to draw the two carcasses into the jackal's cave, and cover them with thorn-bushes, to keep off all carnivorous beasts and birds of prey; we also buried our ostrich eggs in the sand, as their weight retarded our march greatly, and we could leave them here until the morrow.

The sun was set when we rejoined my dear companion and our little Francis, who received us with demonstrations of the most lively joy. A good fire, and a well-cooked supper refreshed our weary bodies, and my little heroes commenced a long narration of the exploits of the day, Master Jack making up for the small share he had had in our victory by boasting and swaggering enough for all. My wife was so frightened at the recital of our danger that she could not restrain her tears; and, although I assured her that the flesh of the bears would make as good provision as the peccaries, she begged me not to return into the desert.

My wife and Francis had not been idle during our absence; they had discovered on the banks of a stream a sort of greasy, white earth, which appeared to me to be fine pipe-clay. They had also collected water enough for the use of our domestic animals, and, by the force of industry and perseverance, had amassed, at the entrance of the defile, a quantity of materials necessary for my projected fortification.

I thanked my good wife for the pains she had taken. We then lighted a large fire to guard us through the night, and our dogs, whose wounds my wife had washed and dressed with fresh butter, lay down beside it. I wished, before retiring for the night, to make a trial of the earth my wife had found, for I suspected that it was porcelain. I made two roughly shaped bowls from it, and threw them into a furnace of hot cinders, and we then all retired to the tent, where sweet sleep soon sealed our eyelids. The next morning it required a strong effort to tear us from our beds, so wearied out had we been the preceding day. I found my two bowls hardened by the heat: they were, as I supposed, porcelain, rather coarse-grained, but well enough for our purposes. We breakfasted in haste; the beasts were harnessed to the cart, and, after a pleasant little ride, we arrived safe and sound at the cavern of the bears.

On approaching, we found the entrance of the cave occupied by a troop of birds, whom, by their conformation, ruffled necks, and the colour of their feathers we should have taken to be turkey-cocks if a nearer examination had not convinced us that they were birds of prey,* occupied in dissecting our bears, as we could see them flying out, carrying away huge pieces of the flesh. I thought, by the immense number of birds, that our work was finished, and nothing would be left but the bones, when suddenly we heard a flapping of wings above us, and a black shadow passed along the ground; we raised our eyes, and beheld an immense bird of prodigious size, whose wings extended full sixteen feet; he came gradually sweeping down toward us, when Fritz fired his gun, and the formidable creature fell dead at our feet: it had been shot in the heart, and the life-blood oozed out from the wound.

The report of the gun had frightened the band of marauders, and they flew away stunning our ears with the horrible discord that they made; we then entered the cavern, and found one of our bears half devoured, and the other completely cleaned of the entrails, a saving of labour, by the by, to us. We loaded our cart with the skins and the remaining meat, and, placing the immense bird, which we had discovered to be a condor* of the largest size, upon the top, we set off for the tent.

49

Preparation of the Bear's Flesh; the Pepper; Excursion of the Boys into the Savannah; the Angora Rabbit; the Antelopes

WE devoted a whole day to the preparation of the bear's flesh. After having skinned them with the utmost care and precaution, I cut off the hams, and then divided the rest of the meat into long strips, about an inch in thickness, and we exposed

the whole to a good current of smoke, as the ancient bucaniers* used to do. The grease was collected in bamboo canes,
and carefully preserved; for, beside its use in the kitchen, my
wife said it was excellent on bread in lieu of butter. We had
about a hundred pounds of fat, together with that which the
peccaries had afforded us a few days before; we abandoned the
carcasses to our dogs, and they, aided by the birds of prey, soon
picked the bones so clean that there remained nothing but two
perfectly white, dry skeletons, which we carried home with us
for our museum. As for the skins, they were carefully washed
with salt water, and rubbed with sand and ashes; and, although
our talents in the art of currying were poor enough, we rendered the skins sufficiently soft for all purposes, without having
recourse to the Greenlanders'* process, who, it is said, chew
them in order to render them soft.

I very much regretted that we had not the leaves of the
ravensara* with us, so that we might impart its delicious odour
to our meat; but the children, while searching the bushes, had
discovered a sort of vine, which, upon examination, I found to
be pepper. I received this new gift with transports of joy, for it
would now enable us to preserve many things that the heat of
the climate had before rendered impossible. The skins of the
bears, the hams, and the strips of flesh received the first
application of our new discovery. The condor came next; after
having taken out every particle of flesh, we rubbed the interior
of the skin smartly with powdered pepper, and, filling the cavity
with moss and cotton, reserved the arrangement of the bird
until we returned to Felsenheim.

Our labours had been too peaceful for the restless, turbu-
lent character of my little boys. I could see that they were tired
and fretful, and I thought the best plan would be to diversify
our work with some amusement. I proposed to them to make
an excursion alone in the desert; my proposition, as one may
suppose, was joyfully received, and the perspective of an un-
checked course rallied the flagging spirits of my little companions. Ernest refused to accompany them, preferring to remain
at home with us. On the other hand, Francis was so eager to
accompany his brothers, that I at last permitted him to go.

Fritz, Jack, and Francis were soon in the saddle, and, after
having gaily saluted us, galloped off through the defile. It was

not without a painful sentiment that I saw them set off alone, abandoned to their own resources; but I felt that it was necessary to familiarize the children to provide for themselves, as some accident might deprive them of their father and mother, and thus they might be prepared to meet the loss. I rested my hopes in the prudence and intelligence of Fritz; I was sure that he would watch over his brothers; and the remembrance of the coolness he had so often evinced reassured me. I turned to God, and prayed to him, in humble assurance that the hand which had brought the sons of Jacob* back to their father would also bring mine back to me.

My wife and I resumed our domestic labours, and Ernest, tranquilly seated in the sand, occupied himself in making cups from ostrich eggs, for we had discovered, by putting our eggs into hot water, that the principle of life was exhausted. Ernest had read of a plan somewhere to separate the eggs by surrounding them with a string steeped in strong vinegar. The action of the acid on the lime contained in the shell forms a circular line, which gradually eats through; but the lining membrane of the egg was so hard that it was necessary to cut it with a knife; it had all the elasticity of parchment.

We soon quitted this occupation to undertake another. While examining a small cavern which we had discovered near the tent, I found several minerals, among others a piece of amianth,* known as being incombustible; and also a superb block of talc,* as transparent as glass, and which I resolved to fashion into window-panes. Ernest aided me as much as he was able, and we soon detached a splendid piece, about two feet in length, and the same in thickness. My wife, who received everything that could recall Europe to her mind with pleasure, was overjoyed at our new discovery, especially when I informed her that this mineral could be divided into leaves no thicker than paper.

We had been thus occupied the best part of the day, and, as evening approached, we gathered around our hearth, where our good housewife was cooking two bear's paws, which had been well soaked in brine, and the smell of which, as it escaped from the pot, promised us a delicious supper, and we sat down to while away the time in conversation until our huntsmen returned. We did not wait long, for the galloping of their steeds was soon heard, and in another moment they were at our sides.

Jack and Francis each carried a little kid on his back, with the feet tied together, and the game-bag of Fritz appeared to me to be pretty full.

'A fine chase, papa,' cried Jack. 'Storm carried me through the desert like a flash of lightning. Fritz has two magnificent Angora rabbits* in his pouch, and also a complaisant cuckoo, who led us to one of the finest hives I have ever seen; we will be able to get plenty of honey.'

'Jack has not told all,' said Fritz: 'we have taken a whole troop of antelopes prisoners, and have driven them into our domains, where we can hunt them and tame them just when we please.'

'Fritz has not told all,' answered I; 'he has forgotten the chief part. He has forgotten that the great blessing of the day is, that God has brought back three little boys, cast out alone in the desert. Let us commence, my friends, by giving thanks to Heaven for this favour.'

After prayer, turning toward Jack, whose face seemed very much swelled, I said, 'What is the matter with your cheeks? Have your adventures been dangerous in any way?'

Fritz interrupted his answer, and began the following narration.

'After quitting you, we took the direction of the valley, and, finding a narrow place where two or three trees had fallen down, we took advantage of this natural bridge, and crossed to the other side of the river. We rode on some time without perceiving any thing, our coursers going at full speed, and the sun not being high enough to be unpleasant. At last we discovered, in the distance, two herds of small animals, of what kind we could not distinguish, but I thought they were either antelopes or gazelles. Our first care was to call our dogs together, and keep them close by our sides, as we knew the animals were more afraid of them than of us. I then divided my forces: I gave Francis the line of the river as his position; Jack occupied the middle, while I, mounted on the onagra, sustained the right wing, and endeavoured to drive the animals to the centre. We effected this movement, and one of the herds passed the river as quietly as if the act had been voluntary. The other herd did not seem to perceive us until we were close to them, when suddenly they raised themselves from the grass where they had been lying, and, stretching out their long necks and little heads, surmounted by short, pointed ears, set off at full speed; and now our chase commenced. We urged our coursers onward, and, giving our dogs their liberty, we soon forced the entire troop over the river, and drove them into the defile which separated us from the savannah. After we had secured them in our dominions, the next thing was to keep them there. Several plans were proposed, but at last the following was chosen. We stretched a long cord from one side of the defile to the other, and fastened to it every light thing we could find, the continual motion of which frightened the animals away whenever they approached it; the ostrich plumes in our hats, our handkerchiefs, etc., furnished us with materials.'

'Admirable!' said I, as the boy stopped as if to see how his stratagem was received, 'admirable! The only thing is, in the night it cannot be seen; but it truly was a bright thought for such a boy.'

'It is not original,' replied he: 'I read it in Levaillant's Voyages* to the Cape of Good Hope. It is practised by the Hottentots in order to prevent their domestic cattle from straying away.'

'I am very glad to see,' said I, 'that your reading has not been lost on you, and you now see how much a good book sometimes assists one. You did not, very likely, think, while reading Levaillant as a mere matter of amusement, that you would ever make any use of his narrations. But about the rabbits,' added I, 'what do you intend to do with them? If they should happen to get in your mother's vegetable garden, there would not be much of it left.'

'No, no; but I thought that one of our two islands would make a good home for them; for instance, Shark Island would make a magnificent warren, and furnish us many a good dish, and fine furs to make caps out of.'

'But how did you come to take them alive?'

'The honour of the capture is due to my eagle; he pounced down upon a troop of rabbits, that were flying before us, and carried off two in his talons. I rescued them before he had injured them, and he consoled himself by killing another, which he soon devoured.'

I could see that Jack was watching every opportunity to put in a word, and I laughingly requested the poor fellow to speak.

'In my turn!' said he, 'in my turn! Only I galloped on with Francis while Fritz was chasing the rabbits; the dogs followed us, and suddenly we saw them jump forward, and run after two little animals about the size of a hare, that fled with incredible rapidity. Away we all went, and, after a hot chase of a quarter of an hour, we captured the two fugitives. There they are,' continued the young narrator, throwing down before us two beautiful little animals; 'I think they are young fawns.'

'And I think,' said I, 'that they are antelopes.'*

'Well, whatever they may be,' continued Jack, 'our dogs behaved admirably, and so, I can say, did their masters. But that was nothing to what happened afterward. We had scarcely commenced our progress when a sort of cuckoo began to fly before us, singing away as if to defy us. Francis, who always is imagining something marvellous, exclaimed that it was some prince, enchanted by a fairy, who wished to conduct us to his palace. "Bah!" said I; "I'll soon break the enchantment"; and I had already levelled my gun, when Fritz requested me to recollect that it was loaded with ball, and that I should only waste the charge. I accordingly slung my gun on my back, and we rode on, the cuckoo flying on before us, when suddenly he stopped just over a bee's nest, artfully concealed in the ground. We now held a council of war about the nest, and discussed the plan of attack. Francis begged to be excused, recalling to our memories the former attack at Falcon's Nest. Fritz was willing to do all the advising part, but would rather leave the execution to somebody else; so you see, at last, the whole affair devolved upon me. Armed with some sulphur matches that I found in my knapsack, I advanced and tried to suffocate the bees by throwing the lighted matches down the hole, when suddenly a rumbling noise was heard, and, in a second, a swarm of bees emerged, attacking me on all sides; my hands and face were violently treated, and it was with the greatest difficulty that I mounted my buffalo and rode away, bearing with me the honourable marks of the conflict. I could scarcely believe,' said Jack, as he finished his recital, 'that so small an animal could cause so much pain.'

'Well, well,' said I, 'you have received a good lesson in natural history; take care that you do not forget it; but go to your mother, and let her put something on your face to allay the pain.'

I then made a basket of willows, covered with canvas at the top, in which to put the rabbits and the antelopes, so that they might be easily carried to Felsenheim. We were undetermined whether we would keep them at the grotto or abandon them on one of the islands of the coast.

My children would have preferred to keep the pretty creatures by us; but I was afraid that they might get into the garden, and do some damage, so it was resolved at last that they should be put on Shark Island, and there left at liberty.

I reflected long upon what Jack had told me concerning the strange bird that had shown them the nest of bees. I easily recognized it as being the 'cuckoo-indicator'* of naturalists; but, thought I, 'how, if this coast is uninhabited, could the bird have known that human beings liked honey, and would be willing to share the discovery with him? Is not such conduct a sign that we are not the first men, who have trod this soil? May not the interior of the country be inhabited?' These considerations were of the highest importance to us, and I was convinced that it would not be prudent to advance into the interior, unless with the greatest caution. I also resolved to build a fortress on one side of the coast, and I chose Shark Island as its situation, as it appeared to me that a strong fortification that would command the coast of Felsenheim, and fortified by our two cannons, would enable us to defend ourselves against all attack from the interior, if any ever took place.

I showed my young huntsmen the splendid block of talc that we found, and the window-panes I had manufactured from it; but the admiration it excited was soon effaced by the welcome call to supper. The bear's paws formed the principal dish; but, although they exhaled a most appetizing odour, yet the shape so much resembled a man's hand, that Jack cried out, like the ogre in the story of 'Hop o' my Thumb',* 'I smell fresh meat.' This sally made us all laugh; I attacked the dish, and we found it was one of the most delicate and delicious things we had ever eaten, and my wife was loaded with praises for her good cookery. After supper we lighted our torches and fires, attended to our smoke-house, and lay down with unburdened consciences to enjoy our night's rest.

50
Capture of an Ostrich; the Euphorbia; the Vanilla

AT the break of day I was up, and awoke my sons; our labours were almost done—our bear's meat was smoked, our fat all tried-out into bamboo vessels; and the rainy season, which was rapidly approaching, warned us to return to our home in the grotto. Nevertheless, I wished to make another excursion into the desert, to see whether a second visit to the nest of ostrich eggs would not succeed better than the first, and I likewise wanted to gather some of the gum of the euphorbia.*

As we wished to accomplish this excursion as rapidly as possible, it was resolved to go on horseback. Fritz gave up his onagra to me, and took the young colt, and Jack and Francis each mounted their respective beasts. As to Master Ernest, he preferred to remain at home; he had become the habitual companion of his mother, and succeeded Francis as assistant in the kitchen.

We took with us Turk and Billy, and set off, following the direction of the Green Valley, tracing over again all the places rendered illustrious by some remembrance of our last excursion—the spot where we had encountered the bears, the turtle marsh, and, at last, the rock from which Fritz discovered the ostriches. To this rock we gave the name of 'Arab's Tower', in allusion to the mistake he had made in thinking the ostriches to be Arabs of the desert.

Jack and Francis galloped off at full speed, and, as the plain was so level that they could not escape from my eye, I let them go on. I retained Fritz by my side to aid me in gathering the euphorbia which had congealed in the sun. I had provided myself with a vessel to put it in, and I soon filled it with the little drops of hardened gum.

This gum is one of the most violent and subtle poisons. The inhabitants of the Cape of Good Hope make use of it to poison the waters where the wild beasts come to drink; but they are obliged to watch their flocks lest they might happen to drink of the same spring; and they do not mind losing a few sheep; as they are richly paid for it by the magnificent furs that they obtain from the lions, tigers, hyenas, etc., that are poisoned by the water. The Hottentots do more—they eat the flesh of the animals thus poisoned, and are said never to experience any ill effects from it. My son asked me why I took so much pains to collect this poison.

'I intend to use it', said I, 'to destroy the monkeys—a cruel means, I will allow, but necessity drives us to adopt it. We can also employ the euphorbia in preparing the skins of birds and other animals; it will preserve them from corruption, and keep out all insects. To whatever use, however, we may apply it, the greatest precaution must be observed, as it is capable of producing the most dreadful results.'

During our work, the two cavaliers had almost disappeared in the savannah, and it was with great difficulty that our eyes could follow them, surrounded as they were by a cloud of dust. They had passed far beyond the ostrich nest, to which we directed our course, wishing to see whether the eggs had been abandoned or not.

We had scarcely come in sight of the nest when we saw four noble ostriches rise from the sand and advance toward us. Fritz's first care was to prepare his eagle for the conflict; and, to prevent it from renewing the former scene of carnage, he fastened its beak so strongly that it was almost harmless. Our dogs were also muzzled, and we stood still, in order that we might not frighten the birds. On they came, with half-extended wings, gliding over the ground with inconceivable rapidity. They seemed to think us inanimate objects, for they came on directly for us until they had arrived within pistol-shot; they were three females and a male—the last a little in advance, with his beautiful tail-feathers floating behind him. The moment of attack was come; I seized my string with balls, and, calling up all my sleight of hand, I launched it against the male ostrich. Unfortunately, however, instead of catching him around the legs, as I intended, the balls of my string took a turn round his body, and I only fastened his wings to his sides. It diminished his speed somewhat, but the victory was not complete; and the frightened bird turned round, and, using his long legs, endeavoured to escape: away we dashed after him, I on the onagra, and Fritz on the colt. But we were nearly exhausted, when, happily, Jack and Francis rode up, and cut off his farther retreat. Fritz then unhooded his eagle, and, pointing out the ostrich to him, he immediately pounced upon his prey; and now commenced an arduous chase. Jack and Francis on one side, and Fritz and I on the other, tormented him, and harassed him without ceasing; but the most useful combatant was the eagle. The presence of this new enemy troubled the ostrich greatly; he felt him on his head, and heard the flapping of his wings, while, on the other hand, the eagle, furious at finding his beak strongly fastened by a ligature of cotton, was so violent that, by a vigorous stroke of his wings, the ostrich fairly tottered. Jack then threw his string and balls so skilfully that the noble bird bit the sand of the desert. A cry of joy burst from the huntsmen, the eagle was recalled and hoodwinked,* and we hastened to our prize in order to prevent his breaking the bonds that confined him; for he was so very violent, and struggled so vigorously, that I hardly dared to approach him. I imagined that by depriving him of light I might reduce his fury, and I threw my hunting-sack, my vest, and handkerchief, over his head. I had discovered the secret: no sooner were his eyes covered than he became as quiet as a lamb. I approached, passed a large band of sea-dog-skin around his body, two other bands were attached as reins to each side, and his legs were fastened with strong cords, long enough to allow him to walk, but which confined him sufficiently to prevent his escape.

'A fine, prize, truly,' said Jack, when our work was done. 'We have got the giant; but how shall we tame him?'

'Wait awhile,' answered I; 'the most ferocious nature yields to education: do you not know that the Indians even tame the enormous elephant, and that by a very simple means?—They place a wild elephant between two others, already tame; they deprive him of the use of his trunk; and, fastening him strongly, leave the tame elephants to teach him better manners.'

'Admirable, papa!' cried Jack, bursting into a fit of laughter. 'According to that, we now want two tame ostriches, and I do not think that either Fritz or I are large enough to supply their places.'

'I thought of that before,' replied I; 'but, instead of ostriches, we have other auxiliaries that will well answer as a substitute for them. The bull and the buffalo, for example, will, I think, do very well on each side of our captive, while you, and your brother, each armed with a whip, can teach him to march in a line with them.'

'Oh yes, papa; that will be fine fun, and is sure to succeed', was the answer to my proposition.

I then attached our two coursers before and behind the ostrich with strong cords; and when all was ready, my two cavaliers jumped into their saddles, and I pulled the covering from the head of the ostrich.

The bird remained some time immovable, as if astonished at the return of light. It soon made a start; but the ropes pulled it roughly back, and it fell down on its knees; again it made the attempt, and again it was foiled. It tried to fly, but its wings were tightly fastened by the band I had passed around them; its legs were also restrained: it threw itself from side to side with the utmost violence, but the patient buffaloes did not pay the least attention to the pulling and hauling. At last the bird appeared convinced of the inutility of its efforts, and, submitting to its two companions, set off with them at full gallop. They dashed gallantly on for half an hour, until the buffalo and the bull, less accustomed to the sands of the savannah than the ostrich, forced it to abate its rapid pace, and adopt a slower system of movement.

While the two young cavaliers were thus occupied, Fritz and I set out in search of the ostrich nest. The cross of willows which we had planted in the ground near it, at our last visit, still remained, and, as we approached, a female bird rose up off the nest and fled rapidly away into the desert. Her presence appeared to us a good augury, as it assured us that the eggs yet retained the principle of life. I had taken care to bring with me a sack and a quantity of cotton, I now took out six of the eggs, and, enveloping them as carefully as possible in the cotton, placed them in the sack, leaving the others in the nest, in hopes the mother would not discover the theft. The sack containing the eggs was carefully fastened on the back of the onagra, whom I led slowly along; Fritz mounted his colt, while Jack and Francis marched before us, escorting the ostrich, and controlling its waywardness by an occasional blow.

We traversed the Green Valley without perceiving anything uncommon, and soon arrived at the tent, where Ernest and his mother received us with an astonishment they could not find words to express.

'What, in the name of Heaven,' cried my wife, as she perceived the ostrich, 'are you going to do with that immense bird? Have we so much more provisions than our other animals can consume, that you must go into the desert to find another pensioner? It is said that the ostrich eats iron; shall I give him some now? Once more I ask, what do you mean to do with him?'

'A post-horse, mamma,' cried Jack, 'a post-horse that I mean to name, "Hurricane", for nothing else can equal him in rapidity. Nobody else shall ride him but me, and I will give you Storm, Ernest, because you have no courser.'

I endeavoured to allay the inquietude my good Elizabeth experienced at the sight of our new captive.

'The ostrich', said I, 'has not such a very voracious appetite as you suppose; on the contrary, it is a very quiet animal, who is contented with green herbage which he can easily find for himself; and, besides, if we are obliged to help him, he certainly is able to help us.'

While I was making this short apology to my wife for the introduction of the ostrich, Jack and Francis were quarrelling, behind my back, for the ownership of it.

'Jack pretends', said Frank, 'to esteem the ostrich as his property. I don't think it is quite fair; I helped to catch him as well as he.'

'Very well,' said I, 'let us divide him. Fritz, you may take the head, for it was your eagle that stunned him; I claim the body, for it was my string and balls that caught that; Jack, you own the legs—your balls captured them; and we will give you, Francis, a feather from the tail, as it was there, I believe, you kicked the bird to make it stand up.'

This distribution of the victim made my little boys laugh, and each one renounced his pretensions, preferring to make the conquest common glory.

Ernest had listened to all our conversation, and, at last, bursting into tears, he exclaimed.

'Why does it happen that I am always absent while you are enjoying yourselves?'

'As to that, my friend,' answered I, 'remember that it was your own wish to stay at home. I do not blame you, my child; for the good God gives to all their particular dispositions: to you he has given a taste for study and retirement; your brothers are more fond of active life. Besides,' added I, 'you have your days of triumph, when you discover some new treasure for us; and, if a European vessel should ever arrive at our shores, you will be our interpreter, and with you the captain must communicate.'

These words healed the little wound that the noisy joy of his brothers had inflicted on the heart of poor Ernest; and the idea of being useful consoled him.

I fastened the ostrich securely between two trees, and the rest of the day was devoted to preparations for our departure on the morrow. We had a number of new riches to collect, and I wished to leave nothing behind us.

The next day we set off early. The ostrich took his place between the bull and the buffalo, as before; he was, at first, inclined to be refractory, and threw himself from right to left, but all in vain: his two conductors were like immovable masses, against which all resistance was unavailing.

Fritz mounted the young colt, 'Rapid', and I the onagra, while Ernest directed the cart, in the middle of which my wife sat in all her majesty, among the provisions. Our march was slow, but it was very picturesque, as may be imagined.

We halted at the entrance of the defile where my sons had suspended the cord with the feathers attached, to keep back the antelopes and gazelles. In the place of the cord, we erected a solid palisade* of bamboo, high enough to keep out all animals that do not climb. We planted a row of thorn-bushes on each side, and sprinkled a layer of sand all around, so that we could discover what sort of animals might frequent it. During the construction of this fence we made a new discovery; it was that of the Vanilla bean,* which I recognized by its brown pods and balsamic odour.

Our labours at the defile detained us a long time, and it was night when we arrived at the cabin of the Hermitage. We found our smoking-hut the same as before, and our provision of peccary untouched. We lighted a fire, and, after a frugal repast, extended ourselves on our sacks of cotton, and courted balmy sleep.

The next day, we discovered a new treasure: our hen-house had received an addition of twenty young hens—the product

of the eggs Jack had brought home in his hat. My wife was
enchanted at this discovery, and caught several pairs to take
home with her.

We were so anxious to return to our dear Felsenheim, and
to resume all our comforts and luxuries, that we resolved not
to stop again until we arrived there. It was long after noon
when our weary journey was finished. We were worn out with
fatigue; the sun's rays had been pouring down on our heads
all day, and our strength was so exhausted we could scarcely
give our animals their evening food.

51
Education of the Ostrich; the Hydromel; the New Hat

THE day after our arrival at Felsenheim, my wife commenced
'cleaning house'. Windows were opened, beds aired, and all
swept and garnished. While she and the two younger boys were
thus employed, I, with the two elder, unpacked and distributed
the riches we had brought home.

We had tied the ostrich, at first, under a tree, and securely
fastened his feet; but we changed his situation, and tied him
to one of the strong bamboo columns that supported the
gallery.

We next visited the eggs, and they were, like the first, submitted to the trial of warm water. Several of them fell heavily
to the bottom; but three or four moved slightly when immersed
in the water, and these were carefully preserved, in order that
we might try the experiment of hatching them by cotton and
artificial heat. For this purpose I constructed an oven, in which
I took care to maintain that degree of heat which the thermo-
meter marked as being the natural heat of the hen.

We then installed our Angora rabbits on Shark Island; we
constructed a burrow in the ground, similar to those of
Europe, and, before putting them in, we combed them, and
removed all the superfluous hair. We also fixed wooden combs
over the entrance of each burrow, so that the rabbits, when
passing in or out, would be deprived of some part of their fine
wool, which I intended to manufacture into hats.

The two antelopes were also transported to Shark Island.
We should have liked very much to have kept these charming

little creatures about us, but the fear of the dogs and beasts of prey forced us to condemn the timid creatures to confinement. In order to render their exile as agreeable as possible, I erected a hut, in the middle of the island, to shelter them, and we took good care to provide them with plenty of provisions. It was a pleasure to see the slender, timid creatures bounding gracefully over the high grass, and we greatly admired their light movements, the rapidity of their course, and the beautiful shape of their bodies. The antelope is of a deep-brown colour, in some places approaching to black; a long line of white hairs extends down the neck and the backbone to the tail, but it is almost entirely hidden by the brown hairs which cover the back; upon each of the cheek-bones is a white spot, and there are also white spots on the haunches. Its legs are very slender, and its feet extremely small; its tail, which is very short, is covered with long brown hairs, and the upper lip is furnished with a black mustache. They are the most graceful, timid little creatures that can be imagined. The antelope carries with it the valuable perfume called musk, which is much sought after by the American hunters; and, it is said, a very cruel method is used to deprive them of this treasure. They beat the antelope with sticks, until lumps and blood-blisters are formed on the skin; these contusions are tied tightly with a string, so that the extravasated blood in these sort of pockets cannot escape; they are then left until they fall off of themselves, and in them is found the perfumed blood which becomes musk, and is worth its weight in gold.

There remained alive but two of the little turtles which we had brought home, and these were transported to our goose-swamp. We had originally intended them for the kitchen-garden, but my wife was afraid that they would do her salads more hurt than good, and at last they were condemned to the mud of the swamp. Jack was sent to deposit them there, and he had scarcely arrived at the swamp, when we heard him call out for Fritz to hurry thither, and bring his stick with him. I supposed that they were going to knock down two or three of the bullfrogs with which the swamp abounded, when what was my surprise to see them coming toward us, dragging with them an enormous eel that they had found in one of the fikes* Ernest had constructed there before we had set out for the savannah. The other fikes had all been broken through by the fishes that had been caught in them. The eel was joyfully received; our good housekeeper instantly cooked us a piece, and the rest was prepared as the sailors prepare the tunny fish.*

The pepper and the vanilla bean, climbing plants, found a place about the bamboo columns which we had erected to support a sort of piazza at the entrance of our grotto.

The vessels of grease we had brought back, and the smoked flesh of the peccaries and the bears, were placed in the magazine of provisions, where they presented a formidable array, by the assistance of which we could brave all attacks of famine.

When these first works were finished, we occupied ourselves in embellishing our habitation and rendering our position more agreeable.

The two bear-skins were placed in the sea-water, being loaded with large stones to prevent the tide from carrying them away.

My wife also took the heath fowl we had brought home under her charge, and she let Master Knips and the jackal understand that they must be respected, or else their lives would pay the penalty. The condor was placed in the museum, and we intended to devote a day to finishing his preparation. The block of talc, the asbestos, and the porcelain earth that we discovered, were also placed in the museum, but not simply as curiosities, as I intended to convert the talc into window-panes, the porcelain into all sorts of utensils, and to make from the asbestos incombustible wicks with which to supply the lamp I had suspended from the roof of the grotto. But all these works were deferred until the rainy season confined us to the house.

I also deposited the gum of the euphorbia in the museum, taking care to envelop it carefully in paper, and write on the outside, in large letters, 'poison', so as to prevent any danger resulting from it.

The skins of the rats Ernest had brought annoyed us by the odour of musk they exhaled. I made them up into a packet; and, remembering what I had read concerning the manner in which sailors bring over the asafetida,* a sort of fetid gum, by hoisting it up to the top of the mast, I placed our rat-skins out in the open air, under the gallery, where their offensive smell could not reach us.

All these operations consumed two whole days. Jack, who was always pleased with change, was quite delighted with them; but Ernest could not reconcile himself to all these goings and comings. He declared that he was much more happy while perusing some book under a shady tree than we were while arranging what we call riches. I tried to remedy the faults of the two boys, if possible. I remarked to Jack that the whole of life could not always resemble a magic lantern,* where one scene quickly succeeds another, and that uniformity should always mark our occupations. As for Ernest, I showed him that an inactive mode of life allowed the noblest faculties to bury themselves in a useless dream, beneficial neither to one's self nor to others.

I was thinking, at the time, of a project that, in employing all hands, would not allow the philosopher to be less industrious than his brothers. I wished, before the rains came on, to prepare a field to receive the seeds we had hitherto confided to the earth without any order or regularity. It was a difficult enterprise, and we felt in all its reality the force of that law which condemned man to gain his bread by the sweat of his brow. Our faithful animals were of much assistance to us; but the sun was so intense that the slightest labour utterly exhausted them. We could work but four hours in the day: two in the morning and two in the evening; we were able, however, to prepare at last about two acres of land, which would furnish us an ample harvest of maize, potatoes, and manioc root. How many groans, how many complaints I heard while we were thus occupied! But self-love, the natural stimulant to action, and foe of idleness, came to the aid of my sons, and even Ernest persevered until the whole work was completed.

'Oh!' said Jack, 'how good this bread will be. I hope I'll have a good appetite to eat it: any way I'm sure we have earned it.'

During the intervals of our fatiguing field labours, we occupied ourselves by beginning the education of the ostrich. It was an enterprise as difficult as it was novel; but I had read that it could be accomplished, and I was resolved to try it.

Our pupil began by putting himself in a terrible passion; he struggled, snapped at us with his beak, and cut up all sorts of capers; but we could find no better remedy for such conduct than to treat him as we had treated Fritz's eagle, that was, by burning tobacco under his nose. This had the desired effect, and we soon saw the majestic bird totter and fall insensible to the ground. We had recourse to this plan several times. Little by little we relaxed the cord which fastened it to the bamboo post, and gave it room to wander about the doorway. A litter of rushes was provided for him; calabashes filled with sweet nuts, rice, maize, and guavas were placed every day before the animal; in a word, we neglected nothing that we thought would consort with the fellow's taste.

During three days all our cares were in vain: our choice dishes were regarded with great disdain, the beautiful captive would not eat, and it carried its obstinacy so far that at last I was seriously afraid of the consequences. At last an idea occurred to my wife which relieved us from all embarrassment. It was to poke down the throat of the bird, willy-nilly, balls of maize and butter. The ostrich made horrible faces at first, but when it got a taste of the balls, all trouble on that point was over, and the delicacies we placed before it were quickly devoured, the guavas, by the by, being especially favoured.

The natural savageness of the bird disappeared more and more every day; it would let us approach it without striking at us, and after some days we thought we could, without much risk, unfasten it, to take a short lesson in the art of walking. We placed it between the buffalo, and the bull and put it through all the exercises of the stable—to trot, to gallop, stop short, trot again, walk slow, etc. I cannot say that the poor bird relished his first lesson very much, but the tobacco-pipe and the whip were two admirable instructors, and when he was disposed to become unruly a whiff of tobacco would set all to rights.

At the end of the month its education was complete, and it had so well succeeded that I now seriously thought of making our new conquest of general usefulness. I wished it to associate with our domestic animals, to submit like them to regular movements, and to stop and march as we wished. The first thing that was to be thought of was a bit; but how could I contrive a bit for a beak? I had never seen one, and I must confess that I felt greatly embarrassed; at last I achieved my task. I had remarked that the absence of light had a very direct influence upon the ostrich; it would stop short when blindfold,

and could not be induced to move until its eyes were uncovered. This discovery was the basis of the new invention that
I constructed. I made, with the skin of a sea-dog, a sort of hood,
like the one we had made for the eagle, which covered the
head, being fastened about the neck. I made two openings in
the side of this hood, one opposite each eye, and covered each
of these holes with one of our little turtle-shells, attached to a
whale-bone spring, fixed in such a manner that it would open
and shut. Reins were fastened to these springs, so that, by their
action, we could admit the light or shut it out, just as we
pleased. When the two shells were open, the ostrich galloped
straight on; when one was opened, it went in a direction corresponding with the eye that received light, and when both
shells were shut, it would stop short. The most fully trained
horse could not have obeyed better than our ostrich did, under
his novel head-dress.

Our first success encouraged us; and, as human vanity always has a share in all the actions of life, it was found necessary
to decorate the hood with all our disposable ornaments.
Accordingly, ostrich-plumes pulled out of its own tail, ribbons,
and tinsel were sewed on to the covering, and made a very gay
appearance as they fluttered in the air.

My children thought that the education of our captive was
now complete; but I was of a different opinion. The ostrich is
a very robust animal, and capable of supporting a great deal
of fatigue. I wished to learn it to carry burdens, to draw a
carriage, and even for horsemanship; I began, consequently,
to fabricate harness for each of these occupations. Of the first
I shall say nothing; but the third, that is to say, the saddle and
all that pertained to equestrian exercises, was a masterpiece of
saddlery. I had brought my saddles and bridles to such perfection that I have not the least doubt that, had I resided at
the Cape of Good Hope, the land of ostriches, I would have
obtained a brevet* of invention and the pompous title of first
saddler to the kingdom. But, whatever was the merit of my
invention, I must confess we had a great deal of difficulty in
making the ostrich submit to our wishes; our hardest task was
to make it submit to our mounting it; but I knew that patience
and perseverance are the two first elements of success in im-
parting education. I was not, therefore, discouraged, and at last

we had the satisfaction to see our new courser galloping between Felsenheim and Falcon's Nest with one of our young cavaliers mounted on his back.

After the training of the animal was achieved, the question of ownership came up again, with all its difficulties. Jack would not give up his pretensions, while Francis and his brother protested loudly against his right of possession. I felt myself obliged to interpose in this case my parental authority.

Jack was lighter and more agile than his two elder brothers; on the other hand he was stronger than Francis, who perhaps might rival him in agility. These two considerations decided the matter in his favour, and he was adjudged the ownership of the animal, but on one condition: that everybody should be allowed to ride him, and that he should be more generally recognized as common property than the other animals.

This decision, restricted as it was, overcame Jack with joy, and the others submitted, contenting themselves with joking the happy proprietor on his new bargain; but they did not disconcert Jack much: he received all their sarcasms as a traveller receives the snowflakes on his mantle.

The artificial nest of ostrich eggs, which we had enveloped in cotton and placed in a stove had succeeded; that is to say, out of six eggs, three had hatched. The young ostriches were the drollest-looking animals that could be imagined: they looked like ducks, mounted on long legs, and they tottered awkwardly about on their slender stilts. One of the three died the day after its birth; the two others survived, and we endeavoured to preserve them by taking all possible care for their comfort. Maize, acorns, boiled rice, milk, and cassava were set before them in rich profusion.

Our next care was to give our bear-skins their necessary preparation. I carefully removed all particles of flesh that adhered to them, rubbed them with vinegar several times, and then with a mixture of ashes and grease, worked at them constantly until they had attained the desired softness, and we thus obtained two superb warm coverings.

We had had nothing to drink but water since our arrival on the island, if I except the barrel of Cape wine that we had saved from the shipwreck; but that had long ago been exhausted, and I now determined to make some sort of drink for the winter.

I had often heard of the hydromel* of the Russians; we had the primary material, honey, from our hives, and I determined to make the experiment. We boiled some honey in a sufficient quantity of water, and after having filled two barrels with the fluid, I threw in a large cake of sour corn-bread, to make the liquor ferment; when that process was finished, we tasted it, and found it was of a pleasant flavour, agreeably acid, and a great resource for our long winter days. We placed the two tuns in our cellar, or, to speak more justly, in the hole we had dignified by that name. We then set to work and made a choicer drink than the first: to our honey and water we added nutmegs, ravensara,* and, in short, a collection of all the aromatic plants we could find. This drink was reserved for extraordinary occasions, such as holiday banquets, anniversaries, etc.

After the hydromel came vinegar: it had become an article of necessity to us in that hot climate, and my wife received this new fruit of our industry with signal marks of favour.

When all our provisions were gathered in, and we felt sure that we could get through the winter without famine over-taking us, we commenced our manufacture of hats. It was a labour as difficult as novel for us. Of course we did not display all the skill and fine taste that those who work for the London and Parisian dandies are obliged to possess; but we only wanted a comfortable covering for our heads.

The first question that presented itself was the form of our hats: each one gave in his opinion, but necessity came into the council, and obliged us to give our new manufacture the form most in unison with our means of execution. It was extremely simple: I cut a wooden head, which divided into two parts, and on which we spread a thick layer of soft paste, composed of rat-skin and the glue of fishes. We let it dry, and as it took the exact impress of the mould, we obtained a sort of cap, of which my readers can form some idea of the shape.

It had cost us a great deal of trouble to produce even this ill-looking affair. My sons were scarcely less satisfied with it than I was; but our European hats were so dilapidated that it became a matter of necessity to procure something to replace them.

'Is it a hat, bonnet, or a cap?' asked Master Ernest, laughing. 'A fine question to submit to the academy of Felsenheim at its first meeting.'

'Hat or cap,' said Fritz, 'it is of a most abominable colour, and I vote for some sort of colouring to dye it with.'

'Yes,' replied Ernest; 'I vote for red, it's the poets' colour.'

'And the cardinals' and the doctors',' chimed in Jack. 'Dye the cap red by all means, and we will have a cardinal's hat for professor Ernest.'

This sally made us all laugh, and the discussion proceeded. Francis preferred grey, Jack green, as being the favourite colour of the hunter, while Fritz—the prudent Fritz—voted for white, he having read that this attracted less heat than any other colour, from which he concluded it was the best for a cap.

'I am sorry,' said I, 'that I can not accommodate you all; but I have not half the colours you mentioned. Fritz's colour showed his judgement; Jack picked out his more for ornament than use; and as for Ernest, whether he was thinking of cardinals or not, his must be the colour, as it is the only one I can procure easily.'

I now had recourse to the cochineal, and I soon gave to our beavers a beautiful brilliant purple tint. The hat looked better; I adorned it with a couple of ostrich-plumes, and it looked better still; my wife passed a ribbon round it, which she had found in her enchanted sack, and the disdain with which my poor beaver had at first been received was changed into anxious requests for its possession.

But its destination had been fixed beforehand: it belonged to Francis by right, as he had lost his old hat a few days before.

Francis was a beautiful child, his face mild and pensive, and the new hat became him much. His beautiful auburn curls falling over his neck, with his infantine face and innocent expression combined to give him the look of Tell's son, such as the chronicles of our country represent him at the moment when his father is about to shoot the arrow. This souvenir enhanced the value of the new hat. Switzerland! William Tell!* These two words brought with them so many sad recollections that we were affected to tears. We thought ourselves transported to the bosom of our mountains. Ernest repeated the legend of the hero of Switzerland: my wife sung some of our mountain songs. Imagination—that magic fairy—pictured to us our huts, our trees, our precipices. We forgot for a time that there was between us and Switzerland an extent of perhaps

3,000 leagues, and we thus passed one of the most agreeable evenings we had experienced since our shipwreck.

52
The Ondatras; the Porcelain Cups

I HAD made but one hat; but I had four heads to cover, and each one wanted a hat like the first. But materials were wanting, and I engaged my little boys to procure as many rat-skins as possible before we proceeded to a new fabrication. I began by making a lot of rat-traps similar to those used in Europe, and, armed with these, we set off for the rat city.

For bait I employed a sort of little fish that we found in abundance in the marsh, and which the rats appeared very fond of. My traps succeeded, and we returned to the grotto with an ample supply of rat-skins. We now had leisure to examine these industrious animals, their construction, form, and habits.

The ondatra* is about the size of a small rabbit; its head is short and thick like that of the water-rat; it has large eyes, short ears, rounding, and covered with hair on both sides; its soft, shining fur is of a reddish brown, and its tail flattened and covered with scales.

These animals much resemble, in the general form of their body and most of their customs, the family of beavers. They construct their habitations of dry plants, particularly of rushes, and cover them with glazed earth. At the bottom of these habitations are different openings by which they pass out to seek their nourishment; for they do not amass provisions for the winter: they have also subterranean burrows to which they retire when their habitations are attacked. These habitations, which are destined to serve but for one winter, are reconstructed every year. Several families occupy the same habitation, which is sometimes covered, in northern latitudes, with a thickness of snow and ice to the depth of eight feet, so that these animals must necessarily pass a dreary existence until the return of spring. During the summer they wander here and there in couples, nourishing themselves upon roots and herbs; they become extremely fat, and acquire that musky odour which has given them the name of musk-rats.

Our hat manufacturing occupied us full ten days, and we obtained, by constant practice, some very nice hats. The cochineal, of which we had a great abundance, furnished us with a brilliant red dye, which made our caps look quite respectable. To see us walking gravely along the coast, with our red hats, anybody would have taken us for four dignitaries of the church of Rome. We left to Francis the privilege of wearing feathers, the border we had put to our hats superseding them with us.

Our success in the manufacture of hats emboldened us to try our hand at other things. We were much in want of kitchen utensils, and I was obliged to pass from the art of hat-making to that of potter.

I did not understand much about pottery; and what puzzled me most was the way in which the earth was to be prepared before using it; and I began my experiments with very little hope of their ever succeeding.

I constructed, in one corner of the grotto, a large stove, divided into compartments destined to receive the different articles; earthen pipes were conducted all around, so as to equalize the heat as much as possible. These preparations occupied me a long time, as I had no idea how the thing should be done, and I can safely say I invented rather than imitated a furnace for pottery.

I next took a certain quantity of the porcelain earth, which very much resembled fine white sand. I carefully removed all foreign particles, such as bits of stone, etc., as I was afraid they would cut our hands while working the porcelain. I also mixed a quantity of the talc we had brought home for window-panes, thinking, perhaps, it would render the mixture more firm and solid. When all was well worked up together, I left it a little while to dry, while I set to work to invent a machine for turning our utensils on. The wheel of one of our cannon-carriages, fixed horizontally on a pivot and surmounted by another wheel, united to it by an axle and turning with it, formed my machine. I first turned out some plates and dishes, cups and saucers, bowls, and other things. I exposed these articles to a very strong heat: a great many broke in pieces, but I completed about half. When baked they were perfectly transparent* and of the most beautiful grain. My wife saw her kitchen apparatus enriched with utensils of all sorts, and, overwhelmed with joy, she promised us, in exchange, numberless good dainties, which, for want of a suitable utensil, she had hitherto been unable to make.

Having overcome the first difficulty, the next thing was to embellish our crockery. Jack wanted to see some of those fine flowers that ornament the porcelain of our country, and refresh the eye with their bright colours. But the art of painting was one that required skill and knowledge far beyond what we possessed. We could not have flowers; but I supplied their places by the following artifice. We had saved from the vessel several cases containing collars and bracelets of coloured beads, intended to serve as a medium of intercourse with the savages. I broke them to pieces with the hammer, and, reducing them to powder, mingled it with the porcelain-dough, and by that means gave to our manufactures different shades of colour, and very much improved their appearance. After the articles which were manufactured on the turning-wheel, came those produced by moulding: I made a quantity of different shaped moulds, and we obtained vases, cups, pots, and several other utensils which, if not equal to the porcelain of Sèvres,* at least looked exceedingly well to our contented eyes. My wife and sons arranged the different articles ostentatiously on the boards that formed our sideboard, and I was happy to see my children exult in an achievement of their own, and regarding as a great victory the advantage industry had gained over necessity.

53
The Kayak*

THE rainy season was now rapidly approaching, and we were soon obliged to give up our excursions. The winds and the rain commenced, the sky that had so long been clear became dark with storm-clouds, terrible tempests announced the approach of winter, and we closed the door of our grotto, happy in having such a comfortable shelter.

The turning-wheel was continually in motion. We improved the quality of our fabrications more and more, and we manufactured utensils that at the outset we had despaired of ever possessing.

We had preserved the shells of the ostrich eggs, and, having divided them by means of a string steeped in vinegar, we converted the halves into elegant vases. I turned some wooden pedestals on which they were placed, and we thus obtained drinking-cups, and vases for flowers in summer.

The condor, which we had heretofore neglected, was now taken in hand. The euphorbia furnished us with the means of preserving the skin; we made eyes of porcelain for him, and the subject of his position in our museum was a matter of serious discussion. At last his station was decided, and he was arranged with extended wings and raised head, his beak half open, and his talons drawn up to his breast. This immense bird, joined to the boa, gave our growing museum an imposing aspect.

Of all the instruments at our disposal, the English turning-lathe was the most serviceable, and my wife made such frequent appeals to its powers that she finished by making me a capital workman.

But these labours were much more interesting to me than to my young family, and I feared that the inactivity to which I saw them reduced would render them indolent. Ernest found occupation enough in his books; but his brothers never entered the library unless when driven by necessity. I felt the urgency of providing some active occupation for them, and one more to their taste than literature; but I could not think of anything, when Fritz came to my assistance.

'We have,' said he, 'in the person of our ostrich, a splendid post-horse, with which to travel the highways of our kingdom; we have carts to transport our provisions; a pinnace, and a canoe, which are riding majestically at anchor in Safety Bay; but one thing is yet wanting: we have need of an equipage that will glide over the surface of the water, as the ostrich does over the sand—we want a light barque that will transport us in the twinkling of an eye to the farthest extremities of our kingdom; coast along the rocks, and bound over the opposing waves. I have read that the Greenlanders have a sort of vessel which they call a "kayak", and which I must have. Why cannot we construct one? We have constructed a canoe—why should we, civilized Europeans, not succeed in that which barbarous savages have attempted?'

I joyfully received the proposition of my son; but my wife, who had an old grudge against the sea and its dangers, disapproved of the project of the kayak. We brought forward all the arguments and inducements we could think of to gain her consent; but all was in vain, and my good Elizabeth was silent, but unconvinced.

However it might be, the construction of a kayak would fill up the vacuum in my sons' occupations, and I resolved to commence it, promising my wife a masterpiece of grace and swiftness.

The kayak, the only embarkation of the Greenlander, is a sort of canoe in the form of a shell, and a piece of walrus-skin, with three or four strips of whalebone, are almost the only requisites for its construction. It is extremely light, and the navigator who has glided in it over the surface of the wave can easily carry it on his shoulder when he has arrived at land. The Greenlander develops in the management of his kayak an address and boldness almost incredible; he attempts with it long voyages, employs it to hunt the walrus and the sea-dog, and, whether the sea be calm or agitated, his kayak is impelled over the waves like a feather; in it the islander never thinks of fear—his legs crossed at the bottom of the boat, and his oars in his hand, he knows not of such a thing as shipwreck.

The Greenlander is no adept in civilization and the arts, therefore his kayak is not a masterpiece of execution; we hoped to improve upon it, and intended that our projected barque should resemble the Greenlander's in no other points but swiftness and lightness.

The strips of whalebone, bamboo-cane, and Spanish rushes, with some sea-dog skin, were the materials that we employed. Two arched strips of whalebone fastened at each end, and separated in the middle by a piece of bamboo fixed transversely across, formed the two sides of our canoe; other pieces of whalebone, woven in with rushes and moss, well covered with pitch, formed the skeleton. The first improvement on the kayak was to arrange it so that the rower could remain sitting, while in the kayaks of the Greenlanders one is obliged to remain with the legs crossed, like a tailor, or else to lie down in the bottom of the boat—both very uncomfortable positions, and depriving the rower of much of his strength.

I shall say nothing of the exterior decorations, of the elongated, and, consequently, more graceful form than the original; but on the whole, this assemblage of osiers, whalebone, and bamboo formed a construction so light and elastic that it would rebound like a ball from the earth; and when we submitted it to the water, although heavily laden it scarcely drew two inches. We were engaged upon our new work more than a month; but it succeeded so well that my sons prophesied it would work miracles.

When the skeleton of our boat was finished, and the interior covered with a coat of gummed moss, we commenced the construction of an envelope. For this I took the two entire skins of sea-calves,* fastened one at each end of the canoe, and then drew them down under it, where they were strongly sewed together, and covered with a gum-elastic coat to render them impervious to water. I also cut out oars of bamboo, and fastened bladders to one end, so that they might be useful in case of accident. I also constructed, in the bow, a place to receive a sail, in case we should decide, at a future period, to put one there.

Fritz, as the author of the idea of the kayak, and being the eldest and the most capable of managing it, was solemnly installed owner of the kayak, Jack and Ernest being but little tempted by so seemingly dangerous a construction.

There was yet an important thing wanting in the completion of our Greenland boat: it was the equipment of him who was to manage it. I had often heard of a sort of apparel well known to those who dwell near the sea, and which consisted in enveloping a person in an airtight dress, lighter than the volume of liquid his body displaced. I described this apparel to my sons, narrated to them how the head of the swimmer was covered with a hood, furnished with a pipe intended to let in air, when it was necessary to breathe under water. My description of the costume and the air-chimney fairly turned the boys' heads, and they would not rest, night or day, until they persuaded me to ask their mother to construct such a suit for them.

My good Elizabeth, to whom our desires were as laws, kindly undertook the work, and so nimbly did her needle ply, that in a few days she had made a complete swimming-costume for Fritz.

A jacket of the skin of the whale's entrails, hermetically sealed and sewed round the borders, so that the air could not possibly escape, was furnished with a flexible pipe closed with a valve, so that it could be inflated or exhausted at the pleasure of its wearer.

The winter had glided insensibly away: reading, the study of languages, and other literary pursuits had been mingled with our domestic avocations, and helped to render the gloomy days we passed in the grotto more pleasant and agreeable.

But our emancipation from the grotto was approaching; the wind calmed, the sea resumed its wonted placidity, the grass sprung up under our feet, and we revisited Falcon's Nest, with its giant trees and its rich harvest of springing grain.

The swimming-costume was the last thing that we had made, and Fritz was anxious to make a trial of it; consequently, one fine afternoon, dinner over, he put on his jacket, which was drawn close round his neck; then his hood, with its pipe for air, was fitted to the jacket, and two pieces of talc inserted in such a manner as to enable him to see.

Our first movement, on seeing him thus accoutered, was to burst into a fit of laughter; but Fritz plunged gravely into the water and struck out for Shark Island. We followed him in the canoe, and arrived about the same time. We unfastened his hood, and found that not a particle of water had penetrated it: everyone was rejoiced at the success of the experiment, and we all persuaded our kind mother to make us one each.

We then set off to explore the island, and endeavour to discover what had become of the colony we had planted there. Our first visit was to the antelopes. They fled at our approach; but we saw with pleasure that they had devoured all the provisions we had provided them with. We strewed some rushes in their little hut, for a litter; and, after renewing the stock of provisions, left the spot, so that the timid animals could return. My sons and I wandered over the island, gathering pieces of coral and beautiful shells to adorn our museum. My wife made another discovery: it was that of a marine plant, unknown to all present but herself: gathering a large quantity of it, she placed it in the canoe, and when she arrived at our home she carefully locked it up in her provision cellar. This conduct astonished me.

'You must have some precious treasure there,' said I, laughingly; 'perhaps it is tobacco, and you hide it away for fear we will get hold of it.'

She smiled, and answered that in a little while I should know the name and properties of the mysterious plant. I was not quite satisfied with this answer, but I was obliged to be content and wait the result.

The earth was as yet too wet to permit of our resuming our excursions, and we spent the last few days of our sojourn in the grotto arranging our shells and coral on tablets in our museum. This occupation suited Ernest, who, at heart, was really proud of the title we had given him of librarian and first director of the museum at Felsenheim. He had studied very hard during our seclusion, and he explained to us the formation of coral, how it is developed amid the waves, and, gradually rising, constitutes large islands. He descanted on the polypi, and neglected no occasion to play the professor; yet I must say we listened with pleasure to his lectures.

'Conchology', said he, 'is one of the most difficult and least explored of all the branches of natural history. The learned man seems to have stopped here as being the farthest limit to which he could push his investigations. There are four sorts of shells: first, those composed of a single piece, called *univalves*; second, those composed of two pieces of different sizes, and sometimes shape, called, *opercules*; third, those of two equal pieces, called *bivalves*; and, fourth and last, those composed of a variety of unequal pieces, and called *multivalves*. Shells are employed in different ways among different nations. That variety which is called Guinea money, or curry, is really used as money in Guinea, the Cape Verde Islands, Senegal, Bengal, and some of the Philippine Islands. In Bengal, bracelets and collars are made of them; the Canadians also use them for the same purpose. In Egypt and Africa the ladies hang shells in their ears, as earrings. The inhabitants of Tyre used to draw from the *murex*, a sort of shellfish, a beautiful purple dye. The Turks and Levantines adorn the harness of their horses with the curry shells, and also cover boxes with them. From the *burgan* shell is procured a beautiful mother-of-pearl, called burgandine, which is set in gold, forming rich articles of jewelry. Sculptured ring-pieces, etc., are made from the *cameo* shell, and these are called *cameos.** Some oysters produce pearls, which serve as ornaments, and are worth an immense sum of money. Some industrious people have made artificial bouquets of flowers from shells, which look so natural that they have often deceived the eye. Among the Romans the shells called *buccins** served as trumpets in war. They are the same shells that the Dutch now call 'Trumpets'. The Savages, a people fond of the song and the dance, join together *buccins, porcelains,* and one or two other kinds, which, being exposed to a current of air produce an agreeable sound. In some countries drinking-cups are made of the *nautilus* shell. Shells have, from time immemorial, been used for voting, in public assemblies. The law of *ostracism** derives its name from a Greek word signifying oyster or shell. This law, it is said, was established among the Athenians, to exile those whom their great riches had rendered suspected by the people. In Corsica, silk is made from the *byssus* shell. In the isle of Ciana they calcine shells to serve as lime. In England shells serve to blanch wax; the English also use them like the farmers of Sardinia and Sicily, to fertilize the land. There are several species of shellfish that are eatable, such as mussels, oysters, periwinkles, etc. The Romans were extremely fond of shellfish; and one of their writers has left to us a long article on the manner of fattening them, and rendering them more agreeable to the taste.'

The ground was gradually becoming more dry, the pools of water which covered it fast evaporated, and we resumed our accustomed excursions to all parts of our dominions.

One evening, when we had returned from Falcon's Nest more wearied than usual, the heat had been so excessive, my wife presented me with a large bowlful of some sort of transparent jelly, of a most delicious flavour and coolness: it was a mixture of sugar and aromatic plants agreeably acidulated; and after having eaten some mouthfuls, we felt ourselves thoroughly refreshed, and we unanimously declared we had never tasted any thing that could be compared with it. In vain we conjectured what it might be. My wife preserved her secret.

'It is ambrosia,'* said professor Ernest.

'It is——it is——' said Jack, scratching his head.

'It is,' replied my good wife, laughing heartily, 'nothing more than an extract from that plant which I gathered on Shark Island.'

'Can it be true?' exclaimed I, with admiration. 'How did you recognize it; for I never remember to have seen its name in books.'

'What strange beings you are, gentlemen,' replied my wife, with all the authority that her new discovery invested her with; 'women, poor creatures! appear to you to be only fit for the kitchen, and you are all astonished if by chance they happen to find some little useful thing which your science had over-looked. Ah, gentlemen philosophers, here your science was at fault. I have surpassed you—me—a poor, feeble woman.'

'It is too true,' said I, 'and we must yield the palm to you. But how', continued I, 'did you happen to think of extracting this delicious jelly from the plant which you discovered?'

'The idea was not original, it is true. The Dutch lady, one of our fellow-passengers, who had lived a long time at the Cape of Good Hope, told me that the inhabitants gather on the borders of the sea a sort of marine weed, which they carefully wash and dry in the sun; they then boil it, adding to it sugar and citron, and obtain a jelly similar to this. In lieu of sugar I used the juice of our canes; for the citron I substituted some leaves of ravensara, some drops of hydromel, and a vanilla bean, the whole combined forming the refreshment that has found so much favour with you.'

We thanked our kind stewardess for this new attention to our wants, and gave her all the credit due, as remembering well how a thing is accomplished is almost equal to inventing it.

A second excursion to Shark Island gave us leisure to exam-ine the different plantations we had made: they had succeeded admirably, and we found several young trees already some feet above the ground. Our rabbits had also prospered, and the family had increased to an enormous extent.

We made, too, a short excursion to Whale Island; our plantations here had also succeeded—all was prosperity around us. Our maritime possessions and those on terra firma afforded a most agreeable spectacle to the eyes of the pro-prietors. Abundance, richness, and a luxuriant vegetation gave promise of an excellent harvest. We stopped a moment upon the rock which overlooked the island, to cast a glance upon the scene around us; the thought of the treasures we surveyed carried our hearts toward the Lord who gave them, and we blessed his name, and sincerely thanked him for his goodness and loving kindness toward us.

One day, when I was occupied by my domestic cares in the interior of the grotto, three of my sons disappeared without saying anything: they carried with them their arms, provisions, and a number of rat-traps. The latter easily explained the secret of their expedition: they had gone for rat-skins in order to make some new hats. I wished them good luck, and thought nothing more of the matter.

Ernest, always fond of home, had remained reading in the library; my wife was occupied in the kitchen, and I resolved to imitate my sons, and attempt an excursion alone. I had need of some large blocks of wood with which to grind the grain we had gathered; but I would not cut down one of the trees around our habitation for fear of disfiguring our residence. I went to the stable for a horse; but all except the buffalo had disappeared, and I was obliged to be content with him. I soon fastened him to the sleigh, and we set off in company in the direction of the Jackal's River. I took with me Folb and Braun; the faithful Billy remained with Ernest, and Turk had gone off in the morning with his young masters.

My intention in choosing the river road was that in passing I might take a look at our plantations of manioc and potatoes which extended along its bank. I had not seen this land which we had prepared with a great deal of trouble, for four months prior, and I expected to find an abundant harvest preparing for us. Judge, then, of my surprise, on approaching, to find the whole plantation a scene of ruin: the roots that had just begun to sprout were all trodden under foot, or scattered over the ground—in a word, it was a scene of utter desolation. I thought at first that perhaps my sons had gathered the harvest; but the prints in the moist earth soon revealed the authors of this devastation: it had been done either by the wild pigs, or else by the family of our old sow. I felt angry about the matter, and said to myself, 'Is it possible that, among all our possessions, these animals should have chosen to destroy that which cost us most pains and trouble?'

But my brave companions, Folb and Braun, had gone off in search of the despoilers, and they soon returned, driving before them a whole herd of pigs, at the head of which trotted our old sow, grunting most melodiously. I was so irritated at the unlucky animals, that almost instinctively I raised my gun, and, by a single shot, brought down two young porkers. The others took to flight, and the dogs would have pursued them, but I called them back, and, cutting off the heads of the two pigs, gave them to them. I then placed the bodies on the sled, and, having marked with a hatchet the trees I had chosen, so that I would know them again, I set off for Felsenheim, with a saddened heart at the devastation I had witnessed.

I recounted the whole affair to my wife, who was deeply grieved, and would hardly look at the two pigs I had brought home. However, I at last persuaded her to regard them as lawful booty, and to prepare them for the table. Ernest aided her, and they commenced operations on the smallest pig, who was soon on the spit, roasting beautifully, while some potatoes, placed in the dripping-pan, received the fat which escaped.

54
Return of the Boys; Harvesting

TOWARD evening we began to grow anxious about the return of the boys, when suddenly Jack appeared in the distance. He arrived at full gallop on his ostrich, having left his brothers far behind. He brought nothing with him, pretending that his courser would receive no other burden than himself. Fritz and Francis coming up, we discovered that each of them carried before him a sack full of game, the products of the chase, in which they had been extremely fortunate; and they brought back with them four of those beasts whom we had christened 'beasts with a bill', twenty ondatras, one monkey, a kangaroo, and two varieties of the muskrat, which they had found in the swamp. The first was the musk-beaver, not much different from the ondatra, excepting in the formation of his snout. I recognized it as being the *Tolay* of Buffon.*

Francis laid down before us a bundle of thistles with extremely sharp points, which would serve admirably to card wool, hair, etc. But each one was anxious to narrate the details of the expedition, and, according to custom, Jack commenced.

'First of all,' said he, 'honour to my courser, honour to the horse with long legs—honour to the hippogriff.* He carried me along so fast that I was obliged to shut my eyes, and the motion really took my breath away. There is one thing yet wanting to complete my equipments: I want a mask with glass eye-lenses; you will make one, papa—I must have it.'

'Ah, my dear cavalier, I am sorry; but I shall not make you a mask.'

'Why not?'

'For two reasons: the first is, that, instead of asking for a thing, you command it; and you seem to have forgotten that *must* is not the word to use to your father; the second reason is, that, instead of having recourse to the industry of another, you should attempt the thing yourself; so, then, if you want a mask, make one yourself.'

'You speak justly, my dear father,' said Jack, extending his hand; 'pardon my impoliteness, I beg of you; I will endeavour to correct it.'

'That is well,' said Fritz; 'everyone for himself is the principle we have been carrying out all the morning; we had nobody but ourselves to prepare our dinner. But, dear papa, what do you think of the fine furs we have brought back?'

'I receive them with all the thanks they merit; but I would rather my huntsmen would let me know when they are going away, and not run off and leave their parents ignorant of where they are.'

'That is true,' said Fritz. 'We thought of that when we were a league away from here; but I promise you it shall never happen again.'

The boys' frankness disarmed me, and I could not reproach them. While my sons were unharnessing their coursers and leading them to the stable, my wife gave the last turn to her roast pig, and we were soon all seated round the table.

'Truly,' said Francis, as he snuffed up the delicious odour of the pig, 'this is a great deal nicer dinner than the one we cooked out in the desert; and, I must confess, I have not much relish for a hunter's life, when he has to cook his own dinners.'

'Admirable!' replied his good mother, laughing. 'I am

enchanted to have so well divined the taste of my little Francis.'
And she took occasion to remark to us the treasures with which
she had loaded our table. By the side of the pig was placed a
plate of nice fresh salad; opposite to that was a dish of the
Hottentot jelly; for dessert we had a sort of fritters made from
guava apples, sweetmeats, of cinnamon preserved in sugar; a
bottle of hydromel completed this dinner, which was laid out
with as much precision and nicety as if we had been at Zurich*
instead of on a desert island.

During the repast, each one recounted his adventures, Fritz
describing their passage through the valley, the attack of the
ondatras and the beavers. 'We also,' said he 'then saw those
"beasts with a bill", coming out of the swamp to partake of a
repast not intended for them. We then caught a fish or two in
the lake, and, relieving our dinner with a plate of ginseng
cooked in the ashes, sat down to our humble meal.'

'Pooh, pooh!' cried Jack the boaster; 'who cares for rats and
fishes, it is to my courser and me that you owe this royal prize,
this noble kangaroo.'

'Oh yes,' added Francis, 'a prize very easy to take, as it
remained quiet until you came up and shot it.'

'For my part,' continued Fritz, 'I have brought home
nothing but a plant; but it is of more value than the kangaroo:
examine these thistles, I beg of you; see their hard, sharp
points. Will they not be excellent to card the hair in manufacturing our hats?'

'Oh, go away with your thistles,' replied Jack; 'my game is
worth twice as much as a wagon-load of them.'

We now had before us the whole of the game which our
adventurers had brought home. The rats were soon passed
over; we were too familiar with them to dwell long on their
probable uses. The *musk-beaver* received a more careful examination; but the kangaroo was an object of special study on the
part of Master Ernest.

'The kangaroo', said he, 'is one of the most curious animals
of the New World; it is sometimes nine feet in length, from the
extremity of the snout to the end of the tail; they sometimes
weigh as much as fifty pounds; the hair is short and thick, of a
reddish-grey colour, lighter on the flanks and belly; it has a
small, elongated head, large, erect ears, and a nose furnished

with a moustache; its neck and shoulders are small, increasing in size gradually toward the haunches; the forelegs of the kangaroo are about eighteen inches in length: they serve the animal merely to scratch the earth and to convey victuals to its mouth; but its hind legs are prodigiously strong; it springs often seven or eight feet high. There are but three claws on each foot, the middle one of which is considerably longer than the others. The tail of the kangaroo is long, thick at the butt, but gradually tapering; the animal uses it for defence and can strike blows with it strong enough to break the leg of a man.'

Each one of our young adventurers had a thousand different stories to relate, each one vaunting his own prowess and extolling his share in the events of the day. I had not time to listen to their boastings, and I turned to examine the products of the expedition and determine their use. The thistles of Fritz, which I recognized as being the 'carding-thistle', were received by me as a precious discovery—one more instrument added to our resources. My sons had also brought home some cuttings of sweet potatoes and cinnamon: their good mother received them with joy, and the next morning they were carefully planted in the kitchen-garden. I complimented my sons on their foresight, and felt happy to see some thoughts of the morrow introducing themselves in such young heads.

The next thing was to find an expeditious and easy manner of skinning the kangaroo, and I invented for that purpose a machine which caused a great deal of laughter among the children.

We had found on board the vessel, in the surgeon's case of instruments, a large syringe. I took it, and fixed in the sides of the cylinder two valves, intended to perform the functions of a pneumatic machine.* Without saying any thing concerning it to my sons, who stood watching me with astonishment, I ordered them to suspend the kangaroo by the hind legs, at such a height that the breast of the animal would be about level with mine. When this preparatory arrangement was concluded I made an incision in the skin, and then took hold of my syringe. The gravity of my sons could contain itself no longer: a hot fire of jokes and sarcasms was opened upon me; but still I looked as grave as a judge.

'Wait a moment,' said I, 'and you may appreciate my work by its results.'

I then introduced the end of the syringe into the incision I had made in the skin, and commenced to work the instrument. By little and little the skin of the animal became inflated, and soon the kangaroo was but a shapeless mass.

'To work, to work,' cried I to the astonished boys: 'beat this blown-up skin with your sticks, and you will soon have it off.'

And really, after having made an incision the length of the belly, the skin peeled off easily.

'Well,' said I to Jack, 'do you understand the efficacy of my procedure, yet?'

'I behold a miracle,' said he, 'but I do not know how it was done.'

'Listen, then. You must know that the skins of some animals are only fastened to the flesh by a tissue of extremely tender and delicate fibres and blood-vessels. These fibres are elastic; but, stretched beyond a certain point they will break, and thus separate the skin from the flesh. Such was precisely the action of my syringe on the kangaroo. It insinuated between the flesh and the skin a certain volume of air, which, distending the skin, broke loose the small blood-vessels, and thus ren- dered the skinning of the animal a very easy operation.'

'Why, papa, you must be almost a sorcerer to imagine such a thing.'

'Not in the least; I have often seen it performed, and every village butcher practises and executes it much more skilfully than I do.'

We now undertook and successively finished a multitude of other domestic works, all calculated to surround us with more of the comforts and luxuries of life. I went over to Whale Island one day to choose from the bones of the whale some that might serve as pestles with which to pound our grain. I found six, which I fitted into mortars that I had hollowed out of blocks of wood. My wife used them for the first time in pounding our rice; my sons aided her, and it was an interesting spectacle to see us all engaged at work. It was for ourselves we laboured; we depended on no market, and, in consequence, we gave our labours all the time they needed. Our chickens and the ostrich clustered around the rice-pounders, and not a grain fell from the mortars that was not instantly snatched up. The ostrich especially, towering high above the common fowls, and stretching out his long neck to pick up the grains, was a curious spectacle. Everything bore the impress of civilization, and demonstrated how much industry could overcome difficulties.

The grain that we had sown before the rainy season I perceived had now come to maturity, although it was not more than five months since we had confided it to the earth. We now had our hands full of business. The herrings would soon arrive, then the sea-dogs would come; and my dear Elizabeth lamented piteously while she enumerated all the labour we yet had to perform. There was the manioc to dig up, the potatoes to gather and sow—in short, a thousand cares to attend to, a thousand labours to undertake, that would occupy more time than the year has days.

I tranquillized my good companion as well as I was able, assuring her that the manioc would not be injured by remaining in the ground; and as to the potatoes, I informed her that she had nothing to fear from this precious fruit, as our soil was warm and sandy and they would keep a great while in the earth.

I decided that our labours should commence with the grain, the chief and best of our resources; but, wishing to effect the harvest in the shortest possible time, and with the smallest expenditure of strength, I resolved to adopt the Italian method rather than the Swiss.

I commenced by levelling a large space before the grotto, to serve as a threshing-floor. We then, after having well watered it, beat the earth, for a long time, with clubs. When the sun had dried it up, the operation was repeated, and we continued it until we obtained a solid, flat surface, without a crack in it, and almost as impenetrable to water as to the sun's rays. When we had finished this, I harnessed the buffalo and the bull to the famous osier basket which we had dignified by the pompous name of 'palanquin', and which had been the innocent instrument of torturing poor Ernest. Jack and Francis recalled the scene to the philosopher, and invited him to take another ride, but Ernest was not a boy that would be taken twice in the same trap, and he respectfully declined the honour.

On arriving at the field we were about to reap, my wife asked me where I would find anything with which to tie up the blades into sheaves.

'We will need nothing of the sort,' said I; 'everything is to be done according to the Italian method. Those people, naturally averse to labour, never use sheaves, as being too heavy to carry.'

'How, then,' asked Fritz, 'do they manage to carry their harvest home?'

'You will soon see,' said I.

At the same time I gathered up in my left hand all the stalks it could contain, and, taking a long knife in my right hand, I cut off the stalks about six inches below the head. I then threw the handful into the basket. 'There,' said I, laughingly, to Fritz, 'there is the first act of an Italian harvest.'

My children thought it was an admirable plan; and in a short time the plain presented but an unequal surface, bristling with decapitated stalks, here and there dotted with a forgotten blade.

'I must confess,' said my good wife, as she cast a look over the field, 'that I do not much approve of an Italian harvest. Why, only look at the fine blades you have left behind.'

'Not so fast, my dear woman, not so fast,' said I; 'do not condemn my new method yet, bad as it seems: what we leave behind now we will drink afterward.'

'That is an enigma I cannot solve.'

'I did not expect you could; but sometimes, if things were not put in enigmas they would be forgotten; however, to explain mine, I will tell you that the Italian drinks what he does not eat of his harvest, only it is under another form. In Italy, grass, hay, and pasturage are extremely rare and dear. The Italian endeavours to provide for this scarcity by converting the refuse of his harvest into forage. He never cuts down the straw of the grain, and the substance that remains in the stalk makes the grass grow, and forms a solid mat. Then he cuts it and gathers an excellent forage for his cattle. The blades that were left about the field renders the supply of milk, from the cow that eats them, much more abundant. And now, you see, I have explained my enigma.'

'Very good,' replied my wife; 'but if the Italian feeds all his straw to his cattle, out of what does he make litter for them?'

'None is needed. The soil of Italy is a healthful one, and does not throw out that humidity which prevents our cattle from sleeping on the bare ground. But we have no time for further explanation: after having gathered the grain like the Italians, we must thresh it like them. Gentlemen, to the grotto, and prepare your coursers—we will have need of them.'

We now hastened to the grotto, taking with us the grain we had just cut. When we arrived there, Ernest and his mother received orders to sprinkle the blades over the threshing-floor I had prepared, while my three cavaliers stood by their coursers' sides, laughing at our new invention for threshing grain.

'Ah,' said Jack, 'my courser will cut capers here, such as she never did in the desert.' 'Threshing grain on horseback!' said another. 'Harvesting at a hand-gallop!' said a third. The laugh and jest went merrily round, and my innovation at least made us all joyful, if nothing else. But I kept a sober face, looking as if I was quite certain my project would succeed.

When everything was prepared, 'To the saddle!' cried I, 'to the saddle!' and I told them they had nothing to do but display their horsemanship among the grain. I leave the screams, the shouts of laughter to the imagination of readers; the bull, the onagra, and the ostrich rivalled each other in swiftness; my wife, Ernest, and I, each one armed with a pitchfork, followed after them, throwing the grain under the feet of the animals. Everything went on marvellously well, when two incidents, which I had not foreseen, rekindled the flame of my wife's irony, who was not yet a sincere convert to the Italian method. The bull and the onagra, unable to withstand the temptation before them, suddenly stopped short in their career, and, stretching out their long tongues, each took up a huge mouthful of the grain.

'How is this?' said Fritz, as his courser stopped. 'Is this also a part of the Italian method?'

'Very economical, truly,' said his mother, ironically.

'As to the bull helping himself,' said I, 'it is one of those events that are not foreseen in time to be prevented, and, besides, does not Holy Scripture say, "The ox shall feed on the produce of the mill that he turneth." '*

The aptness of my quotation re-established the honour of the Italian method, which but a moment before had tottered on its base.

When the grain was all threshed, we set to work to clear it of the straws and dirt that had become mixed with it. This was the most difficult and the most painful part of all the labour. We laid the grain on close hurdles, and with wooden flails we endeavoured to disengage the dirt; but this was not to be effected, except at the expense of our eyes, mouth, or nose. The poor little workmen coughed terribly, and we were obliged to desist every few moments to clear our throats.

Our feathered colony, which had taken care to keep clear of the heels of the animals, now that they were gone, flocked around us, picking up, by single grains, as much as the onagra and the bull had taken at one mouthful. 'Let them alone,' said I to my sons; 'what they take now we will recover some other time; and if the corn diminishes, the fowls will grow fatter.' But I had spoken too late, for my wife had already frightened them, and they were all far away.

We were several days engaged in these works, and we wished to see exactly how much we possessed. We found ourselves rich enough to defy all attacks of famine; we had sixty bushels of barley, eighty of wheat, and more than a hundred of maize, from which I concluded that the soil was more favourable to this last than to the barley and the other European grains we had sown at the same time. We had not prepared the maize as we had the other grains; but, after having dried the stalks, we detached the grains by beating them with long, flexible whips; we took this care because we wanted its soft and elastic leaves to stuff our matresses. My wife also burned the stalks, the ashes containing an alkaline quality very useful in bleaching linen.

I had not lost sight of my intention of obtaining a second harvest before the end of the season, and we now set to work to clear our fields of the straw; but we had scarcely commenced when we beheld an innumerable swarm of quails and partridges start up from the dried stalks, where they had been enticed by the few blades of grain we had left behind. As we were unprepared for them, they all escaped, save one quail, which Fritz brought down with a stone; but the presence of these birds after the harvest was a precious discovery for following years, and we anticipated with pleasure the superb chase of quails and partridges we should have after our harvests.

When the land was all cleared I sowed it anew; but remembering what I had learned in Europe, not to exhaust the soil, I varied my original mode of operation, and contented myself by sowing, for the second crop, wheat and oats.

Our agricultural labours were scarcely finished when the bank of herrings appeared off Safety Bay. Our winter provisions being so abundant we did not take as many as customary of these, and we contented ourselves with preparing two barrels, one of salted and one of smoked herrings; we also preserved some of the fish alive, which we put in the Jackal's River, so that at any time we could obtain them.

The sea-dogs then had their turn: my pneumatic syringe did wonders; and, thanks to its assistance, our labour on these animals was but trifling. The skins, the bladders, all were put to use, experience having taught us the value of these riches. We had not been able to finish our kayak until this time, and we now provided it abundantly with bladders, so that it might float more lightly on the surface of the water. When this work was finished, preparations were made to launch the boat and try her power over the winds and the waves.

55
Trial of the Kayak

THE trial of the kayak was a grand holiday fête; all were anxious to join in it, and when Fritz appeared, clad in his maritime costume, he was formally invited to take his place in his boat of skin. I had forgotten to say before that the kayak was furnished with two little wheels of copper, so that it could be used as well on land as on sea. This advantage enabled my sons to arrange the ceremony with all possible pomp. Fritz was installed upon his bench, as proud as Neptune or any other marine god setting off on a distant voyage. The form of the kayak was not a bad resemblance of those immense shells that fable has assigned to the sea-gods as having been used for chariots. The gravity of the hero who sat enthroned, holding an oar in his hand in place of a trident, the efforts of his brothers, who pushed the kayak behind, sounding some sea-conches they had found, with all their might, all this presented a spectacle as animated as it was picturesque. I laughed heartily; but my good Elizabeth, always preserving her hatred against the ocean, could not hide the big tears that rolled from her eyes when she thought of the dangers that menaced so frail and fragile a skiff. To reassure her, I united the canoe and held myself ready to start at a moment's notice, if any real danger should threaten our Greenland sailor. When all these precautions were taken—'To the sea!' cried I to Fritz; 'to the sea!' 'Goodbye', repeated his brothers, and the kayak glided into the water with inconceivable rapidity. The surface of the bay was calm and tranquil, and soon the Greenlander was dancing gaily over the waves: then, like a skilful actor, he began executing a series of evolutions, each more adroit or more audacious than the other. Sometimes he would shoot off far out of our sight, then suddenly he would disappear in a cloud of foam, to the great terror of his mother: in another moment we saw his head above the floods, and an oar that he had elevated to signalize his triumph.

The address and the audacity of our young sailor provoked, as one can easily imagine, loud and frequent applauses on our part: on his part, not content with acting on the surface of the bay, he turned his frail bark toward the Jackal's River, and attempted to mount the current; but this proved too strong for him, and threw him back so violently that he disappeared from our sight. To jump into the canoe and fly to the assistance of the poor Greenlander was the affair of an instant. Jack and Ernest went with me, and we left Francis on the shore with his mother, who abandoned herself to all the terrors that maternal love could inspire. The wheel of the canoe appeared to us too slow, and while I exerted all my force in turning it, my two sons took each an oar. We scarcely touched the surface of the water; yet we could not perceive anything, our cries had no echo but the rocks, and our sight was lost in the foaming waves that boiled up around us. I felt my heart beating violently, and I had not the courage to express my uneasiness to my sons; when suddenly, in the direction of a rock just visible through the foam, I saw a light cloud of smoke issuing forth, and, putting my hand on my pulse, I counted its beat four times before that smoke was followed by a report.

'He is saved!' cried I, 'He is saved! Fritz is there in the direction of the smoke; before a quarter of an hour he will rejoin us.'

I then fired my pistol, which was instantly answered by another report in the same direction. Ernest drew out his watch. After a hard row we perceived Fritz, and in a quarter of an hour we reached him.

We found the young hero of the sea established on the rocks. Before him lay a morse,* or sea-cow, which he had killed with his harpoon. I commenced by reproving my son for his imprudence.

'My dear father,' answered he, 'it was the current that swept me away in spite of myself: my oars were like straws before the impetuosity of the Jackal's River, and I found myself thrown back into the sea, at such a distance as to lose sight of land altogether. But I had not time to fear: a company of sea-cows passed along, almost under my nose. To throw my harpoon and strike one of these animals was the work of an instant; but the wound I had inflicted was not mortal, and, instead of weakening him, it seemed, on the contrary, to inspire him with new strength. He dived down; but the traces of blood he left behind, and the bladder of air fastened to the end of the rope of the harpoon, served as guides to follow him. The second time I was more successful, and I launched a second harpoon direct in his side. This last blow was decisive, and, after some struggles, the monster extended himself on this rock. Remembering our precaution with the boa, I fired two pistols at the head of the animal, and probably those were the reports you heard.'

'You have achieved a truly heroic action, and the combat was a perilous one. The morse is a redoubtable monster; and, instead of flying, he would have turned upon you, and God knows, my poor child, what would have become of you if your frail boat of leather had been torn by the terrible teeth of the morse. But, God be praised, you are safe, and that is better than the capture of ten such animals, which are not very precious game. I do not know what use this will be to us, notwithstanding it is fifteen feet long.'

'Well, then, if it is good for nothing,' answered Fritz, 'I will keep the head myself; I will prepare it and fasten it to the bow of my kayak: its long, white teeth will have a fine effect, and I will now call my kayak "The Morse".'

'The teeth of the morse', said I, 'are the only things worth preserving. They are as white and hard as ivory: but make haste, for the sky gives sure token of a storm.'

'This head will be a splendid ornament, Fritz, on the bow of your kayak', said Jack.

'Yes,' replied Ernest; 'rotten flesh will be a very agreeable perfume, certainly.'

'Rest in peace, doctor,' answered the sailor; 'rest in peace. I will prepare the head of my walrus so that its odour will not be more disagreeable than that of the stuffed animals in our museum; but I must be quick and finish my work.'

'I thought,' said Ernest, while Fritz was thus occupied, 'that the morse, walrus, and such animals did not inhabit these southern climes. How is it possible that we find them here?'

'Without doubt,' said I, 'these amphibious animals appertain to the northern seas; but their presence in such a burning climate is easily explained. A tempest, or a terrible storm might have transported them; besides, there is a species of this animal found at the Cape of Good Hope, and called dugon;* and perhaps our animal is one of these. There are, however, some slight differences between the kinds, but their food in all regions is similar, viz., the marine plants and the shellfish that they detach from the rocks with their long teeth.'

But Fritz had now finished his labour, and, while I was engaged in cutting off some strips of the animal's skin, he entreated me to add three very useful things to the furniture of his kayak: a compass, that he might guide himself, if at any time he should be blown off the coast; a lance and a hatchet, so that he could attack an enemy or defend himself, if necessary. I could see no objection to his demands and as we had more than one compass I promised my son to place one in the prow of his boat, and also to furnish him with a lance, and hatchet.

I wished to take Fritz and his kayak into our canoe, but he refused, and dashed on, saying he would announce our return to his mother. I let him proceed, and he soon passed us. While we were rowing quietly on, Ernest, who always wanted to know the whys and wherefores of everything, asked me how I managed to calculate the exact distance that separated us from his brother.

'In a very simple manner,' said I. 'All that is necessary is a little information, possessed by almost everybody at all acquainted with the phenomena of nature. It is known that light passes through space with extreme rapidity, and that in a second after it is evolved it has traversed space to the extent of eighty leagues. Sound, on the contrary, is much slower in its transition, and in the same time passes through a space of 1,032 feet. I knew that my pulse, like that of every man in good health, beats regularly sixty times per minute: I counted four beatings between the smoke and the report, from which I calculated Fritz was distant from us about 4,160 feet, or about a quarter of one of our leagues. I then immediately knew that it would occupy us a quarter of an hour to row to him. The wind, the rain, the state of the atmosphere might counteract these effects, but generally it can be calculated pretty exactly.'

'Another secret that I did not know before', replied my little philosopher, in a tone that denoted the pleasure he felt at his acquisition of knowledge—'another one of those marvels that appear so impossible to him who does not understand them. But', continued he, 'can celestial light, and the time it takes to reach us, be also determined?'

'Certainly it can. Astronomy has ascertained, with the greatest exactitude, the distance which separates our globe from the sun and stars above us. It has been estimated that the solar rays take eight minutes to descend to the earth, and that the light of Sirius, the dog-star, cannot reach us in less than six years. Likewise, if a cannon was fired in that star, we could not hear it in less than 6,000 years after the detonation.'

'My Goodness! Astonishing! Why, one's head turns with the thought!'

'It almost confuses one when we employ calculation with regard to the fixed stars, they being thousands of times farther from us than Sirius. It is there, my child, it is in that immense book, in that sublime assemblage of wonders, we should study to know the sovereign Author of all things. It is in the presence of this majestic concert of harmonies that man feels his little-ness and humiliation; for, probably, all the stars which stud the blue vault above us are inhabited worlds, in comparison with which our globe appears but as a grain of sand in space.'

56
The Storm; the Drawbridge

THE storm, however, came on quicker than I had anticipated.
We had scarcely accomplished a third of our course, when the
thick, black clouds that brooded over the horizon burst forth
in torrents of rain. The wind, the lightning, the waves were
confounded in horrible confusion. Fritz was too far from us to
allow of his joining us, and I repented of not having taken him
into the boat with us. I desired Jack and Ernest to put on their
swimming-corsets, which we were always careful to take with us,
and to lash themselves fast to the ropes of the canoe, so that
they would not be carried away by the floods that broke over
us. With a mind full of anxiety, I lifted up my thoughts to
Heaven in prayer, that God would watch over and protect us.

The tempest increased, and my anxiety increased with it;
the waves elevated themselves like mountains: at one moment
we would be high in air, and at another precipitated to the
bottom of an abyss, where it would seem we were lost forever.
But the violence of the tempest prevented its lasting a great
while. The waves subsided, and, after a hurricane of a quarter
of an hour, the wind fell, and the storm, for the time, was over,
although black and angry clouds rolled over our heads. Our
canoe had breasted the storm bravely, and had not bent under
the waves that dashed over her, and whirled her round like a
feather on the surface of the water.

Our first sentiment was that of gratitude, and we thanked
the merciful God who had once more preserved us. Fritz and
his kayak were ever present to my mind, his embarkation was
so insecure and his barque so frail for an occasion so trying.
But all that I could do was to appeal to the Throne above, and
ask for strength to support the terrible blow I almost expected.

We redoubled our efforts at the oars and the wheel, and
soon arrived within sight of Safety Bay. We entered the well-
known harbour, and the first objects which greeted our sight
was Fritz, Francis, and their mother, kneeling on the beach:
they were praying for our preservation. The heart of my poor
Elizabeth was almost broken with anxiety, and she needed to
put all her trust in him who alone can comfort.

We leaped from our canoe, amid the cries of joy and the embraces of our dear ones, who rushed to our arms. My wife had not strength to articulate a single word of reproach for the great imprudence we had displayed: her only thought was a feeling of thankfulness to our Almighty Preserver.

We all united in prayer, and retired to the grotto to exchange our dripping garments for dry ones.

'At last,' said Fritz, who spoke first, 'we are again united. I had given up all hopes of ever seeing you again, when a huge wave swept over my little barque; but I held my breath, and the wave passed on, and I found myself still alive. But it was not my exertions that brought me to the shore: there was a stronger hand than mine that sustained my kayak among the waves—the hand of God,' added the young man, 'and to him have I rendered homage.'

'What a day, my father,' said Ernest, still pale with terror. 'How terribly we have suffered.'

'For my part,' said Jack, 'I swallowed more than my share of seawater, and I can assure you it is the most detestable drink that was ever poured down human throat. I was much amused with looking at the doctor here, who sat with his mouth shut, and making most horrible faces.'

'Oh, indeed!' exclaimed Ernest, a little nettled; 'I am really very glad that I have been able to divert Mr Jack at a moment when diversion was not so easy to procure. But I must beg leave to say that, if I was afraid, I kept it to myself.'

''Tis true,' added I, in my turn. 'If Ernest was frightened, he did not make it known, as probably he remembered that cries and screams but render a danger more embarrassing. But let us congratulate ourselves upon the solidity of our equipage; our bark canoe breasted the storm as gallantly as if she had been a ship of the line, and I will not fear now to go to the succour of any ship in distress.'

'Oh, doubtless,' said Fritz, 'the canoe did very well; but my kayak has also some claims to the honours of the day. Twice or thrice it was submerged, and not a single thing is broken. Only we must take care, when we go out to succour any ship, that we do not go too far.'

'Ha ha,' said Jack, laughing; 'on condition, then, that vessels will agree to be shipwrecked in fine weather and near shore, you will agree to save them.'

'Why,' said Fritz, pursuing his idea, 'why can we not construct on Shark Island a sort of fort where we can place a signal gun: its echoes when fired could be heard through the wind and rain, the unfortunate people could answer, and we could fly immediately to their assistance.'

'Oh, if we could see men once more!' replied my young sons, carried away by that instinct of sociability which binds all the members of the human race—'men on our shore—men like ourselves—oh, what happiness it would be!'

'Without doubt it would be very fine. If I had the enchanted cap of Prince Fortunatus* I would take a cannon under each arm and carry them over there. I cannot see how else we could transport them thither. Ah, gentlemen, if the work could be done in reality as easily as it can in imagination, what a fine fort we would have there. But how can a man and four boys, aided by his wife—a fine cook, it is true, but what could she do in military affairs—construct a sea-fort with cannon, walls, portholes, etc.?'

'I think,' said my wife, ironically, 'that every new difficulty that presents itself should be attempted, not despaired of; for it displays how much your ingenuity can effect.'

'Good, good,' replied I, jestingly. 'We will adjourn the commencement of our fort for the present, and occupy ourselves with securing the boats.'

The canoe was then drawn up on the sand, the kayak carried to the grotto, and the walrus's head to the work-room to be prepared, so that it could be attached to the kayak.

The rain had been so abundant that the Jackal's River had overflowed its banks and damaged some of our constructions, which demanded instant restoration.

While we were occupied in considering these ravages, chance caused us to make a new discovery; it was a kind of small pear, about the size of a plum, with which the sand was strewn. They looked so nice that my little gluttons hastened to taste them; but they had scarcely touched them with their teeth than they threw them down in disgust. I wished to know what kind of fruit it was, and, taking one up, I recognized it as being the fruit of the clove-tree,* another addition to our stock of spices.

The clove-tree grows in the Molucca Islands, situated near the equator; it is of the form and size of a laurel; its trunk is about a foot and a half thick, hard, branching, and covered with a bark like that of the olive. The branches are of a reddish colour, and produce very many leaves; the flowers grow in bunches at the end of the branches: they are rose-coloured, with four blue petals of a very penetrating odour. The middle of the flower is occupied by a number of purplish stamens; the calyx is cylindrical, divided into four portions, of the colour of soot, of an aromatic smell; after the flower is dried, it changes into an oval-formed fruit like the olive, first light-coloured, then reddish, then a blackish brown, and containing a hard nut divided by a deep furrow. If left on the tree they will not drop until the second season; they are then planted, and in eight or nine years will bear fruit. The Dutch preserve the new cloves with sugar, and in sea voyages they eat them after dinner to render their digestion better and to prevent scurvy. The cloves are gathered before the flowers open; the season is from the month of October until February, and the fruit is mostly gathered by hand. The rest are knocked off with sticks, on linen cloths placed on the ground. At first the cloves are reddish, but they blacken in drying; they are then exposed, on hurdles, to the action of smoke, which gives them the colour we know them by. None understand the preparation of cloves better than the Dutch of Ternate;* they are almost the only persons who gather, cultivate, and prepare the cloves for the use of the whole world.

We now employed ourselves in building protections against any other storms that might visit the coast of Felsenheim. During our labours we received the visit of a superb company of salmons. We captured a number, which were salted and smoked according to the customary manner; we preserved some alive by passing a strong cord through the gills and fastening them to stakes. Sturgeons* are conducted, living, to Vienna, up the whole course of the Danube,* in this manner.

'How', said Francis, with an air of simple astonishment, 'is the salmon a freshwater fish here? I always thought it belonged to the sea.'

'Little one,' replied doctor Ernest, pompously, 'the salmon is both an inhabitant of the sea and the river; it is a superb fish, and well repays the care taken to secure it. It has, as you see, a sharp, small head in proportion to the size of its body; the opening of its mouth is very large, and the upper jaw is much the longest; its nose is pierced with two holes, placed near the eyes. These are round, situated in the side of the head, with a silvery iris and a deep-black pupil. The entire length of the salmon is from 28 to 30 inches. A naturalist whom you do not know, called Peyceres,* has made some very curious anatomical observations on the entrails of this fish. They are most generally found in the neighbourhood of the Baltic, and the rivers which empty into that sea. The salmon has one peculiarity that distinguishes it from other fish: it is that of continually endeavouring to contend against the currents of rivers: it leaps with much agility, doubling its body into a complete circle. Its great enemy is the leech,* which torments and wearies it by continual bites, and it is to escape these that the salmon throw themselves up. Salmon have been taken which weighed 30 or 40 pounds. Its flesh is not very fat, white when raw, but the salt and the action of the fire gives it a beautiful red tint.'

Jack interrupted the lesson, by remarking to the doctor 'that he was as good a cook as a philosopher'; but Ernest contented himself with smiling disdainfully, and saying that he pitied the poor wretches 'who, not being able to understand science, take pleasure in deriding it'.

We had resumed the peaceable course of our domestic avocations, when, one clear moonlight night, I was suddenly awakened by barks and cries, as if all the jackals of the country, the bears and tigers of the savannah, had made an invasion into our domain. I rose in a great fright, and arming myself with a gun, I walked to the door of the grotto, which we generally left open on account of the fresh air. Fritz had also heard the noise, and I found him half dressed, ready to face the danger.

'What do you think it is, papa,' said he; 'a new invasion of jackals?'

I dissembled the real fear I entertained, and assured my son that doubtless it was our pigs, who were making us a nocturnal visit. I did not think my supposition would be true. We ran out, and found that our dogs and the jackal had captured three large hogs. Our first movement was to laugh: we tried to call off our dogs, but in vain; they had the poor pigs by the ears and they would not let them go, and we were forced to open their mouths with our hands. The pigs never waited to see who were their liberators, but scampered away and were soon across the river.

I attributed this invasion to negligence on our part, and thought we had forgotten to take up the planks from Family Bridge; but, upon examination, I found that they had been all removed, and that the audacious pigs had come across on the beams of the bridge.

This occurence convinced me that Family Bridge was not sufficient for our security: instead of a barrier, it was only a means of entering our domains. I had long contemplated the erection of a drawbridge, and now appeared the proper time for constructing it. To be sure, a drawbridge was not a little thing to undertake; but after having constructed two vessels, attempted and executed a thousand other things which required more skill than the simple art of carpentering, we could not recoil before the idea of constructing a bridge.

I understood the turning-bridges; but as I had neither vice nor windlass, I was obliged to adopt the simplest kind of drawbridge. I constructed, between two high stakes, a sweep that could be easily moved, and by the means of two ropes, a lever, and a counterpoise, we had a bridge which could be easily raised and lowered. It would only insure us against the invasion of animals, the river being too shallow to oppose any obstacle to a more serious attack. Whatever it was, our domains were enriched with a new masterpiece, and my young people exerted themselves in a thousand gymnastic exercises about the stakes of the drawbridge; it was lowered, raised, and for a few days it was a great source of amusement for them.

57
The Hyena; the Carrier Pigeons

THE drawbridge suffered the fate of all new inventions, admiration evaporates so quickly! and at the end of several days, if anyone climbed the stakes, it was that they might have the pleasure of seeing the antelopes and gazelles bounding over the plain of Falcon's Nest.

'Behold,' said one, 'how graceful and light those animals are! they scarcely touch the earth. What a pity we cannot tame them, or, at least, approach them without scattering the whole flock as the wind does the dust.'

'To take them,' said Ernest, 'you will have to adopt the plan resorted to by the Georgians in capturing buffaloes.'*

'Tut, tut,' said Jack; 'cannot you find an example nearer home than Georgia?'

'For the world of thought,' replied the professor, gravely, 'there is no limitation, and it would be as well to become acquainted with the Georgian method before rejecting Georgia as being too distant.'

'Well, then, doctor, give us a lesson.'

The professor, who willingly forgot the sarcasms and pleasantries that were showered upon him, whenever he availed himself of an opportunity to display his scientific knowledge, now began to explain his former remark.

'In the savannahs of North America, in some places, beds of marl are found which contain salt, of which the animals are very fond: the buffaloes especially flock in great numbers to this luxury which nature has provided for them. The natives of the country lie in ambush for them there, and numbers fall victims to their avidity. In the absence of salt marl,' continued the professor, 'we can, if we wish, prepare artificially a representative of it, where the graceful antelopes will fall into the snare. We can, for that purpose, mix together the porcelain clay and some salt.'

'Adopted! adopted!' responded all the little boys unanimously. 'Long live philosopher Ernest, first professor of the academy of Felsenheim, doctor, librarian, manager of the museum, naturalist, and so forth!'

To plan an excursion and ask my permission was the work of an instant, and my hare-brained youngsters promised themselves so much pleasure that I had not the disposition to deny them.

'Oh do, do, papa,' was the general cry; 'an excursion is much more fun than constructing bridges.'

'I will make some pemmican,' said Fritz; 'we have bear's meat enough left for it.'

'And I,' said Jack, with a mysterious air, 'I will take two pigeons with me. I have got an idea in my head.'

'And I', added little Francis, 'will take care of the coursers; and if Fritz will take my advice he will take the kayak along—it will sail so nicely on the lake; and, perhaps, we can capture some of the black swans. Oh! how beautiful a pair of those swans would look in the basin at Falcon's Nest.'

The weather was calm and serene, and everything promised a pleasant excursion to the adventurers.

Fritz walked up to his mother, who was occupied in her kitchen-garden, and, saluting her with all the grace of a polished cavalier, asked her if she would give him some pieces of bear's flesh to make a pemmican.

'Tell me, if you please,' answered his good mother, 'what a pemmican is.'

'It is a very common and highly esteemed preparation in North America. The Canadians make it almost their only food. It is composed of bear's or goat's flesh, which they chop and beat up until it is reduced to a very small volume.'

'And what do you want to do with such stuff? Bear's flesh beaten cannot be such very desirable eating.'

'Oh, mamma, do give me some; we are going on an excursion, and the pemmican is excellent for food on such an occasion.'

'How,' said his mother; 'another excursion without consulting me! Did you suppose I had no objections?'

Fritz brought to his aid all the arts of flattery and address, and he soon returned with the bear's meat. The fabrication of pemmican was commenced immediately, under the inspection of Fritz. The meat was pounded and crushed, until, after two days of hard work, it was reduced to half its former size. I tasted the meat of which Fritz boasted so much, and I did not think it bad.

Baskets, sacks, and all utensils necessary for the excursion were collected together; even our old sled was brought down, and it was mounted on cannon-wheels, and loaded with all that the young adventurers intended to carry with them. The kayak, arms, provision for the mouth and for war—nothing was forgotten, anything that came into their heads they piled on, and a caravan in the desert could not have made more preparation.

The morning of departure arrived. Everyone was awake before day; and Jack, without saying anything to anybody, climbed up into the dove-cot, and took out several pairs of pigeons. They were that species called by Buffon, Turk Pigeon.* 'How is this?' said I, as I saw the youngster placing his pigeons into a basket. 'It appears that you gentlemen are not content with your pemmican, etc., and take precautions to provide a variety for yourselves. I am only afraid that those old pigeons will be pretty tough eating.'

The fellow looked at me knowingly, for a moment, but did not answer. When they were about to set off, I saw him conversing mysteriously with Ernest; but I could discover nothing, and contented myself with waiting a surprise of some kind, as I knew they intended one.

At last they were ready to set out. My wife enjoined my sons to be prudent, we embraced them, and they soon disappeared in a cloud of dust, with their coursers and the sled. Ernest alone remained with his mother and me, and we employed ourselves in constructing a sugar-cane press, which my wife had much need of. The machine, which was composed of three cylinders, placed upright, differed very little from the ordinary presses, with the exception that it was arranged so as to be moved by animals. These labours naturally led us to speak of sugar.

'A few more improvements,' said Ernest, laughingly, 'and we will have a regular refinery soon at Felsenheim.'

'It will be a great while yet,' said I. 'There is a great deal of difference between a sugar factory and our contrivance.'

'I must confess,' replied the philosopher, 'that I have but very undefined notions concerning the way in which the liquid we extract from these canes is transformed into the hard, white, shining substance we call sugar.'

This acknowledgement coming from Ernest was equivalent to a formal demand to explain to him all I knew about sugar-making. I therefore began, 'Sugar,' said I, 'as you well know, is procured from the sugar-cane. The sugar-cane is easily propagated: all that is necessary is to bury the canes in furrows, and from each knot buds a shoot, which becomes the germ of a new plant. It requires nine months to arrive at maturity. It is then cut, the leaves are thrown away, and the stalks crushed under very heavy wooden rollers: the liquor which is pressed out is called sugar-honey. The first thing done with the sugar-honey is to boil it, which must be done immediately, for in twenty-four hours, it becomes sour, and soon changes into vinegar. It is boiled an entire day, throwing in water from time to time; the foam is skimmed off very carefully; and, to purge the sugar more, wood ashes and quick lime are thrown in, the skimming being continued. The liquid is then passed through straining-cloths. The dregs, in some places, serve to feed the hogs, in other places they are mixed with water, and, being left to ferment, make wine. The liquour is then boiled over, and the violence of the boiling is appeased by throwing on the top some oil. The smallest quantity of acid will prevent the sugar from crystallizing and assuming a solid form. The warm liquour is then emptied into earthen moulds, shaped like deep cones, and open at both ends; the little hole at the point is stopped by wood, linen, or straw. The whole art of refining consists in depriving the liquid of a honey-like sap which prevents it from assuming that whiteness, and the requisite solidity and brilliancy. This sap passes out through the little hole. In about forty days the sugar becomes hard and dry; it is of greyish-white colour, and from this the different sorts of sugar are made. When the same preparation has been repeated it forms "brown sugar", the best of which is nearly white, and exhales an odour of violet. The brown sugar, purified by the whites of eggs, or the blood of cattle, forms the refined, or royal sugar, so named on account of the pure, glittering colour of the grain. This sugar, when very dry, will produce a sound if struck with the fist, and if you strike it with a knife, in a dark place, it will emit phosphorescent sparks. The liquid which escapes from the moulds is about the consistency of honey, and it is most commonly known by the name of "molasses". Candied sugar is nothing but sugar melted and crystallized; it can be bought either white or red. A considerable commerce is carried on by the Hollanders in sugar of all sorts, especially with the East Indies, Brazil, Barbados, Antigua, St Domingo, Martinique, and Surinam. The sugar of Brazil is less white, of less consistence, and of a more oily nature than that of Jamaica or St Domingo.'

While we were thus tranquilly conversing, our young adventurers were pursuing their course toward the savannah. I will relate their adventures here, as they were recounted to us on the return of the party.

They had passed over the tract of land that separated Family Bridge from the country which we had called Waldegg, or the Hermitage, and where they intended to pass the day, when, on approaching the farmhouse, they heard cries like that of a person in distress. It was a sort of wild, maniacal laugh, and the animals stopped in terror; the dogs barked and howled fearfully; and the ostrich, more frightened than the others, fled in the direction of the Lake of Swans with such rapidity that all the efforts of its master could not check it. The bull and the onagra trembled so violently that Fritz and his brother were obliged to dismount.

'There is some ferocious animal here,' said the eldest to Francis; 'our coursers will run off if we do not hold them; and I think that the beast must be either a lion or a tiger. You go forward a few paces, while I hold the coursers, and if you perceive anything, return in haste to me, and we will concert a plan to capture it, or else mount and fly as fast as possible.'

Frank seized his gun, put two pistols in his belt, called Folb and Braun, and calmly walked on in the direction of the strange laugh. He had not gone more than thirty paces when he perceived, through the bushes, an enormous hyena,* who, after having killed one of our sheep, was devouring it, while ever and anon that strange laugh of joy would echo from its blood-stained lips. The presence of the little hunter did not disturb the monster in his horrid repast. While rolling his flaming eyes, he tore the poor sheep in pieces. But Francis wanted neither courage nor presence of mind: he placed himself behind a tree, and, taking good aim, he discharged both barrels of his gun, and was so fortunate as to break both the forelegs and pierce the breast of the hyena. The dogs then rushed on; their terror changed into rage. The most terrible combat now ensued between them and the furious monster; growls and cries resounded through the air, and the blood flowed in torrents.

Fritz, who had succeeded in attaching the onagra and the bull to a tree, now ran up at the sound of the double explosion and the noise of the dogs. They would have fired again and terminated the combat, but the dogs were so close to the hyena that they were afraid of hitting them, so that they were obliged to await the issue of the combat. Folb took the hyena by the throat, and Braun by the muzzle, and there they held him until he dropped down dead. My sons uttered a cry of joy, and, calling off the dogs, dressed the wounds they had received by rubbing them with hydromel and bear's grease, which they had brought with them to eat.

Jack soon returned. He could not stop the ostrich until it had arrived at the middle of the rice-field, and it required the greatest exertion to force it to turn back.

On seeing the monster that his brothers had so courageously attacked, Jack did not the less withhold his admiration because he had not been a participant in the act.

The hyena, with its yellow fur, striped with coarse, black hairs; its paws armed with sharp nails; its muzzle elongated like that of a wolf; and its small, round, red eyes, is one of the most ferocious, savage animals one could imagine. The hyena is about the size of a wild boar, but its body is shorter and more compact; its head is better set and more round; its ears are long and erect; and its legs, especially the hind ones, are very long; the eyes are placed like those of a dog; the hair is long, of a dark-grey colour, mixed with yellow, and with transverse stripes of black. It is, perhaps, the only species of quadruped that has not four claws on the hind feet. This savage and solitary animal dwells alone in the caverns of the mountains, and in holes which it digs for itself underground. Nothing can exceed the ferocity of its nature, and even if taken very young it can never be tamed. It lives like the wolf, but it is stronger and hardier, and it sometimes has attacked men. It will follow a herd a long distance. It often breaks open the doors of stables and kills the cattle; its eyes sparkle like diamonds, and it is asserted that it sees better in the night than by day. The hyena will defend itself against the lion, does not fear the panther, and will kill the ounce.* When driven by hunger, it will dig up the ground in graveyards, and eat the bodies it disinters. It is found in almost all parts of Asia and Africa. The capture of this animal was decidedly the most heroic action we had performed since the shipwreck.

When my sons had established their tent, etc., at Waldegg, they set off with the sled to bring the hyena thither. The following day was entirely devoted to skinning the animal and preparing the hide. While they were thus employed, we were calmly conversing under the vault of the grotto.

'I wonder where my brothers are,' said Ernest. 'I think we shall very soon have news from them.'

'What put that idea into your head?' said his mother.

'Oh, I dreamed it,' said Ernest.

'Bah! A great confidence your dreams will induce,' replied my wife.

While we were thus talking, a bird, whose genus we could not discover on account of the obscurity, fluttered in at the open door of the dove-cot.

'Shut it, shut it!' cried Ernest; 'Tomorrow morning we will inspect our new guest. Who knows, perhaps it is a courier from New Holland, and bears dispatches under its wing from Sydney, Port Jackson,* and so forth.'

'Why, how is it your thoughts run on dispatches and news this evening, Ernest?'

'Ah, it is nothing,' answered he, with indifference; 'only the arrival of that pigeon recalled to my mind something I have been reading today, concerning the correspondence the ancient Greeks and Romans carried on by means of carrier pigeons. Is not the story true?'

'Certainly it is,' answered I. 'Of all the inhabitants of the air, none can fly to such an immense distance. Pigeons are sometimes trained to be carriers; but there is a particular kind called carrier pigeons, and I believe I have a book in the library that gives a very clear description of the wandering pigeon.'

I rose, and taking the book from the library, read as follows: 'Ornithologists have given to this sort of pigeon the name of *"Columba migratoria"*, that is to say, traveller pigeon,* and its habits fully justify its name. Sometimes visiting the Gulf of Mexico, other times fixed at Hudson's Bay, it will pass over in its excursion more than 700 leagues easterly; but it never flies farther westward than the chain of Rocky Mountains. Some individuals, more enterprising than the rest, have, however, crossed the ocean, and sometimes reached Scotland. Their power of flight and the extent of their vision is astonishing; and although they soar very high, they can perceive the fruits that they feed on, such as juniper berries, wild fruits, etc., and they will fly down to obtain them. They fly in such dense, numberless bodies, as sometimes to intercept the light of the sun, and it has been calculated that they fly twenty-five post leagues per hour. If human industry could train these rapid coursers, telegraphs* would be useless, and a message could be conveyed from Zurich to Berlin in a very few hours. The structure and form of their bodies is admirably adapted to the long voyages they undertake. Their wings are proportionately longer than those of any other sort; their long, flat tail is a rudder which serves to assist the force of their wings. There is a very great difference in the colour of the plumage of the two sexes; the modest exterior of the female beautifully contrasts with the dazzling plumage of the male, who is not only handsomer, but larger than the female; from the extremity of his beak to his tail, his length is about two feet. The head is of a bluish slate-colour, the wings and back the same, only dotted over with black and brown spots, the breast is a reddish hazel, the neck is adorned with the most beautiful colours—green, purple, gold-colour, and scarlet glitter in all their beauty; the belly is pure white, and the legs and feet are of a beautiful red, and a large band of shining black passes across the tail. The distinctive and predominant character of this kind of pigeon is the love of society; there are no isolated individuals in their excursions: all fly in one solid mass. A celebrated naturalist estimated the number of a troop he encountered on the banks of the Ohio at hundreds of millions, and his calculation was far from being exaggerated. This cloud of birds was three hours in passing over his head, and its length was 65 leagues; counting two birds to the cubic foot, this band was composed of 1,200,000,000 of birds, and they flew so closely that they quite obscured the ground beneath them. These immense columns are formed by the reunion of a great number of distinct parties, but all having a common purpose, executing the same manœuvres in the same space; they have also the singular habit of all choosing the same roosting-place, where they assemble at evening, and disperse the next morning in search of food. The weight of the birds breaks down the branches of the forest, and gives it the appearance of having been visited by a severe storm. The nourishment consumed each day by this enormous quantity of pigeons has been calculated, allowing a moderate ration to each individual, although they eat often and much, and it has been found they consume more than the most populous of the European capitals. At the break of day they disperse to lay under contribution, a space equal to several of the Swiss cantons. Some divisions of the great band fly very far, but always return punctually to the breeding-place. This spot is always selected as retired as possible, and concealed from the natural enemies of the birds; but all precautions are insufficient against man, the most dangerous of their foes. As soon as a breeding-place is discovered, the whole country for miles around make preparations to visit it. The carts are hastily laden with a few household utensils, a quantity of salt, empty barrels, etc., and all the family set off for the pigeon-roost. When all have met, a sort of police guard is established to keep the peace, and then the work begins. At evening the firing commences, and continues as long as they can see the game. After the birds have dispersed in the morning, they commence gathering up the game, and, during the whole day, are laboriously employed in picking, cleaning, and packing away the immense number of pigeons. During the harvest, the pigs are driven in, and they almost visibly fatten, so nutritious are the pigeons. A brig has been seen at New York entirely laden down with this singular merchandise, and the cargo had a rapid sale. The life of these unhappy pigeons is but a continued series of dangers and perils. Those that survive the carnage commence building their nests, and these cover an immense space. In the state of Kentucky a breeding-place has been seen 1 league broad and 16 leagues long. The nests are occupied about the commencement of April, and toward the end of May the young birds learn to fly, and the whole band commences its grand voyage. The female sits thrice a year, and often has three different nests to construct; for about eight or ten days before the time of their departure the hunters arrive at the forest, armed with hatchets and everything necessary for a long stay; the trees, laden with nests, are cut down, and the cries of the victims, the noise of the wings of the parent birds as they flutter around their unhappy offspring, mingled with the repeated blows of the hatchet, and the crashing of the trees, make a deafening clamour. The young pigeons are extremely fat, and they are killed for the purpose of obtaining the fat. One large tree laden with nests and young birds will furnish a family with grease enough for several months.

'The traveller pigeon of America can only preserve its manners in the immense forests west of the Alleghany mountains; the bands which venture east of them find their passage obstructed by numerous enemies. When hunger compels them to seek the cultivated plains, they are taken by hundreds in nets; everybody that owns a gun uses it against the unhappy birds, and they are the accompaniments of every repast; but the time is approaching when pigeon-hunting will be less productive; as the population increases in the interior of the continent, the pigeons, finding their movements so contracted, will lose their social habits, the persecuted race will gradually diminish, and it will insensibly change its manners, and become disseminated and confounded with the other kinds of the same genus.'

I stopped here; Ernest made several remarks upon the instinct of the traveller pigeon, but he was reserved and guarded in his words, and I asked him what was the matter.

'Tomorrow you will see—tomorrow', was the only answer he made; and soon we were all buried in sleep.

58
The Pigeon Letter-Post; the Black Swans; the Bird of Paradise

THE next morning, Ernest rose before me, and paid a visit to the dove-cot; I said nothing; and, after breakfast, I saw him coming in, holding in his hand a piece of paper, folded and sealed like a government letter, which he presented to me on bended knee, saying, as he did so, 'Noble and gracious lord of these lands, I beg you to excuse the postmaster of Felsenheim for the delay that the dispatches from Sydney and New Holland have experienced; the packet was retarded, and did not arrive till very late last evening.'

His mother and I burst into a laugh at this ridiculous speech.

'Well,' replied I, continuing the jest, 'what are our subjects in Sydney and New Holland engaged in? Will the secretary open and read the dispatches?'

At these words, Ernest broke the seal of the paper, and, elevating his voice, commenced,

'*The Governor-General of New Holland, to the Governor of Felsenheim, Falcon's Nest, Waldegg, the field of the Sugar-canes, and the surrounding country,*

'GREETING,

'Noble and faithful ally! We learn with displeasure that three men, whom we suppose to be part of your colony, are making inroads into our savannahs, and doing much damage to the animals of the province; we have also learned that frightful hyenas, have broken through the limits of our quarter, and killed many of the domestic animals of our colonists. We therefore beg you, on one part, to call back your starving huntsmen, on the other, to provide measures to purge the country of the hyenas and other ferocious beasts that infest it. Especially I pray God, my Lord Governor, that he will keep you under his holy protection.

'Done under our hand and seal at Sydney Cove, Port Jackson, the twelfth day of the eighth month of the thirty-fourth year of the colony.

'PHILIP PHILLIPSON, Governor.'

Ernest stopped in laughter at the effect the letter produced on us. I felt that there was some mystery, and I was anxious to get at the bottom of it. Ernest enjoyed my evident embarrassment, and, jumping up and down as children do, he let fall a new paper from his pocket. I caught it up, and was going to read it, when he laid his hand on my arm, saying, 'Those also are dispatches; they came from Waldegg, and, although less pompous than General Phillipson's, perhaps they are more truthful. Listen, then, to a letter from Waldegg——'

'Oh do explain to us', said I, 'this prolonged enigma: did your brothers leave a letter before they went? Is the news of the hyena true? Did they act so rashly as to attack the animal?'

'Here is a letter from Fritz,' replied Ernest; 'my pigeon brought it to me last night.'

'Thanks, many thanks, my dear child, for your idea,' said his mother; 'but this hyena! oh, quick, read me the letter.'

He opened the paper and read the following words:

'DEAR PARENTS, and you, my good ERNEST—I will inform you of our arrival at Waldegg; we there found a hyena, who had devoured several of our sheep. Francis alone has all the honour of having killed the monster, and he deserves much praise for his intrepidity: we have passed the whole day in preparing the skin, which is very fine, and will be very useful. The pemmican is the most detestable stuff I ever tasted. Adieu, we embrace you tenderly in spirit.

'FRITZ.'

'A true hunter's letter,' cried I. 'But this hyena, how could it have found its way into our domains? Has the palisade been overturned?'

'My poor children!' said my wife, with tears in her eyes; 'may God watch over them, and return them safely to my arms.'

'We shall, probably, receive another letter this evening,' said Ernest, 'and that will give us further details of the expedition.'

After dinner a new pigeon was seen to enter the dove-cot. Ernest, who had not remained quiet one moment during the day, immediately shut the door of the dove-cot, removed from the wing of the aerial messenger the dispatch he had brought, and delivered it to us; it read as follows:

'The night has been fine—the weather beautiful—excursion in kayak on lake—capture of some black swans—several new animals—apparition and sudden flight of an aquatic beast, entirely unknown to us—tomorrow at Prospect Hill.

'Be of good cheer;

'Your sons,

'FRITZ, JACK, AND FRANCIS.'

'It is almost a telegraphic dispatch,' said I, laughing; 'it could not be more concise. Our huntsmen would rather fire a gun than write a sentence; nevertheless, their letter tranquillizes me. I only hope that the hyena which they killed is the only one in the country.'

My wife was more contented, and we retired to rest, praying that God would continue to watch over and protect our absent ones. We received other letters from them at intervals; but they were so concise that I will continue here the narration the boys made on their return.

Delivered from the terrible neighbourhood of the hyena, they had undertaken to explore the marsh around the Lake of Swans. Fritz embarked in the kayak, and his brothers followed, as near as possible to him, along the shore. The black swans* afforded a fine chase to our young huntsmen. A loop of wire fastened to a long bamboo was the means they employed; but they captured only three young swans, the old ones being too strong, and defending themselves with their powerful wings.

After the swans came a bird of a new kind, who, by his majestic walk and noble appearance, seemed the king of birds. The boys threw the wire loop over his head, and, drawing him to the shore, fastened his feet and wings, and laid him along-side the swans.

While they were occupied in examining their magnificent prey, which Ernest afterward pronounced to be the 'heron royal',* an extraordinary animal rushed out from the reeds, and, passing close to their sides, struck them with terror. It was an animal about the size of a young foal, of a form like that of the rhinoceros, only it had not the horn on the nose which that animal has; the upper lip was very prominent, and the whole body of a very dark brown colour. My three huntsmen were not very distinguished naturalists, and the best name they could find for the beast was the tapir* or anta of South America.

The tapir is an animal which is commonly found in Guiana and Brazil. The form of its body resembles that of the pig, and its upper lip is much longer than the lower; its throat is armed with four teeth, its eyes are small, its ears round and drooping; its tail is short, pyramidically shaped and destitute of hair. It has legs and feet like the wild boar, and has four nails on the hind, and three on the fore, feet. The hair of the tapir is short, and, when the animal is young, spotted with white; but as it grows older it becomes a dark, uniform brown. The tapir is a most excellent swimmer; it will dive down and pass a great distance under the water, thus defying all attempts of the hunters to take it. Naturalists say that the tapir sleeps all day under water, and at night roams through the forest in search of food. The Portuguese were the first that gave it the name of 'anta'. The savages think the flesh of this animal equal to beef; they also cover their shields with the skin, which is very hard and durable.

Fritz was unacquainted with its characteristics when the animal appeared. He began, however, to pursue it in his kayak; but the tapir swam away so rapidly that he was soon obliged to desist. During this time Jack and Francis had set out for the hut, carrying with them the black swans and the beautiful heron royal. In their way thither they encountered a flock of cranes, which hovered around their heads, uttering piercing cries. A great number were soon brought down, not by fire-arms, but by the bows which the boys carried. These were provided with long, triangular-pointed arrows. They captured in this way two beautiful birds called 'Numidian Girls',* who were in company with the cranes.

Fritz, on rejoining his brothers, felt a little piqued when he perceived, from their trophies, the fine chase that they had made; and, on the other hand, his unsuccessful pursuit of the tapir made him a little ashamed. He wished to retrieve his honour and repair the damage which his reputation as a good hunter had sustained; so, calling the dogs to him, and accom-panied by his eagle, he directed his course toward the wood of guavas. He had not been there more than a quarter of an hour before his dogs started a flock of the most beautiful birds Fritz had ever seen. He cast off his eagle, and, while it was pursuing one, a second fell down from fright into his hands. He also captured a third, which had become tangled in a shrub; this last was a magnificent one: its tail was more than two feet in length, and two of the feathers were longer than the others and glowed with the most beautiful shades of gold, green, and brown, terminated at the end by a spot of black, exactly like velvet. Ernest recognized it afterward as being the Bird of Paradise,* the *Manu cordiata*, which is the most elegant in form and plumage of all the birds of New Holland. This bird has always been associated with fable; even its very name has but lately been ascertained. People have imagined that there was no spot on earth worthy to receive this beautiful bird, and that it would repose in no other place but Eden. Some persons have even asserted that it had no feet at all, and never nested, but slept on the wing; and, what was most ridiculous, that the female carried her eggs into the air and hatched them while aloft, never resting but for a few moments, when she would suspend herself by the two long feathers of the tail, to some tree. The nourishment of the Bird of Paradise agreed with its constitution; therefore it existed on perfumes and vapours, and drank nothing but dew-drops. Marvellous qualities were also attached to the bird, and the man who was so fortunate as to ensnare one was thought to enjoy the peculiar favour of Heaven, and to possess a safeguard against all maladies; consequently every means was employed to capture them, and the chase of Birds of Paradise became very profitable. But science came, and with its magic wand dispersed the clouds of superstition which hung round the Bird of Paradise. Natural history solved the mystery. It was discovered that the Bird of Paradise had feet, fed on solids, and that its plumage alone made it more valuable than other kinds. The Bird of Paradise flies very high, and roosts in the tops of the highest trees. Its real size is about that of the jay; the abundance of its feathers, however, makes it appear much larger. The feathers that surround the base of the beak are of a beautiful velvety black, which appear to change into deep green, which colour extends over the head and throat, falling over on the yellow and metallic green feathers of the neck; the rest of the plumage is of a deep maroon on the belly, and a light maroon on the back. The two long tail-feathers of the male measure two feet nine inches, but those of the female are shorter. In this species of bird, as in all others, the plumage of the male is much more splendid than that of the female.

Our huntsmen, after all their exertions, had acquired a most ferocious appetite; and although their repast was frugal, yet they did it ample justice. The cold meat of the peccary, guavas, cinnamon-apples, and potatoes cooked in the ashes, were all devoured with thankfulness. The pemmican alone was disdained, and declared unworthy of its reputation.

Before night, our young adventurers filled a sack with ripe rice; they also gathered a quantity of cotton, which they intended to carry the next day to Prospect Hill, the end of their proposed excursion. Fritz, who had brought with him some of the euphorbia to poison the monkeys, needed some coconuts, divided in two, to hold the baits for the mischievous animals. My sons did not like the trouble of climbing a tree, so they chose out one that was loaded with fruit, and cutting it down,

as the Carribees do, they obtained a supply of coconuts and
two enormous palm cabbages.

When they told me of this, I reproved them for cutting
down the tree, and forbade them ever doing the like again.
The palm is one of the most beautiful trees in the country, and
unites in itself the most precious vegetable riches; and I tried
to impress it on their minds that it was wrong to cut down
anything that could be rendered useful in its natural position.

My sons quitted Waldegg and directed their course toward
Prospect Hill; but I will let Fritz tell the story himself.

'On entering the wood of pines,' said he, 'we were greeted
by a concert of sharp cries which proceeded from every tree;
they came from the monkeys, who sat among the branches,
grinning most horribly at us. From grimaces they passed to
blows, and we soon felt a shower of pine cones,* which might
have hurt us severely if we had not discharged our pistols, and
frightened away the whole tribe. This reception did not
increase my benevolent feelings toward the monkeys, and
confirmed me in my project to chastise them. We found, on
the borders of the wood a sort of millet, about ten feet in
height which I easily recognized as being the doura,* or black
millet. This field was very large; but in different places the
stalks were broken down, as if a hail-storm had passed over it.
We hastened to reach Prospect Hill, and on approaching it
we easily perceived that the monkeys had been there. Our
plantations had been ravaged, and our little farmhouse
devastated and infected by the ordure the villainous animals
had left behind them. We cleaned out the interior with a
broom of millet. I can not express our rage and disappointment. We passed the afternoon in endeavouring to construct
a place to protect us against any attack from the horrid animals
during the night. I must here, my dear parents,' continued
Fritz, 'beg pardon for a fault of which I must plead guilty. I
took the gum of the euphorbia without asking your permission,
for I was afraid that you would not trust me with it. We commenced the preparations for our work before night. The
coconuts and calabashes which we had brought were all put
into requisition; we filled them with rice, guavas, palm-wine,
and all sorts of enticing ingredients. To each one of these
articles I added a portion of the gum of the euphorbia, and

then scattered those snares all over the forest, and awaited the issue of the morrow. We were about to retire to our matresses of cotton, when suddenly a brilliant light lighted up the horizon, and we thought at first that it was a ship on fire. We immediately quitted the hut and ran as fast as possible to the summit of Cape Disappointment. It appeared to be a mass of fire, perfectly round, which arose from the bosom of the waves and elevated itself above their surface. It was the moon. I do not think I ever saw a more marvellous sight. The sea was calm, the waves were dashing gently against the foot of the cape; the evening breeze scarcely ruffled the smooth surface, on which the beautiful moon shone down, and in which she reflected all her splendour. We remained some time gazing on this scene, in religious silence, our souls lifted up toward the Lord, and we thanked him instinctively for the mighty exhibitions of his power which he unceasingly presents to the admiration of man. But the calm contemplation in which we became absorbed was suddenly interrupted by a discord of the most diabolical sounds I ever heard. It was a confused mixture of bellowings, groans, and piercing screams. In answer to these formidable accents, our dogs responded by long-drawn barks, and the jackal yelled in sympathy with the cries of his fellows, that grated on our ears in the direction of the savannah. We heard the piercing cry of a wild horse, mingled with a deep, sonorous growl, which I felt sure proceeded from a lion or a tiger. This singular amalgamation of sounds lasted about a quarter of an hour. We were about to descend when we heard the galloping of a horse in the distance, and we regained our hut, full of apprehension that there was an elephant or a hippopotamus in the neighbourhood. We found everything tranquil about the hut; but we had scarcely laid our heads on our pillows when another outburst commenced in the pine wood. It was at first a solo; but voices gradually joined in, and the noise was so shrill and piercing that it nearly split our ears; and our dogs, whom we had fastened to the stakes of the hut, kept up a most deafening clamour in response. It was impossible for us to sleep a wink: the howling and screaming continued all night, and I felt heartily glad when morning arrived. We arose eager to ascertain the results of the night. Alas! we found all our musicians asleep on the earth; but it was an eternal sleep.

There lay the monkeys, who had partaken of the fatal meal we had prepared for them. The earth was strewn with their dead bodies, evincing that the euphorbia had produced a most terrible effect. We threw the poisoned monkeys and the utensils we had used into the ocean, and felt rejoiced that our disgusting and hideous work was accomplished. It was then that Jack composed the famous letter which you were so unfortunate as not to receive. Here it is, a splendid specimen of the sublime and beautiful:

" PROSPECT HILL,
" the 11th, 12th, and 13th inst.

"The caravanserai* of Prospect Hill has been cleaned and again rendered habitable; pain and trouble did the labour cost us; but the guilty ones have paid for it with their blood. Nemesis* has poured poison into the cup of vengeance, and Ocean rolls its waves over the dead bodies of the traitors. The sun smiled benignly on our departure this morning, and he will bid us adieu this evening at the defile of the savannah.

"Valete, valete." '*

I will now resume the narrative myself, and explain the effect of Jack's letter, which we really had received. The passage about Ocean and the dead bodies puzzled me, as I did not then know what I have just now written, and which was afterward told me by Fritz. Another dispatch arrived which added to my anxiety. It contained the following words:

'The palisade of the defile which leads to the savannah is destroyed; the sugar-canes have been all trampled down, and we have discovered large footprints, like those of the elephant, in the sand. There are also the prints of the hoofs of wild horses. Come quickly to our aid, dear parents; there is much to do for the safety of the colony. Lose not an instant, we beg of you.'

I leave it to the reader to imagine the inquietude into which this letter threw me. I saddled the onagra without losing a moment, and, leaving Ernest and his mother to follow me on the next day, I set off for the defile. There was a distance of six leagues between my sons and me; but I accomplished it in three hours.

My children were surprised to see me arrive so promptly, and they received me with transports of joy. The idea I had entertained of the devastation was but faint in comparison with the reality. The sugar-canes were irretrievably lost: they had been trampled down, and the leaves torn off, by some animal that I was sure must have been an elephant. All our trouble in erecting the palisade had been wasted; the stakes had been all torn up, the trees nearby deprived of their bark, the bamboos had been treated no better than the sugar-canes, and every young shrub I had planted had been torn up. I examined attentively the footprints in the sand, and was convinced that the larger ones were those of the elephant, and the smaller ones those of a hippopotamus; but I could discover no traces of the hyena. I surrounded our tent with dry branches, and amassed an abundant provision of combustibles, so that we might keep off any beasts by fires at night. We each took our turns in watching and replenishing them: but nothing troubled our repose. We were all seated quietly round the fire, talking of the ravages that had been committed, when my sons asked me to give them some information concerning the elephant. I willingly complied.

'The elephant', said I, 'is the most singular of all quadrupeds. In considering his form, with regard to our ideas of proportion it is very ungraceful; his body is large and thick, his legs clumsy and ill formed, his feet round and crooked, his monstrous head is covered with a thick skin, and the skull in front is seven inches in thickness; his ears fall over his cheeks like withered leaves; his trunk, his feet, his tusks, are as ungraceful to the eye as they are necessary to the animal. The warm climates of Asia and Africa are the special habitations of the elephant: those of India are much larger, and consequently stronger than those of Africa. When the elephant is deprived of his skin and flesh, the hind legs appear shorter than the front, because they are less disengaged from the mass of the body; these legs resemble more those of a man than those of a quadruped; the under surface of the foot is furnished with a horny bone, about an inch thick, and resembling the sole of a shoe. The strength of the elephant's legs is proportioned to the heavy weight of his body; but it is said he will go very fast, and easily keep up with a man running: he also swims very well. Some authors have said that the want of suppleness in the limbs of the elephant prevents him from rising when he lies down; but this is an empty fable.

'The most peculiar organ of the elephant is his trunk, in which we find movements and uses not in any other individual animal. The trunk is very long, and the animal lengthens and contracts it at pleasure. This organ, which is properly his nose, is furnished with veins, nerves, etc., and is hollow, like a tube. The extremity of the trunk is enlarged like the mouth of a vase, and it has a membrane like a finger at the end. By means of the trunk the elephant executes as much as we can do with our hands. The neck of the elephant is too short to allow of its putting its head to the earth to crop the grass or appease its thirst; it dips the end of its trunk into the water, and, sucking it full, turns it under, and empties it into its throat. The young elephant sucks with its trunk and empties it into its mouth. The elephant is able to tear down trees and break branches with this powerful organ, and it can squirt water from it to an immense distance. The mouth of the elephant is situated in the very lowest part of the head, and is furnished with eight teeth, four in the upper and four in the lower jaw. The elephant has also two very long tusks proceeding from the upper jaw, and which serve the animal as defence; these tusks furnish ivory. The elephant has very small eyes, and it has eyelashes, which distinction is confined to it, man, monkeys, the ostrich, and the great vulture. The body of the elephant is covered with a wrinkled skin, very dirty and disgusting in appearance. The wild elephants live on grass, herbs, branches of trees, etc., and in the month of August they make incursions into the rice-fields, and do much damage, unless they are kept off by continual fires. These enormous eaters can exist seven or eight days without food; their drink is water, which they always render muddy before tasting. The wild elephants sometimes enter the tobacco-fields, which they ravage. If the plant is young it does them no hurt; but when it is old and strong, it puts them to sleep, and then the negroes fully avenge themselves for the damage the monstrous beasts have done. The elephant has much instinct and docility. It is susceptible of attachment, affection, and grief. It is very easily tamed, and one is surprised to see so powerful a beast so docile.'

Ernest and his mother arrived after dinner, bringing with them the wagon, the cow, the ass, and all necessary utensils for our encampment, which was likely to last a good while.

We immediately began the construction of a solid fortification across the defile, one that would effectually keep out all intruders. I will spare my readers the details of this tiresome work, which occupied us constantly for more than a month. My good Elizabeth shared in our toils and inspired her sons with ardour and perseverance. Sometimes we would relax from our labours; Fritz would make excursions in his kayak, and the other boys would wander off, and always bring us home something useful.

59
The Redoubt; Different Discoveries; Jack's Fright; the Fort on Shark's Island

OUR next labour was to construct some sort of a fort to shelter us whenever we should visit the defile. We had not strength enough to build a regular fort, and, besides, our knowledge of fortification was very limited. At last Fritz thought of a plan of a Kamschatdale* fort, which he had read of somewhere, and which I thought, with a little improvement, would answer admirably.

The Kamschatdale fort simply consists of four high stones, upon which are laid planks and boards, forming a platform upon which a hut of bark or branches is constructed—not a very formidable fortress certainly, but yet capable of defending us in case of an attack from wild beasts.

Instead of four stones for the foundation, we chose out four trees to answer the same purpose. We did not cut the branches off close but left them as rests for the beams of our platform; these trees resembled the plane tree of Europe, and were adorned with the vines of the vanilla bean, which were climbing up the trunk. We surrounded our platform with a high and strong network of rushes and branches, leaving an opening for entrance, and we covered the roof with the waterproof leaves of the Talipot palm.* These leaves grow so large that ten men can be covered by one of them. We were glad to have discovered a tree of this palm, as it afforded us many facilities. Our fort bore a strong resemblance to Falcon's Nest, and, surrounded as it was by green trees and flourishing verdure, it did not look much like a military construction.

To ascend to the platform, we employed one of the simplest means I could imagine: it was by a beam which descended perpendicularly to the ground, and notched deeply into steps. We also arranged this so that it could be raised and lowered at pleasure.

Fritz and Jack promised themselves wonders from our new fort, which overlooked the savannah for a great distance, and we could see the river running like a silver thread through the immense plain; and, by means of spy-glasses, we could discern troops of buffaloes and other animals feeding around the brink.

Our labours at the fort were diversified by some important discoveries. One day Fritz made an excursion to the river of the savannah, and found among the rich vegetation there some unknown shrubs, of which he brought me specimens to examine. One kind bore, in large clusters, a beautiful green fruit, tipped at the end with violet, and shaped like a large gherkin;* the others were covered with quantities of small flowers, interspersed with large fruits like cucumbers. On examination, I recognized them as being two of the most precious productions of the tropics; the largest of the fruits was the cocoa bean, of which chocolate is made; the other was the banana, that forms an article of food for the inhabitants of several countries in America.

We tasted these far-famed fruits, but did not find them very excellent. The beans of the cocoa, are filled with a sort of viscous matter, like thick cream, but of an insipid taste and an odour like that of an over-ripe pear.

'How singular it is,' said I, laughing, 'that this fruit (I spoke of the banana), reputed to be so delicious, should fall so far behind our expectations; probably it wants some preparation before we pass our judgement on it.'

But the chocolate that could be made from the cocoa-bean was much more interesting than the banana to my sons.

'Cocoa! chocolate!' cried they all, eager for a taste. 'Papa, we *must* make some chocolate.'

'Very well, gentlemen,' said I, with less enthusiasm; 'but before you make it, you ought to know something about it, and be able to tell how this bitter cocoa-bean is transformed into delicious chocolate. Come, now, let me see which one of you can give any information concerning the origin and preparation of this luxury.'

These words were followed by a moment's silence on the part of my sons, and then the doctor commenced:

'The cocoa-tree,' said he, 'is one that varies in height according to the soil in which it grows. The wood of this tree is porous and very light; its leaves are about 9 inches in length, and 4 in breadth; the leaves succeed each other so rapidly that the tree is never destitute of them, and it is also always covered with an immense quantity of flowers resembling very small roses. The fruit, when it has attained perfection, is about the size and form of a cucumber, and furrowed, like a melon, on the sides. The fruit grows along the stalk and on the dry branches. The cocoa-bean is an object of considerable commercial importance: in the new continent, and on the coast of Caracas, it is extensively cultivated. The trees are planted 12 or 15 feet apart, so that they will grow much better; they thrive best in low, wet soil, and are propagated from cuttings. When the cocoa is considered to be ripe the gathering of it is entrusted to the most skilful negroes, who, with long, thin poles, knock down the ripest pods, taking good care not to disengage those that are not matured, or are in flower. In a good season a gathering is made every fifteen days, otherwise once a month. The pods are not allowed to remain in heaps more than four days; if they remain longer they will germinate: they are, therefore, never gathered until just before they are wanted for use or shipment. The kernels of the cocoa are deprived of their bark by the aid of fire; they are then moderately roasted, pounded up in a heated mortar, and formed into cakes by mixing with them an equal weight of sugar.'

'Oh, what a thing science is!' interrupted the blockhead Jack. 'I have drank chocolate often, and never once thought of asking about its origin. As long as I could get a cupful, I did not care where it came from, and I humble myself before the science of professor Ernest, and move, that he drinks the first cup of chocolate from our factory at Felsenheim.'

'Adopted!' was the general cry; and I saw the doctor's fine eyes radiant with triumph.

The banana was the next object of our discussion.

'Is it not strange,' said I, 'that we should so little relish a fruit that a celebrated writer* has declared to be one of the greatest blessings given to us by Providence. '"The banana",' says he, "is all that was necessary for the first man. It produces the most healthful food in its farinaceous, succulent, sugary, aromatic fruit, just the diameter of the mouth and grouped like the fingers of the hand. The tree resembles a magnificent parasol, the top formed by an assemblage of long, large, shining green leaves. These leaves droop at the extremities, and thus form a charming cabin, impenetrable to the sun or rain. The leaves are very flexible, and the Indians cover their huts with them, make thread from them, and use them as shrouds for corpses; so the banana tree is able to nourish, lodge, clothe, and bury a man. There are many sorts of bananas and of different heights, from that of an infant to double the size of a man. There are in the Isle of France* dwarf bananas, and also gigantic ones, originally from Madagascar, the long and curved fruit of which is called *ox-horns*. A man can easily climb these trees, the stalks of the old leaves serving as rests for the feet; one of these bananas is enough for a meal. The taste of bananas is enough for a meal. The taste of bananas is different in different kinds; the dwarf kind has a flavour of saffron; the common sort called *fig-banana*, has a rich, sweet, mealy taste, and has the consistency of butter in winter, so that there is no need of teeth to bite it. The banana has also other peculiarities: although it has no skin, yet it is never attacked by insects or birds, and it will ripen if preserved in the house, and will retain its flavour for more than a month. Bananas are found throughout the torrid zone, in Africa, Asia, the two Americas and the islands of their seas. Travellers have justly called the banana *king of vegetables*, for there are numberless families between the tropics who live entirely upon it. It is under its delicious shade, and by partaking of its invigorating fruit, that the Indian Brahmin* is able to exist so long. A banana tree on the bank of a stream suffices for all his wants."'*

During this discussion, my wife, who had opened several bananas, sought in vain for some seeds which she might take home to plant in her kitchen-garden. I told her that the banana contained no seeds, and was always propagated from cuttings, which will easily grow, if planted in rich, wet earth. My wife also wished to plant some of the cocoa-beans in her garden; but she was obliged to renounce her project, as Ernest told her that unless the seeds were put in when the fruit was gathered they would be useless. It was resolved, in consequence, that Fritz should set out the next day in his kayak, and go in search of the elements necessary for the reproduction of these two precious plants. My wife never forgot her kitchen-garden, and whenever she came across a useful addition, immediately planted a specimen there.

The next day Fritz embarked; and, fearing that his kayak would not be large enough to hold the cargo he intended to bring home, he fastened a raft of rushes behind it. He was ashamed, he said, to go for some banana cuttings only, and he intended to bring home something else. We occupied ourselves during the day in preparing to set out for Felsenheim, and Fritz did not return till late in the evening, when we saw him coming toward us, the kayak and the raft loaded down to the water's edge.

'Bravo! bravo!' cried his brothers, as they saw Fritz advancing, laden with green branches. The cargo was soon unloaded and dragged up to the hut with as much contentment as if it had been the galleons of silver that Admiral Anson* captured. Fritz also gave into Jack's hands a wet sack that seemed to contain something alive; the fellow took it, and, looking in, exclaimed, 'Good! Fritz has executed my commission', and then ran off and hid his treasure in the bushes.

Fritz now came up, holding in his hand a superb bird, the feet and wings of which he had fastened, and which he presented to us as the principal booty of the day.

It was the Sultan Cock* of Buffon, the king of water-fowls, so called from its beauty of form, and the brilliancy of its plumage. I easily recognized its long red legs, and its beautiful green and violet plumage, with a red spot in the forehead. My wife wished to add it to the inhabitants of the farmyard, and as it was very gentle it soon became as tame as the rest of our domestic fowls, who appeared jealous of the newcomer.

Fritz now recounted to us the details of the day. He informed us that he had ascended the river for a great distance, and that he had been astonished at the majestic forests which bordered it, and threw a sombre shade over its waters. He had encountered several families of turkeys, pintados,* and pea-cocks, whose cries and screams imparted an air of life to the sombre river. Farther on, the scene had changed: there were enormous elephants feeding along the bank, in troops of twenty or thirty; some were playing in the water, and squirting the cooling fluid over the heated bodies of their companions; tigers and panthers, too, lay sleeping in the sun, their magnificent fur contrasting strangely with the green bank upon which they reclined; but not one of these animals paid the least attention to the young navigator.

'I felt my inability and weakness', said Fritz, 'on finding myself face to face with these terrible enemies; my gun, my balls, and my skill would have been of little use, and I thought I had better retrace my steps. I commenced to turn my kayak round, when what was my surprise to see, at about the distance of two gunshots before me, a long and large mouth, armed with rows of formidable teeth, and the whole apparatus moving directly toward me. I cannot say how I found strength enough to escape, I felt so frightened at the apparition. I took a lesson then in natural history that I have no desire soon to repeat.'

'What animal was it,' asked Francis, 'the mouth and teeth of which Fritz saw coming out of the water?'

'An alligator, probably,' said Ernest; 'or, if you would prefer using a name more familiar to you, a crocodile.'

'A crocodile! What, the same animal that the Egyptians adored as a god?'

'The very same,' replied the doctor, enchanted at the opportunity of displaying his science. 'The crocodile is the largest and strongest of the family of lizards to which it belongs, and it is thought that this is the animal alluded to in the Holy Scripture by the name of leviathan. The crocodile, which at the Antilles* is also called cayman,* is a monster of the greatest voracity; it is produced from a very small egg, but it attains a length of more than 20 feet: it is covered with a hard, scaly skin, of a bronze colour mingled with green and dirty white spots; its mouth opens as far back as the ears, and the jaws are furnished with a great number of canine teeth, long and round, white and pointed, and which fit exactly in one another; its eyes resemble those of the pig, and it makes a noise very similar to the grunt of that animal; its paws are armed with

sharp claws; its tail is rounding, and as long as all the rest of
its body. Crocodiles are found in the Ganges, the Nile, the
Niger, in Asia, in Africa, and in several of the great rivers of
America. In Egypt they are very plentiful: they dwell in the
rivers, and remain immovable in the mud until their hunger is
aroused, and then they are extremely watchful. They feed on
fish, and are excessively fond of human flesh. Crocodiles are
taken with iron fish-hooks, their scaly skin being ball-and
arrow-proof. Crocodiles have been found 33 feet in length.'

It was evident, from the narrative of Fritz, that the environs
of the defile were peopled by wild and savage animals, and that
particular precaution should be taken to secure our domain
against any invasion on their part.

We finished our preparations for departure, and set off at
break of day the next morning for Felsenheim. Fritz asked my
permission to allow him to make the journey by water, in his
kayak, and to return home by doubling Cape Disappointment.
I readily consented, as the ease with which he managed his
little boat gave me nothing to fear, and, besides, I was anxious
to know more about the cape.

We both set out at the same time, and both arrived home
safely. The sailor, in doubling the cape, had made two new
discoveries: among the bushes which covered the rock he
remarked two shrubs, one of which was covered with very
highly scented, rose-coloured flowers, and had long, narrow
leaves; the other had numerous small white flowers, and in its
whole appearance very much resembled the myrtle. He
brought home to us specimens of these two shrubs, one of
which my wife recognized as the caper tree, used in pickling;
the other was a sort of Chinese tea-plant, which was received
with marked distinction.

The hope, although feeble, which we entertained, that one
day some ship would visit our coast, was never forgotten, and
we collected all that the country afforded as precious or useful,
so that in case a vessel should arrive we could exchange with
them, or, if an occasion offered to quit the island, we would
have something wherewith to pay our passage. For that purpose
we gathered, every year, a quantity of cotton, dried different
sorts of fruits, preserved others in sugar, such as cinnamon,
ginger, vanilla etc. One can suppose, then, that the discovery

of tea was of the highest importance; and, as I examined the shrub, I related to my sons all I knew concerning the tea-plant. 'This shrub, which grows in China and Japan, is cultivated with the greatest care, especially that destined for the consumption of the imperial family; the fields in which it grows are divided into compartments, like a vast garden, intersected by canals of running water, and straight walks which are carefully swept every day. Those who gather the imperial tea, which is composed of the first and smallest leaves, are obliged to cover their hands with gloves; they must abstain from eating fish and certain meats, and must bathe twice a day, so that nothing impure can mingle with the precious harvest, over which is placed a guard of soldiers who watch with the strictest scrutiny. In China, and, generally, in India, the tea is prepared by the hands of women. About the month of May, the mothers of families, the children, and the female slaves, leave their homes and visit the tea-plants every hour of the day, so that they may gather the leaves before they are fully developed; at evening the leaves they have gathered are taken home and spread on plates of polished iron, heated to different degrees of temperature; they are stirred continually with the hand until they begin to curl, and are then spread on rush mats, fanned till cool, and then again submitted to the heated plates. These operations are repeated four times, the women rolling the tea-leaves until they take the form which we see them under. When the tea is perfectly dry it is enclosed in porcelain vases hermetically sealed, but more commonly in boxes lined with lead. The consumption of tea increases considerably every year; and in Europe, where it is almost universally used, between 8 and 10 millions of pounds are annually consumed. The Dutch, English —all the people of the North—use considerable quantities; and in France, where, forty years ago, tea was used but as a medicine, it is now an article of almost general consumption. But nowhere will bear comparison with the United States as to the quantity consumed. The Americans have always had a sort of passion for tea, and the trade in this article was one of the indirect causes of their Revolution.'

These details vividly interested my young family, and it was determined that, the following year, we should have a regular gathering of the tea, and preserve it for our use.

Jack arrived at the drawbridge half an hour before the rest of us, the long legs of his ostrich giving him clearly the advantage. The first care of the young scamp was to run to the marsh of ducks and choose a convenient place to deposit the mysterious sack. We arrived and unpacked with all the tranquillity of good householders who return to their homes after an absence of several months. Fritz arrived some time after us.

Our first care was to dispose of all the fine birds we had brought home. The heath cocks, Canada fowls, and the cranes (one of which had a wing broken), were placed on one of our islands. The heron royal, the Sultan Cock, and the elegant 'Girl of Numidia', were placed in the marsh with the ducks and geese, and shared with them the crumbs of our repast.

These first cares occupied us a good part of the day, and, while awaiting supper and listening to Fritz's tale of his voyage, suddenly we were surprised by hearing a most horrible noise, resembling distant thunder. These strange sounds appeared to come from the marsh. The dogs began to bark, the animals seemed frightened, and I rose immediately to ascertain the cause of our ears being assailed by such a hideous uproar.

'Jack,' cried I, 'bring me my gun; we will see who's the musician; and you, Fritz—how! do you sit still in time of danger?'

Fritz smiled, and made me a sign to reseat myself, telling me, while Jack was gone for the gun, that the noise was the croaking of two monstrous frogs that were confined in the bag Jack had put in the marsh.

'Admirable!' cried I. 'Let us all get up, and when he reappears, exhibit signs of the greatest inquietude; thus the fellow will fall into his own trap.'

Jack soon returned with the guns—he brought two.

'Very well,' said I; 'you are a brave fellow, Jack, and brought the other gun so that you could share the danger with me.'

Jack answered nothing; but, turning to Ernest, who feigned the greatest fear, asked him if he knew what animal it was.

'Yes, and we are going to attack him right away. I can very well distinguish his form through the rushes.'

'And what do you call him?'

'A jaguar', answered Fritz.

'But what is a jaguar?'

'A jaguar', said Ernest, 'is the most beautiful tiger of America; its fur is superb. The naturalists call it "felis cincalor"; it has——'

'It has—it has—' interrupted the coward. 'I know very well what it has; but I do not want to have to fight one'; and the fellow set off on a full run to the grotto, and he soon appeared on the exterior gallery. We began to laugh heartily, and then Master Ernest revealed the secret.

'It was your sack,' said he; 'it was the two frogs we heard that was the jaguar—the tiger with the rich fur, the monster that you fled from; a grand soldier you would make, truly.'

Jack became the star of the evening. He was called knight of the jaguar, hero of the frogs; and all the compliments he was so fond of showering on his brothers were returned with interest.

When we were a little rested from our fatigue, my wife recalled to mind Falcon's Nest and its aerial chateau, which we had almost forgotten since our discovery of the salt cavern.

'It is wrong', said she, 'to let that beautiful habitation go to ruin. Although Felsenheim offers us a sure protection in winter, yet Falcon's Nest, with its gigantic branches and pleasant verdure, is the most agreeable habitation we could possess.'

My wife spoke reasonably, and I promised her that I would do as she wished. We left Felsenheim and took up our residence in our old habitation. The roof that we had made over the roots was new plastered with gum and resin; the staircase was repaired; we substituted a bark roof for the old linen one over our chamber in the tree: we made a balcony all around it, and repaired everything, so that it was a clean, agreeable habitation.

But the embellishments at Falcon's Nest were but a prelude to more considerable and difficult works. Fritz had not renounced his idea of fortifying Shark Island, and making that a sort of rallying-point in case of danger. He teased me so about it, and his head was so full of plans and projects, that it was impossible to resist him, and the work was at length begun. One can easily conceive how great were the obstacles that a man and four boys had to contend with, in order to transport two cannons to the island, and level them on a platform more than 50 feet in height. It cost us immense labour to transport the cannons to the island. I placed on the platform we had built a large capstan; and, to shorten the time and reduce the labour in passing round the rock. I let down a rope, made into loops, so that we could easily ascend and descend. The cannons were attached by strong ropes, and then hauled up by the capstan. This work cost us a whole day of hard labour; but at last the cannons were landed on the platform, and established with their mouths toward the sea. We placed a long pole in the rock, with a string and pulley, so that we could hoist up a flag at any time. How glad we felt when our work was done, and how proud we were of our ingenuity. When we had crowned this military construction with a flag, a cry of joy was uttered; and, as economical as I felt we must be in powder, six times we fired our cannons, and the echo of the rocks repeated the noise over a vast extent of ocean.

60
A General Review of the Colony after Ten Years of Establishment

IT is with dismay that I cast my eyes over the number of pages I have filled, and which every day grow more numerous.

Although I would like to mention the minutest details of our domestic life, yet I have some consideration for my readers, who would throw down the book in disgust and grow weary of the monotony of the design; therefore, I must content myself with merely describing our principal occupations.

Ten years have passed away since we were thrown on this coast, each year resembling the preceding one in the similarity of its works: we had our fields to sow, our harvests to gather, and our domestic cares to attend to. These formed the almost unbroken circle of our existence. My only desire is, that the end I intended in writing this journal may be fulfilled, and that my readers, if I ever have any, may learn how to provide for their necessities should they ever be cast on a desert island as I have been.

Providence had willed that the theatre of our exile should be in one of the most favoured quarters of the globe; and every day we offered up our thanks to him for his goodness and beneficent kindness toward us.

The ten years we had passed were but years of conquest and establishment. We had constructed two habitations, built a solid wall across the defile, which would secure us against invasion from the wild beasts which infested the savannah. The part of the country in which we dwelt was defended by high mountains on one side, and the ocean on the other; we had traversed the whole extent, and rested in perfect surety that no enemy lurked within it. Our principal habitations were beautiful, commodious, and especially very healthy. Felsenheim was a safe retreat for us during the storms of winter, while Falcon's Nest was our summer residence and country villa; Waldegg, Prospect Hill, and even the establishment at the defile were like the quiet farmhouses that the traveller finds in the mountains of our own dear Switzerland. My good Elizabeth made the comparison, and, pointing to the mountains in the distance, she would say, 'Do you see the Alps, and their white summits? Those tall trees that seem to touch the skies are the firs of the Black Forest; and there, behind the farmhouse, extends the Lake of Constance, with its clear, calm surface.'

The remembrance of our native land is never obliterated from the mind; the love of one's birthplace is a love that survives youth, and exists in all its ardour in the bosom of the old man.

Of all our resources, the bees had prospered most; experience had taught me how to manage them, and the only trouble that I had was to provide new hives each year for the increasing swarms; and, in truth, so great was the number of our hives that they attracted a considerable flock of those birds called *merops*, or bee-eaters,* who are extremely fond of these insects.

The plumage of this bird is brilliant and charming; but we were obliged to put an end to the devastation they wrought on our hives. We constructed some snares with glue and resin, and captured a number of these marauders, the brilliant colours of whose plumage were a rich addition to our museum of natural history. The study of this science was one of our greatest delights: we possessed in our library several fine works which could guide us in the different branches of this interesting science, and nature each day spread before us a rich treasury of wonders on which we could investigate and experiment; the bees especially, their intelligence, their sagacity, their love of labour, their curious manners, enlisted our close attention. But the mind of man cannot penetrate the secret of the intelligence that is displayed by so frail a creature, and we felt ourselves constrained to say that the grandeur of God does not especially evince itself in the glorious luminaries he has placed above to light us, or in the terrible animals with which he has peopled the desert, but in the smallest productions of his will. The bee in its hive is not less admirable than the lion which ranges the forest, or the immense whale which sports on the waves of the ocean.

Our dove-cot had also succeeded well; but it was not large enough, and we had been obliged to suspend baskets on the adjoining trees, where our pigeons might build their nests.

We also finished the gallery which extended along the front of our grotto; a roof was made to the rock above it, and it rested on fourteen columns of light bamboo, which gave it an elegant and picturesque appearance; large pillars supported the gallery, around which twined the aromatic vines of the vanilla and the pepper, and each end of the gallery was terminated by a little cabinet with elevated roofs, giving them the appearance of Chinese pavilions, surrounded by flowers and foliage. These steps led up into the gallery, which we had paved with a sort of stone so soft when dug out as to be cut with a chisel, but hardening rapidly in the sun.

The environs of our habitation were rich and agreeable; our plantations had perfectly succeeded, and between the grotto and the bay was a grove of trees and shrubs, planted in tasteful confusion, which gave the spot the aspect of an English garden.

Shark Island no longer was an arid bank of sand: palm and pineapple trees had been planted everywhere, and the earth was covered with a carpet of vivid green; while far above the trees towered a staff upon the top of which the Swiss flag floated gaily in the breeze. The scene around us was always animated and gay; the swans clothed in mourning colours mingled with geese white as the driven snow and the heron royal with his silvery crest, or the flamingo in his robe of rose-colour, would stand by the marsh and capture the frogs with which it abounded. Under the shade of the beautiful trees our little troop of ostriches reposed, unmindful of the clamour raised by the troops of cranes and turkeys that clustered around them; the Canada fowls and the heath fowls, joining together and disdaining the society of their fellows, crossed to the other side of Family Bridge.

One could not recognize in this beautiful spot, surrounded by so much that was grateful to the eye and ear, the desert, sandy plain we had found on our first coming. It had for boundaries on the right Jackal's River, which was bordered on our side by a strong and impenetrable hedge of thorn-palms, aloes, Indian figs, karatas, and other plants of the same sort, all so close together that a mouse could scarcely penetrate it; on the left, inaccessible rocks, among which was the grotto of crystal, barred all entrance on that side. Before us, as I have said, extended the blue-waved sea, losing itself in the distance. Behind us the mass of rocks, in which our grotto was situated, was so high and steep that I feared nothing from that side. The only outlet from our little elysium* was Family Bridge, for which we had made a drawbridge, and, that it might better be defended, we built a parapet of stones before it, and mounted on that two small 6-pounder cannons, which could sweep the whole bay, while two mortars armed our ship of war, the celebrated pinnace.

The space comprised between the grotto and the stream contained our gardens; a palisade of bamboos surrounded them and added to the number of our defences. This palisade ran in a right line from our habitation to the Jackal's River, forming a triangle, and containing a small field of grain, a plantation of cotton, one of sugar-cane, some cochineal plants, and several varieties of garden plants, and the kitchen-garden of my wife, with an orchard of all sorts of European fruits. All these plantations were irrigated by tunnels* of bamboo, which conveyed their supply of water from the river, and distributed it over the ground.

Our European trees had grown with a strength and rapidity of vegetation almost incredible; but their fruits had lost their savour: and whether because the soil or the air was unfavourable, the apples and pears became black and withered, the plums and apricots were nothing but hard kernels surrounded by a tough skin; on the other hand, the indigenous productions multiplied a hundredfold: the bananas, the figs, the guavas, the oranges, and the citron made our corner of the island a

complete terrestrial paradise, where all the riches of vegetation
were assembled. But the abundance of fruit brought on
another plague; multitudes of pillagers, in the shape of birds,
flocked to the spot. We kept our bird-snares always ready, and
it sometimes happened that an unknown animal would be
taken in the trap; for example, the great squirrel of Canada,*
remarkable for its beautiful tufted tail and lustrous red skin,
attracted hither probably by our almonds and chestnuts; paroquets, in all their diversity of colours, would sometimes be
caught; blue jays, thrushes, yellow loriots,* abounded plentifully to the great prejudice of our cherries, figs, and native
grapes. Besides the birds by day there were other destroyers by
night, and we had a great deal of trouble to dislodge a nest of
flying squirrels that had taken up their residence in the topmost branches of one of our finest trees.

While our trees were young, and bore fruit but sparingly,
we used every means to capture these winged thieves; but they
seemed to laugh at all our attempts, and at last we gave up the
hopeless task.

Our beautiful flowers also attracted numerous guests: these
were the humming-birds; and it was one of our greatest amusements to watch these little birds flying around us, sparkling like
precious stones, and hardly perceptible by the quickness of
their motions; it was an amusing spectacle to see these passionate, choleric little fellows attack others twice their size, and
drive them away from their nests, and at other times they would
tear in pieces the unlucky flower that had deceived their
expectations of a rich feast. These little scenes diverted us, and
we endeavoured to induce the birds to remain in our neighbourhood by fixing little pots of honey on the branches and
planting the flowers we observed they preferred. Our cares
were recompensed; several couples suspended their little nests,
lined with soft cotton, to the branches of the vanilla which
wound around the columns of the gallery, or on the vines of
the pepper, the perfume of which is very enticing to the
humming-bird.

Our spices, as I have said, prospered well; the nutmegs
flourished beautifully on the lawn before the grotto; and when
at evening we reposed under the portico, their grateful perfume would add to the delights of the evening; but they

attracted such quantities of birds of paradise, that we were compelled to set snares in order to reduce the number of the beautiful thieves.

Our olive trees were the only ones that escaped these invasions of the feathered tribe. As we had two kinds of olives, we gathered the largest before they were ripe, and, after putting them in lye-water, as they do in Provence, we preserved them with salt and spices, as a sort of relish; the other sort we allowed to remain on the trees until perfectly ripe, and then converted them into oil, for the fabrication of which I had constructed a mill that rendered the labour much less fatiguing.

The fabrication of sugar was an object of our special attention, and we gradually improved our manufacture, not that I would say we crystallized it as done in the refineries; but we obtained a very satisfactory result. We had saved from the wreck of the ship many articles intended for a sugar-factory; among others, three metal cylinders with which to press the sugar-cane, three great kettles to boil the liquid in, and ladles and skimmers in abundance.* The press was fixed under a perpendicular screw, working in connection with the cylinders, the whole turned by a lever passed horizontally through the screw, and moved by one of our beasts of burden. We also made another machine of the same sort, to answer three purposes: first, to break our hemp more readily and quickly; secondly, to crush our olives; and, lastly, to bruise the cocoa-bean pods and other substances of the same nature. The bottom of this press was formed by a large stone, with a deep furrow chiselled in it, to let the juice pass off; this stone had a border nine inches in height, and below it, was placed a furnace, so that we might heat it, as is necessary in pressing out certain substances.

These two presses were at first established in the open air, between the drawbridge and Herring Point; but afterward we erected a slight wooden building over them, and it formed a very commodious workshop.

Whale Island had not been neglected: we embellished it with trees and shrubs; but it was here that we always performed our less cleanly avocations, such as the preparation of fish, the melting of fat, the tannery, and the fabrication of candles. The materials for these works were fixed under an overhanging rock, which protected them from the sun and storm.

Our cares were divided between these different establishments, without neglecting those that were more distant from us, and which we called our colonies. At Waldegg we transformed the swamp into a superb rice-field, which repaid our labour by plentiful harvests; we also planted cinnamon, which yielded us an ample return.

Prospect Hill also had its share of attention; for each year, when the capers were ripe, we made an excursion thither, and gathered a large quantity, which my wife preserved in spice and vinegar; and when the tea-plant began to put forth its leaves, again we set out, and, gathering enough for our use, we took it home to my wife, who, with her youngest son, occupied herself in rolling, drying, and preparing it for use. We also gathered the black millet or doura, of so much assistance in fattening our fowls. We used our canoe for all these excursions, reserving the pinnace for the sea.

We made, from time to time, an excursion to the defile of the savannah, so that we might see whether any elephants, or other hurtful beasts, had penetrated into our plantations. Fritz then made an excursion in his kayak up the river of the savannah, and brought back to us a rich cargo of ginseng, cocoa, and bananas.

As Fritz had one day discovered in the woods near the defile traces of birds which, from their noise and form, he judged to be heath cocks, we resolved one day to have a grand hunt after the manner of the Cape colonists. For this purpose we constructed a large quadrangle of the enormous bamboo-canes I have spoken of, piled upon one another until the edifice was 10 feet long and 6 high, and exactly resembled an enormous bird-cage; the top was covered with a lattice of canes, and the door formed of the same. To induce the birds to enter, we dug a deep ditch, which led, like a mine under a city wall, into the centre of the edifice; we covered this ditch with sticks and earth, and placed in the exterior entrance, all along the passage, different sorts of grain: we then retired, and the birds precipitated themselves on the food; the more they ate the deeper they buried themselves in the ditch, until at last, when they arrived at the end, they found themselves captured, and in vain they beat their heads against the trellis-work. We entered and soon took them all prisoners.

During one of our excursions beyond the defile, we captured a superb species of fowl, which we mingled with our European poultry, and saved them from degeneration. These birds had magnificent plumage, and much resembled the turkey,* only they were so tall that they could easily take food from off a table.

The family of Turk and Flora had each year been increased by a certain number of young dogs, which, notwithstanding the brilliant qualities they displayed, we were obliged to throw into the water, as to have allowed them to live would have been our own destruction. To this rule there was but one exception, and, on the earnest entreaty of Jack, I permitted the canine family to retain one new member, which we called *Coco*, 'because', said Jack, 'the vowel *o* is the most sonorous, and will sound so fine in the forests.'

The female buffalo and the cow had each year produced us a scion from their race; but we had only raised one heifer and a second bull. We had called the cow Blanche, on account of her pale yellow colour, and the bull, Thunder, as his voice was so powerful. We also possessed two more asses, which we named Arrow and Alert on account of the swiftness of their course.

Our pigs were as wild as ever. The old sow had been dead many years; but she had bequeathed to her posterity a spirit of savage independence that all our exertions could not modify. Our other beasts had multiplied in the same proportion, so that we could often kill one without any fear of impoverishing ourselves. Such was the state of the colony ten years after our arrival on the coast: our resources had multiplied as our industry increased. Abundance reigned around us; we were as familiar with our part of the island as a farmer with his farm. It was a perfect paradise. It would have been an Eden, but there was one great void—oh! if we could but have looked upon men, our brothers.

For ten years had we watched both by sea and land for some traces of man's existence, but all in vain; and yet, we hoped on, hoped ever, and still gathered up all our treasures of cotton, and spices, and ostrich-plumes, etc., in earnest hope that some day we would again see the blessed face of man.

My sons were no longer children. Fritz had become a strong and vigorous man; although not tall, yet his limbs had been developed by exercise: he was 24 years of age.

Ernest was 23 and, although of a good constitution, he was not as strong as his brother; his reflective mind had ripened; reason now aided his studious disposition; he had conquered his habit of idleness, and was, in a word, a well-informed young man, of a sound judgement, and unquestionably the light of the family.

Jack had but little changed: he was as headlong at 20 as at 10; but he excelled in corporeal exercises.

Francis was 18: he was stout and tall; his character, without any predominant trait, was estimable. He was reflective, without being as deep as Ernest; agile and skilful, but without surpassing Jack or Fritz. In general my sons were good and honest men, with sound principles, and a deep sense of religion.

My dear Elizabeth had not grown very old. As for me, my hair had become whitened by age, or, to speak more justly, there were but a few scattering locks left; the heat of the climate and excessive fatigue had taken them all away, although I still felt young and vigorous.

There was one bitter, sad thought that always haunted my mind, and, turning my eyes to heaven, I would often say, 'My God, who didst save us from shipwreck, and hast surrounded us with so many blessings, still watch over us, I pray thee, and do not let those perish in solitude whom thy hand has saved.'

61
Excursion of Fritz; Result

ONE can easily imagine that my young family was not as easy to govern now as it was during the first few years of our stay.

My children would often absent themselves whole days, hunting in the forest, or clambering over the rocks; but when they returned at evening, fatigued and wearied, if I had intended to reproach them for their wandering life, they would have so much to tell me concerning the rare and curious things that they had seen, that I never had resolution enough to scold them.

Fritz one day went off in this manner, and caused us the greatest inquietude. He had taken with him some provisions and—as if the land was not large enough for him—also his kayak, and gone out to sea. He had set out before daylight, and night was approaching, but nothing could be seen of him. My wife was in a state of the greatest suspense, and, to alleviate her distress, I launched the canoe, and we set out for Shark Island. There, from the top of the flag-staff, we displayed our flag and fired an alarm-cannon. A few moments after, we saw a black spot in the far distance, and, by the aid of a spy-glass, we discovered our beloved Fritz. He advanced slowly toward us, beating the sea with his oars, as if his canoe was charged with a double load.

'Fire!' cried Ernest, in his capacity as commander of the fort, 'Fire!' and Jack touched off the cannon. We descended to the shore and were soon in the arms of our adventurer Fritz. His boat was loaded with different things; and something heavy and dark, which looked like the head of a large animal, was towing behind.

'It appears,' said I, 'my dear Fritz, that your day has not been an unprofitable one, and blessed be God that he has returned you safe and sound.'

'Yes,' replied Fritz, 'blessed be God; for, besides the booty which you see, I think I have made a discovery which is worth more to us than all the treasures of the earth.'

These words, half-whispered in my ear, excited my curiosity; but I thought I would say nothing until the voyager had taken breath. When we had brought on shore his sack, filled with large oysters, as it appeared to me, and the marine monster which served as a counterpoise, we drew the little kayak, with its master seated in triumph in it, up to the grotto. The boys then returned to obtain the remainder of the cargo, while we sat down quietly in the gallery, to listen to Fritz's narrative. He commenced the recital of his adventures by begging us to pardon him for running away, as he had resolved to visit the eastern part of our country, of which we as yet knew nothing.

'I had long ago intended to make this expedition,' said he. 'I furnished my kayak with provisions, and two skins, one full of water, the other of hydromel. I had placed a compass in my boat, a fish-net and a harpoon were on the right, a gun and an anchor on the left; I also put a pair of pistols in my waist, and slung my ammunition-bag around my neck. This morning, before you awoke, I softly arose and ran, as is my custom, to the borders of the sea. The weather was so beautiful, the waves so tranquil, that I could not resist the temptation. I called my eagle, and, seizing a hatchet, jumped into the kayak, and falling into the current of Jackal's River, was hurried out toward the shoals where our vessel had been wrecked. I saw there, in passing, and at not a very great depth either, a great quantity of bars of iron, cannons, and balls, which we may one day get up, if we discover the means of plunging to that depth. I directed my course toward the eastern coast, among shoals and rocks covered with the nests of seabirds, who flew around me uttering piercing cries. Whenever the rocks offered any surface, you would see great marine monsters, extended in the sun, while others were playing and bellowing frightfully in the neighbouring waters. These were sea-lions, and elephants, and walruses of all sorts, who, holding on to the rocks by their long teeth, let their hinder parts rest in the water. It seemed that this was the general rendezvous of these monsters; for I saw, in coasting along the shore, several places strewed with their bones and ivory teeth; and perhaps we can procure some fine carcasses there for our museum.'

'Oh, how delightful!' cried out all the auditors; 'we can make handles for our knives out of those fine ivory teeth.'

Francis, whose reflective spirit had always a remark to make, asked me what use these enormous teeth were to the animals, as they could neither bite nor wound with them.

'All teeth are not for that purpose,' said I. 'Some are arms for attack or defence, such as those of the elephant, rhinoceros, the morse, and the narwhal;* others, like the tusks of the boar, and the curved teeth of the babiroussa,* are a sort of implement with which nature has provided the animal, either to dig up roots, to detach shells from the rocks, or to pull down the branches of trees, so that they can eat the foliage. The hippopotamus alone has such strong and singular teeth as to render it uncertain for what they are designed, as the animal feeds on grass and herbs. Besides, the tusks of the morse and the hippopotamus, being less porous than those of the elephant, are more valuable; and as they are less liable to turn yellow, dentists employ them in the fabrication of artificial teeth.'

Fritz now resumed his narrative. 'I must confess,' said he, 'that when I saw myself encompassed by these monsters, I did not feel very safe, and I endeavoured, as far as I was able, to pass through the shoals unperceived, which I effected after a hard row of an hour and a half. I stopped my course before a magnificent portico of rocks, which nature seemed to have fashioned into the most imposing forms: it was like the arch of an immense bridge, under which the sea flowed in like a canal, while the rock on each side of the entrance advanced out into the sea, like an immense promontory. I did not hesitate to enter this sombre vault, from the other extremity of which issued a feeble light. A delicious coolness filled the cavern: on all sides numbers of the little coast-swallows were flying about; and, on my entrance into the cavern, a swarm of these birds surrounded me, uttering piercing cries, as if they wished to prohibit my farther approach. I tied my skiff to an angular stone in the cavern, and began to examine the inhabitants. I found I was mistaken in considering these birds swallows: they were about the size of wrens, their breast of a pure white colour, their wings of a light grey, the back of a lustrous black. Their nests appeared like those of other birds, made of feathers and dry leaves; but they were placed on a singular sort of a support, resembling a long spoon, made of greyish, polished wax. Some of these nests were empty; and, having examined them with more attention, I discovered that they were made of a substance resembling fish-glue. I disengaged a few of them to bring home to you, and I now beg you will examine them, and see whether they are good for anything.'

'Certainly they are, my son. If we carried on commerce with China or India, we could sell these nests for their weight in gold; for they eat them by millions, and esteem them one of the greatest delicacies.'

A general cry of disgust burst forth from my wife and children, at the idea of eating birds' nests. I explained to them that the feathers and moss lining the inside were not eaten, but only the covering, which is carefully cleaned and cooked with spices, making a transparent, savoury jelly. The word jelly reconciled my wife to the birds' nests, and she consented to an experiment with them.

'Is not one of the greatest delicacies made out of sharks' fins? And why not out of birds' nests?' said I.

Fritz promised that he would clean them the first thing in the morning; then, turning toward me, he asked where the birds procured the gummy matter with which they make their nests.

'It is not positively known,' answered I. 'Some pretend that it is the foam of the sea, which this little bird, called *Salargane*,* gathers in its beak, and that this foam, in drying, takes the appearance of wax or fish-glue; another opinion is, that it is derived from a shellfish of which the birds are very fond. This last opinion appears the most probable, as it possesses animal qualities. But let us leave this discussion and listen to the narrative of our traveller.'

'I advanced boldly through the passage,' said he, 'and came out into a magnificent bay, whose low and fertile shores stretched out into a savannah of vast extent; trees and shrubs everywhere varied the beauty of the scene: on the right, a vast mass of rocks rose up, being a prolongation of those that I had passed through; on the left rolled a calm and limpid river, and beyond this was a thick swamp, which terminated in a dense forest of cedars. While I was coasting along the shores of the bay, I perceived, at the bottom of the transparent waters, large beds of shells resembling large oysters. "Here", said I to myself, "is something that is much better than our little oysters at Felsenheim; if they taste good, I will take some home with me." I detached some with my hook and threw them on the sand, without getting out of my canoe and set to work to obtain more. When I returned with a new load, I found that the oysters I had first deposited on the sand were opened, and the sun had already began to corrupt them. I took up one or two; but, instead of finding the nice, fat oyster I expected, I found nothing but a hard, gritty meat. In trying to detach this from the shell, I felt some little round, hard stones, like peas, under my knife; I took them out and found them so brilliant that I filled a little box with them which I happened to have with me. Do you not think, my father,' added Fritz, 'that they are really pearls?'

'See, see!' said the boys, catching hold of the box. 'How beautiful, how brilliant, how regular!'

I took the box in my hand. 'They are really pearls,' cried I, 'oriental pearls of the greatest beauty. You have, in truth, discovered a treasure, my son, which one day will be, I hope, of immense value to us. We will pay a visit to this rich bay as soon as possible; but continue your story.'

'I pursued my course,' resumed Fritz, 'along the coast, indented with creeks, and covered with verdure and flowers. I came up to the mouth of the river, the calm waters of which floated on tranquilly toward the sea; its surface, overgrown with aquatic plants, resembled a verdant prairie covered with different sorts of birds. I gave to this river the name of St John, as it put me in mind of the description I had read of a river of that name in Florida.* Having renewed my provision of fresh water, I then directed my course toward the other promontory, opposite the arch by which I had entered. This bay, which I named The Bay of Pearls, was about six miles from one promontory to the other, and a chain of rocks separated it from the sea; one entrance only is practicable, for all the rest are covered with shoals and sandbanks, forming a natural harbour, which wants but the neighbourhood of a city to render it perfect. I endeavoured to leave the bay; but the tide had risen so high that it filled the vault, and I was obliged to await its ebb. I stepped on shore, as I saw on all sides, popping up out of the water, the heads of marine animals, which appeared about the size of a calf, and they plunged and frisked about in such a manner that I was afraid they would overset my kayak; so I secured it to a point in the rock, and, taking my eagle in my hand, I stood ready to attack the first game that came near me; for I wished to procure one of the animals, which resembled a stuffed valise, as I thought its thick skin might be of use to me. A company of them soon came, plunging and diving, close to the shore. I cast off my eagle, who seized on the largest and best, and soon blinded him; I jumped on a projecting rock, and, catching hold of the animal with my boat-hook, drew it to the shore. All the others fled as if by enchantment. I had to remove the entrails of the animal, as the weight was too heavy for my little skiff; but, while I was thus occupied, a prodigious number of seabirds clustered around me: gulls, sea-swallows, frigates, and half a dozen other kinds. They came up so close that I whirled my staff around to keep them off, and in doing so knocked down a very large bird, an albatross,* I think. After this operation was finished, I fastened my sea-otter* to the stern of my boat; and, taking a sack full of oysters, made preparations for my return. I soon passed through the arch, and sailed quietly along, until I saw your flag and heard the report of the cannon.'

After this narrative, and while my wife and the younger boys had gone to the kayak, the one to examine the edible birds' nests, the others to look at the pearl-oysters, my son drew me aside and confided to my ear an important secret.

'A very singular circumstance', said he, 'happened on my voyage. In examining the albatross which I had knocked down, judge my surprise when I saw a piece of linen around one of its feet. I untied it, and read the following words written upon it in good English: "*Save the poor shipwrecked sailor on the smoking rock.*" I cannot express to you, my father, what I felt on seeing this linen. I read and reread the line to assure myself that it was not an optical illusion. "My God," said I, "only grant that it may be true. From this moment my only thought shall be to search the coast in quest of the smoking rock, save the sufferer —my brother—my friend. Oh! once more perhaps I may see a human being." An idea occurred to me to attach the linen again to the foot of the albatross, and to write upon a second piece, which I fastened to the other foot, the following sentence in English: "*Have confidence in God: succour is near.*" If the bird returns to the place from whence it came, thought I, the person can read the answer: at all events there will be no harm in trying this experiment. The albatross had been stunned, and I poured some hydromel down its throat to reanimate it. I attached my note to its foot, and let it go, earnestly praying that its mission might be successful. The bird flew up, hesitated for a moment, and then darted rapidly away in an easterly direction, which decided me to take that route in my search. And now, my father,' continued Fritz, with emotion, 'what do you think of this event? If we could find a new friend, a new brother—for certainly we will go in search of the stranger, oh yes, we will go—what joy! what happiness! But, alas! what despair if we should not succeed. The reason I did not communicate this to my brothers and my mother was to spare them the agonies of a hope which, after all, might never be realized.' My son pronounced these last words with sadness.

'You have acted very prudently,' said I, 'and I am glad that you have sufficient strength of mind to resist the temptation of immediately flying to the assistance of the sufferer. Only consider for a moment into what terrible anxiety we would all have been thrown if you had not returned as you did. As for the result of any expedition of discovery I cannot say much; the albatross is a traveller-bird, and it flies extremely swift: the linen might have been put on its foot thousands of miles from here; and even if near, perhaps years ago, and now succour may be too late. But continue to keep the secret, and I will try to imagine whether some way cannot be devised to save the poor unfortunate, if in our vicinity.'

These cold, positive words were dictated by a desire to appease the ardent imagination of the young man, and prevent him from rashly undertaking any precipitate enterprise. I knew that pirates often made a smoke behind some rock to deceive, and I was afraid that this might be a case of the kind. We now returned to the rest of the family, who were yet occupied in looking at the pearls.

'We have a large fortune there,' said Ernest to his brothers. 'Europe pays their weight in gold for the fine pearls that are brought from the Levant; and in 1804 the English government sold the right of fishing in a bank of pearl-oysters on the coast of Ceylon, for 3 million of francs. The pearl-fishing commences in the month of March, and gives occupation to a great number of persons. The Orientals endeavour to throw a veil of mystery around it, and they go through a variety of ceremonies before undertaking it, which they think will render their success secure. They set off in the night, as they think it is essential to be at the pearl-bank before sunrise. About 7 o'clock, when the water is warm enough to permit the divers to plunge in, the fishing commences. It is carried on in the following manner: a sort of scaffold is made which hangs over both sides of the boat; to this is suspended a stone in the form of a sugar-loaf, which is sunk 5 feet in the water, and is called the plunging-stone. The cord which sustains it is joined to a stirrup, intended to receive the foot of the plunger, who remains standing a few moments, until a net-like basket is thrown to him, in which he places his other foot; this net has attached to it a cord, which the diver holds in his hand. Everything being arranged, the diver with one hand compresses his nostrils, so that the water shall not penetrate them; with the other he jerks the rope attached to the stone, and then plunges into the water. Arrived at the bottom, he draws his foot from the stirrup; those on the boat immediately draw up the stone and fasten it to the scaffold; then the diver commences his work, gathering, as fast as possible, the pearl-oysters, which he detaches from the rock with a pair of pincers: he fills his basket in a minute and a half with 150 oysters, if he is an expert diver. When he has finished, he pulls the cord attached to the basket; those on board draw it up immediately, and the diver rises to the surface of the water and swims round the boat until his turn for again plunging comes on. The inhabitants of Ceylon, and the coast of Coromandel are great lovers of the pearl-fishing; and, painful as it is, those men who are engaged in it esteem it a delightful amusement. They will fish steadily for six hours without uttering the least complaint. After the fishing, the oysters are heaped up in large enclosures, and carefully guarded for six days, during which time the shells open, so that the pearls can be extracted. Seawater is then let in on them, and, after a space of twelve hours, they are washed out, and passed to the clippers, who detach the pearls from the shells with pincers.'

After this explanation, given by master Ernest, each one made his particular remarks on the beauty, size, and number of the pearls which Fritz had brought home.

In answer to the questions of Francis, who demanded whether all pearls were of this brilliant colour, I added some additional information to the details given by Ernest.

'Pearls', said I, 'depend for their purity on the state of the water in which they are found; in muddy water they are clouded; in sandy and gravelly places, clear and brilliant. They are also differently coloured in different places. There is a pearl-fishery in the Gulf of California which produces pearls of a bright orange colour; those on the coast of Africa are nearly black; and green ones have been found, which are much esteemed by the Arabs. In Scotland and Lorraine large mussels are found which also contain pearls of a bluish tint and irregular form.'

'And how are pearls formed?' asked Francis.

'Their formation has long been regarded marvellous. It has been attributed to a sort of dew which fell from heaven, from whence comes that beautiful apalogue* of the drop of water which fell into the oyster-shell, and, to console it, Jupiter turned it into a beautiful pearl. As to its real origin, naturalists have discovered that it is in diseased oysters the largest pearls are found, especially in those which have been pierced by a little worm, called urille, which can pierce the hardest shell; to defend itself, the oyster covers the hole with a substance which hardens and becomes the pearl. It is also stated that they are formed from grains of sand or other foreign particles which have introduced themselves into the body of the oyster, and the fishermen always throw a little stone in those oysters they perceive open.'

After the pearls, the edible birds' nests had their turn; but the sea-otter especially excited the attention of my young naturalists.

'What an ugly animal it is,' said Frank, as he turned it over; 'and you call it a sea-otter, do you?'

'Yes,' replied Ernest, 'it is really a sea-otter, one of the most inoffensive animals of the whole marine tribes; it possesses many good qualities, especially a love for its offspring, which surpasses belief, and it will die of starvation if its little ones are taken away. If attacked it makes no resistance, but seeks to escape by flight; if that is impossible it will lie down and cover its face with its forepaws, as if waiting the mortal blow. The otter is a very desirable prey: besides the skin, which forms an excellent fur, its flesh is also much esteemed, and is said to resemble mutton in flavour.'

We then looked over the other articles that Fritz had brought; and when the excitement was over, I suddenly turned to my family, and in a grave tone made the following speech: 'My dear wife, and you, my sons, this day shall be an era in the history of our family. Fritz no longer is a child. In the last excursion that he made he conducted himself with so much prudence and courage that I here resign my parental authority, and declare, before you all, that from this day he is free from all subordination; that I shall consider him as a man, and as a friend, who will aid our counsels, and mingle in our labours.'

This unexpected scene was followed by a moment of silence. Fritz was greatly embarrassed; his mother tenderly embraced him, and the tears of affection flowed from the eyes of my good Elizabeth.

'It is the ceremony of the "toga virilis",* my dear Fritz,' said Ernest at last. 'You are now a man; but take care you do not have to come back to leading-strings again.'

I said nothing concerning the secret which Fritz had confided to me, as I wished to think it over, and see what could be done.

The pearls were too important an object to be forgotten, and my sons importuned me to start immediately on our projected expedition to the newly discovered fishery.

'Softly,' said I; 'before riding, you must saddle your horse; and if you wish that your enterprise should succeed, you must take with you the necessary implements. Let each one of you try to invent something useful for our purpose, and then we will start.'

This proposition was received with joyous acclamations, and each member of the party set his ingenuity to work. I forged for myself two large iron rakes and two small hooks of the same metal; I fixed wooden handles to the first-named, with iron rings attached, so that I could fasten them to the boat and drag them over the banks of oysters; with the hooks I intended to loosen the oysters which the rakes were insufficient to detach. Ernest made a sort of butterfly-net, with scissors attached, intended to receive the birds' nests. Jack constructed a sort of ladder, made by piercing a long bamboo at regular distances and fixing in sticks crosswise; the machine looked like the stick in the parrot's cage. To the top the young man fixed a hook of iron, and a spike at the bottom, so that it should rest firmly in the rocks. Francis, very adroit in making nets, made several very strong ones to hold our oysters.

During this time Fritz worked in silence at his kayak, endeavouring to construct a second seat in it. I alone knew his intention; but I dared not encourage him by evincing my knowledge of his object.

We next prepared our provisions for the voyage: two hams were cooked, cassava cakes, barley-bread, rice, nuts, almonds, and other dry fruits; and for drink we took a barrel of water, and one of hydromel. These stores, with our tools and fishing implements loaded down the boat.

62

The Edible Birds' Nests; the Pearl Fishery; the African Boar

WE had spent an entire day in preparing our cargo. A fresh and favourable breeze and a slightly ruffled sea induced us to embark immediately. Francis and his mother were left to guard the shore, and we gaily put off, amid their prayers and wishes for our safe return. We took with us some of our domestics: young Knips, the successor of our good old monkey, Jack's jackal, Flora, Braun and Folb, all found a place in the boat. Jack occupied the second seat in Fritz's kayak. Ernest and I conducted the canoe loaded with our provisions and animals.

The kayak led the way, and we followed, steering our course through the shoals and rocks with the greatest difficulty. We did not encounter any marine monsters; but the rocks were covered with the whitened bones of morses and sea-horses,* and Ernest made us stop several times, at the risk of bruising our boat against the rocks, in order that he might collect some of these osseous remains for our museum of natural history.

The sea was as calm and brilliant as a mirror, and was covered with the little boats of the nautilus,* a sort of shellfish which much resembles a miniature gondola, and from which it is pretended men learned the art of navigation.

When the nautilus wishes to sail, it elevates its arms, and extends, like a sail, the thin, light membrane between them; it also puts out two other members which serve as oars, and a fifth as rudder. At the approach of an enemy it lowers its arms, draws in its oars, fills its shell with water, and down it sinks in perfect safety; when arrived at the bottom it turns over and empties out the water; and when it wishes to reascend it inflates some air-bladders with which it is provided, and rises to the surface. The shell of the nautilus is of a delicate white, and as thin as paper; the animal is a *polypus* of eight feet, and has a fringe-like substance covering the mouth, which serves as a means of seizing the food of the animal and conveying it to the mouth.

My young naturalists could not behold these beautiful little boats, dancing over the surface of the waves, without wishing to capture some; they threw out a net, and we soon had half a dozen fine ones, which we carefully preserved for our cabinet at Felsenheim.

We soon attained the promontory, behind which, Fritz said, was the Bay of Pearls. This promontory was singular and imposing. Arch rose above arch, column above column; in a word, it resembled the front of one of those old Gothic cathedrals,* embellished with a thousand grotesque carvings and antiquated decorations, with the only difference that, instead of a pavement of marble, we had the blue sea, and the columns were washed by the waves. It resembled a temple elevated to the Eternal, in the middle of immensity. We penetrated into the vault; it was sombre and gloomy, like an old cathedral, and only lighted by a few apertures in the rock. We searched every place in order to see whether there were any of those terrible marine monsters concealed there, but we could only discover a few bones scattered here and there over the rocks.

The noise of our oars frightened the peaceable salarganes, and they flew about in such numbers as almost to render it impossible to guide the boat; but when our eyes became habituated to the darkness, we saw with pleasure that every niche and corner was filled with their nests. These nests resembled white cups, were as transparent as horn, and filled, like the nests of other birds, with feathers, and dry sticks of some sort of perfumed wood.

The trial which we had made of this substance, after boiling it with salt and spices, convinced us that it was a delicate and wholesome food; besides, we knew how highly they were valued in China, and we were so possessed with the idea that, some day or other, a vessel would arrive on our shores, with which we could trade, that I resolved to gather a considerable number of these nests, only taking care to leave those which contained eggs or young ones. Fritz and Jack climbed like cats along the rocks, and detached the nests, which they gave to Ernest and me, who placed them in a large sack we had brought. It was soon filled, and I was glad of it, as the boys were tired, and I could not bear to see them suspended on the ladder over the water. Before we went on, however, I set Jack and Ernest to work to clear the nests of the feathers, etc.

'Really,' said master Ernest, who did not much like this sort of work, 'what is the use of our gathering up this dirty provision; just as if a ship would ever come here to buy them: ten years have already passed and——'

'Hope, my son,' answered I, 'is one of the greatest blessings that Heaven has vouchsafed to man; it is the daughter of courage and the sister of activity, for the courageous man never despairs, and he who works confidently seldom fails in his desires. Hope on, hope ever, should be a maxim engraved on the heart of everyone.'

I now gave the order for departure. Fritz had assured me that the canal which flowed through the vault was navigable, and that, by following the passage, we would soon arrive at the bay. The flood-tide carried us rapidly forward, toward the other extremity of the cavern, and we could not help admiring the magnificence of the passage: the roof was covered with stalactites, wrought by the hand of nature into a thousand fantastic forms, columns and capitals rose in rich profusion, and it seemed as if nature had assembled here specimens of all the styles of architecture—the Doric, the Ionic, the Corinthian,* all had their representatives. At length we issued into a beautiful bay; we were struck with surprise, and remained resting on our oars in silent admiration. The water was so calm and pure that we could see the fish far below us. I recognized the white fish, the shining scales of which are used as false pearls. I showed them to my sons; but they could not understand how a little stone would be worth so much more than the fish-scales, when the latter were full as brilliant as the former.

'It is not the object itself,' said I; 'it is the difficulty in procuring it which costs so much. If every river in Europe abounded in pearls they would be worth nothing.'

'Ah, yes,' replied Ernest, 'it is what is called *"pretium affectionis."* '*

We joked the doctor on his Latin, and while discoursing we arrived at the rocky bank where Fritz had found the pearl oysters. The coast presented a most beautiful prospect; forests which lost themselves in the distance, and high mountains covered with the rich vegetation of the tropics. A majestic river flowed into the bay, and cut the green prairies like a band of silver. We all landed safely, except master Knips, who could not make up his mind to leap the narrow space which separated him from terra firma; twenty times he rose on his hind legs, and twenty times he shrunk back as if he had the ocean to cross. At length we took pity on him, and threw him a rope, by which means he safely landed.

The day was too far advanced to commence our pearl-fishing, and we appeased our hungry stomachs with some slices of ham, fried potatoes, and some cassava cakes; and, after having lighted up fires along the coast to keep off wild beasts, we left the dogs on shore and went on board the canoe, Knips being installed on the mast as vidette.* We drew the sail over our heads, and, wrapping ourselves in our bear-skins, soon sunk to rest. Nothing disturbed our repose save a concert of jackals, who regaled us for about an hour with a most horrible charivari,* to which Jack's pupil responded by most ear-piercing howls.

We rose at daylight, and, after a frugal breakfast, commenced our labours in the pearl-fishery, and, with the aid of the rakes, hooks, nets, and poles, soon brought in a large quantity of the precious oysters: we heaped them all up in a pile on the shore, so that the heat of the sun would cause them to open.

Toward evening the coast appeared so beautiful, and the vegetation so rich and glowing, that it was impossible for us to resist the temptation of making an excursion to a little wood, where we had heard turkeys gobbling all day. Each took with him one of our faithful servants, and we separated. Ernest entered first into the wood, accompanied by Folb; Jack soon followed him, while Fritz and I remained a moment to fix our guns. A few moments after we heard a report, then a scream from Jack, followed by another report. Fritz unhooded his eagle, I snatched up my gun, and we ran in the direction of Jack, who was screaming 'Papa! papa! Quick! I am killed! Quick! Come!'

The poor boy had exaggerated matters a little, for he was not even wounded; but then he lay face to face with an enormous boar, with formidable tusks, who had knocked him down so rudely that he thought himself lost. Jack, notwithstanding his 20 years, was as cowardly as ever.

His brothers ran quickly up; two shots well fired freed him from his terrible enemy, and we loaded him with sarcasms and jokes upon the poltroonery he had displayed. The fellow, however, was so terribly frightened that I feared for the consequences; so I gave him a glass of hydromel, and, after having rubbed the contusions on his head and back with the same liquid, I sent him on board the canoe, where he soon fell asleep on his matress.

Ernest recounted to me the manner in which this affair happened. 'I had entered into the little wood', said he, 'with Folb, when suddenly the brave dog quitted me, and set off in pursuit of a wild boar, who had come out of the forest and was sharpening his tusks against the tree with a terrible noise. At that moment, Jack came up, his jackal perceiving the boar, sprung furiously upon him, while Folb attacked him on the other side. I approached cautiously by passing from one tree to another, until I was near enough to fire; the jackal, however, had received such a terrible blow from the boar's foot that he lay senseless on the grass. Jack then fired, but missed; and the boar, turning round, set off in pursuit of his new assailant, who fled like a Hottentot before him. Without doubt, he would have soon escaped if a projecting root had not tripped him up. Down he fell; I fired but missed, and the boar began to butt poor Jack with his head. He, however, had not time to do him much harm, as Braun and Flora rushed in, and, seizing the animal by his ears, held him so firm that he could not stir. Fritz's eagle now joined the fray, and, flying on the head of the boar, who fairly frothed with rage, began picking out his eyes. Fritz now fired, and hit the animal directly in the throat; it fell right across Jack's body, who could not disengage himself. I then ran up and helped him; he groaned dreadfully, and at first I thought he was seriously wounded; but I found that I was mistaken. He took Fritz's arm and walked away, while I remained by the boar. It was not without some surprise that I saw master Knips with some large black tubercles,* with which the ground was covered; I gathered two or three which I put in my game-bag. Look at them.'

So saying, the young naturalist presented me with six tubercles, resembling potatoes, the odour of which was very penetrating. I opened one, and, having tasted it, I discovered that they were excellent truffles,* of a perfumed, delicate flesh, marbled with white.

'It appears,' said I to my son, congratulating him on his discovery, 'that the boar, who is very fond of these things, was eating them when he was disturbed.'

My son now asked me to give him some details concerning this singular production.

'Naturalists', said I, 'have agreed to place the truffle in the family of mushrooms; it grows without roots, leaves, or stalks of any kind. It would not even be discovered unless its perfume betrayed it, and even that is so delicate we are obliged to use animals whose sense of smell is much more acute than ours to discover it. These animals are dogs and pigs; the former merely scratch the ground where the truffles are, but the pigs dig them up, and would eat them if iron rings were not put around their noses.'

'But', said Ernest, 'is there no way of determining where truffles grow?'

'There is', said I, 'one index that is pretty certain: it is the presence of certain green flies which are produced from worms that feed on the truffles. Truffles are found in nearly every quarter of the globe, but especially in temperate countries. France and Piedmont furnish a prodigious quantity, the flavour and perfume of which are much esteemed by connoisseurs. The truffle is round, of irregular form, and presents a black or grey exterior; its interior substance is a firm, compact flesh, brown, intermingled with white veins. The truffle is classed among the *criptogames*. The secret of reproducing truffles was long undiscovered, but, I believe, it has been at last found out.'

While talking in this way, night came on, and it was necessary to seek repose. We lighted our watch-fires, swallowed a morsel of meat, and then retired to our canoe: the dogs were again left on shore. We were soon asleep, and dreaming of the absent ones at our beloved Felsenheim.

63
The Nankin Cotton; the Lions; Another Excursion by Fritz;
the Cachalot Whale

OUR first care, on rising the next morning, was to set about the preparation of the boar. Jack had recovered from his fear, and, accompanied by the dogs, we set out to look for the dead boar. He was enormous—between a boar and a buffalo in size, and his head was indeed frightfully large. While we were examining his gigantic proportions Fritz cried out, 'Zounds! we can obtain some fine hams here; better, I think, than those we procured from the peccaries.'

'For my part,' said Ernest, 'I claim the head: it will make a fine addition to our museum. But how will we ever get this enormous body to the shore?'

'As to that,' replied Fritz, 'I think I can manage it, if papa will let me.'

'Willingly; but I fear that the flesh of this old African is not better than a European boar. My advice is, that instead of dragging the immense carcass away, we cut off what we want, and let the rest alone.'

My sons agreed with me, and we began to cut off the hams and head of the wild boar. Some branches of trees furnished us with sleds to put the pieces on, and we made the dogs draw them to the shore.

While we were occupied in disposing of our hams, chance made known to us an important discovery. Ernest remarked on the branches we had cut to make our sleds a sort of nut; he opened one, but, instead of a kernel, it contained a beautiful fine cotton, of a deep yellow, which I recognized as being the real Nankin cotton.* This cotton owes its name to a province of China where it grows abundantly, and is cultivated with much care. We made a large provision of these precious nuts, and dug up two young trees to carry to Felsenheim.

Jack shrunk with fright from the head of his terrible enemy, and appeared quite overjoyed that it was going to figure in our museum; but, on the observation of Ernest, that it would be very difficult to prepare, and having heard that boar's head was very fine eating, we resolved to cook it with truffles, in the Otaheitan manner; consequently Fritz and Ernest set to work, and dug a deep ditch, while I cleaned the head and heated some stones. When these preparations were finished, we placed the head, stuffed with truffles, and seasoned with salt, pepper, and nutmeg, in the ditch, and covered it with red-hot stones and a thick layer of earth. While our supper was cooking we suspended our hams over the smoke of the fire, and tranquilly sat down to talk over the events of the day, when suddenly a deep, prolonged cry rung through the forest. It was the first time we had ever heard such unearthly tones: the rocks echoed

it, and we felt seized with sudden terror; the dogs and the
jackals also commenced howling horribly.

'What a diabolical concert,' said Fritz, jumping up and
seizing his gun; 'some danger must be near. Build up the fire,'
continued he, 'and while I try to discover the danger in my
kayak, you retire to the canoe.'

This plan appeared the best we could pursue, and I adopted
it. We threw on the fire all the wood we could find ready cut,
and, without losing time, we regained the canoe. Fritz jumped
into the kayak, and was soon lost in the obscurity.

During all this time the roarings continued, and they
appeared to approach nearer to us. Our dogs gathered around
the fire, uttering plaintive moans. Our poor little monkey
seemed to suffer painfully from fear. Imagined that it was a
leopard or a panther, which had been attracted, by the remains
of the wild boar in the wood. My doubts did not last long, for
we soon discovered, by the pale light of our fires, a terrible
lion, infinitely larger and stronger than those I had seen in the
royal menageries of Europe. In two or three leaps he bounded
over the space which separated the wood from the shore; he
stood immovable for a moment, and then commenced lashing
his flanks with his tail and roaring furiously, every moment
crouching down as if to spring on us. This frightful pantomime
did not last long: every moment he would run to the stream,
lap up some water, and then return. I remarked with mortal
anguish that the animal came nearer and nearer to the shore;
and at length he lay crouched down, his flaming eyes fixed
directly on us. Half in fear, half in despair, I raised my gun, and
was about to fire, when suddenly I heard a report; the animal
bounded up, gave a tremendous yell, and fell lifeless on the
earth.

''Tis Fritz,' murmured my poor Ernest, pale with terror. 'Oh
God! protect my brother.'

'Yes, it is he,' I cried, 'our brave Fritz; he has saved us from
a terrible death; let us go to him.' In two strokes of the oars
we were on shore; but our dogs, with an admirable instinct,
began to bark terribly: I did not neglect this indication; we
threw more wood on the fire, and again jumped into the boat.
It was time; for scarcely were we secure when a second enemy
rushed from the forest: it was not as large as the first, but its

roar was frightful. This time it was a lioness, probably the female of the superb animal which we had just killed. How grateful I felt that both had not appeared together; what could we have done? The lioness ran straight up to the corpse of her companion, smelled it, and licked up the blood which had flowed from the wound; and when she was convinced that he no longer lived, she set up a howl of rage that pen could not describe; she lashed her sides and opened her enormous mouth, as if she would devour us all.

Again Fritz fired, and the shot, less fortunate than the first, only broke the shoulder of the animal. The wounded lioness commenced rolling on the sand, foaming with rage; but all three of our dogs rushed upon her. It was a repetition of the combat with the bears in the savannah; the obscurity of the night, the formidable voice of the lioness, the horrid howlings of the dogs, struck us dumb with fear. Braun and Folb seized the animal in the flanks, and Flora caught hold of the throat. Another shot would have put an end to the combat, but I was afraid of wounding the dogs; so I jumped from the boat, and, running up to the animal, who was held fast by the dogs, I plunged my hunting-knife direct to her heart: the blood spouted out, and the lioness fell; but the victory had cost us dear, for there lay our poor Flora, dead under the terrible wounds she had received from the fangs of the monster.

Fritz now ran up and threw himself into my arms, as did Ernest and poor Jack, who trembled with mortal agony. We lighted our torches, and directed our course toward the field of battle; we found poor Flora, with her teeth yet clutching the throat of her enemy, while the royal couple lay majestically extended on the sand, and we could hardly suppress a sentiment of fear which struck us as we gazed on the terrific beasts.

'What a terrible range of teeth,' said Ernest, as he raised up the head of the lion.

'Yes, and what frightful claws,' said Jack; 'wouldn't they make nice holes in your skin?'

'Yes, my friends,' observed I, 'let us thank God that he has saved us from the danger; thank him for the wisdom with which he has endowed men so that they may be able to conquer such terrible beasts.'

'Poor Flora!' said Fritz, as he detached the dead body of our dear dog from that of the lioness; 'she has done for us today what our old ass did in the case of the boa. Come, Ernest, see if you cannot induce your muse to fabricate an epitaph.'

'Ah! my muse, I must confess, has been too terribly frightened to make any rhymes.'

'Pshaw! Go and meditate while we dig the grave of our poor hunter, and be sure to be ready when we are done.'

Flora received the honours of a funeral by torch-light; we dug a grave and reverently placed in it the remains of the faithful animal, and a flat stone served to mark her resting-place. Ernest composed the following legend, which he read to us, saying that he was too frightened for poetry, and Flora must be contented with prose:

Here lies
FLORA, A DOG
remarkable
for her courage and devotion.
She died
under the claws of a lion,
on whom
she also inflicted death.

'Admirable,' said Fritz. 'It must be confessed, Ernest, you write full as good poetry* as prose.'

Jack, who did not care much for poetry or prose, remarked that the night was waning and, as it would be impossible to sleep, he thought that we had better have something to eat.

'I suppose', said he, 'the poor boar's head must be done by this time: anyway, I mean to go and see.'

So saying, Jack began to dig away the covering of earth and cinders, while Ernest and I dressed the wounds of the dogs. But, instead of the juicy meat poor Jack expected, he found nothing but a carbonized mass of bones and burned flesh. He was going to throw it away in disgust, when I stopped his hand, and, cutting off the burned flesh with my knife, we found underneath some most delicious meat, saturated with the perfume of the truffles in a manner that every epicure knows how to appreciate.

When we had eaten, we tried to snatch a little sleep after the fatigues of the night. At sunrise we were up, and our first care was to deprive our noble prey of their superb furs. My pneumatic syringe, which I had taken the precaution to bring with me, did the business effectually, and we soon obtained two of the most splendid skins that can be imagined; the fur was as soft as silk and of a most beautiful colour. This operation naturally led us to speak of the lion, and I tried to dissipate some errors which my sons entertained.

'Of all the animals of the creation,' said I, 'there are few more generally known than the lion; and there are also few who have had so many fabulous reports attached to them. He is commonly called the king of animals, and his grandeur and clemency are highly extolled. This is an error: the lion is neither clement nor magnanimous, but simply a ferocious animal, who devours his prey like the tiger and the panther, only, when his appetite is satisfied, he is less savage; but the same quality is shared with the other animals. This error has been in general acceptation from time immemorial; the lion has always been represented as the emblem of nobleness and courage, and modern naturalists have decreed him the sceptre among animals.

' "The lion", says Buffon,* "has an imposing figure, a fixed look, firm step, and terrible voice; he is not as large as the elephant or the rhinoceros, nor as heavy as the hippopotamus or the ox. He has not the leanness of the hyena, nor its elongated body; on the contrary, he is so well formed and proportioned that the lion is a perfect model of agility, solid and nervous, not overcharged with flesh and fat, and containing nothing superabundant, but all nerves and muscles. His great muscular force is demonstrated by the prodigious bounds which the lion makes—by the quick movement of his tail, which is strong enough to knock down a man—by the facility with which it moves the skin of the face, thereby adding much terror to its appearance, and, lastly, by the faculty it has of moving its hair, which, when in anger, shakes and bristles in all directions." Doubtless all this is true! the picture is faithful; but it does not prove the magnanimity of the lion, and I cannot understand why all these noble qualities should be heaped on his head. You have heard his roars and have been witness to his terrible anger.'

'Oh, papa,' said Ernest, laughing, 'are you not ashamed to try to dethrone the poor king lion, who has ruled so many centuries; but I uphold him for the sake of our victory. We have conquered the lion, and we have laid the king of animals prostrate at our feet.'

Jack wanted to make a mantle of the lion's skin, such as Hercules wore after his victory in the Nemean forest;* but I adjourned all arrangements about their appropriation till a more convenient time.

The heat of the sun had commenced to corrupt the oysters heaped upon the bank, and the effluvia which they exhaled induced us to return to Felsenheim, and early in the morning we set sail. Jack did not feel much inclination to take his place again in Fritz's kayak, and pretended that the exercise at the oars was too fatiguing for him, and he embarked with us in the canoe, more hands making lighter work there.

Fritz set off before us, as if to serve as pilot; but when he had conducted us through the vault, and over the shoals, he rowed up to our canoe, and, handing me a letter, shot off again like an arrow. I opened the paper quickly, and imagine my surprise when I found that, instead of having forgotten the albatross and the smoking rock, he informed me in the letter that he was going in search of the unfortunate being. I had a thousand objections to make to this romantic project; but Fritz rowed so fast, I could barely halloo through the speaking-trumpet—'Return soon, and be prudent', before he was out of sight. We gave to the cape where he left us the name of the 'Adieu Cape'. We prayed that our adventurer might return safe, and I begged my rowers to redouble their endeavours, so that we could arrive early at Felsenheim, for I suspected that my good Elizabeth would be worried at our long absence of three days.

We finally arrived without accident, and the different treasures we had brought were joyfully received; the truffles, the lion-skins, the pearls, the Nankin, became the objects of a thousand questions, but they could not drive away the thoughts of Fritz; and my wife said she would willingly give up all our cargo of pearls, etc., if she could only see her beloved son.

I had not yet spoken to my wife concerning the reason for Fritz's absence, as I did not like to give any hope which was so likely to be deceived; but in the present case I thought that it was my duty so to do. I therefore confided to her the secret of the albatross; and the good woman, to my surprise, was calm and resigned, and only prayed that he might be successful.

I now undertook the preparation of our lion-skins, and I carried them, for that purpose, to our tannery on Whale Island, where, as I said before, our dirty work was done; we also occupied ourselves in storing our provisions and with the necessary household avocations. Five days had thus passed away and still Fritz had not returned, and his mother was so anxious and worried that I proposed to launch the pinnace and make a new excursion to the Bay of Pearls; she received my proposition with demonstrations of the utmost joy, for she thought that Fritz would return in that direction, and that we should certainly meet him. We lost no time; the pinnace was prepared, and early the next day we bid adieu to Felsenheim, and soon were in sight of the promontory of the bay, when suddenly the vessel ran against a black mass, and was nearly thrown over by the shock. My wife and sons uttered a cry of terror; but the boat soon righted, and I perceived that the obstacle was not a point of rock, as I had thought, but a marine monster of the family of blowers, for we soon saw him throw up into the air two spouts of water mingled with blood. I instantly pointed the cannons of the pinnace, and a discharge of artillery prevented the huge monster from overturning us, which he certainly would have done if the blow had not stunned him. We saw with pleasure that the waves carried the enormous body to a sandbank a little distance from the shore, and there it lay like a stranded ship.

'After the whales,' says one naturalist, 'there is no class of cetaceous fishes as large as the cachalot,* and it even can dispute its supremacy in the ocean with the whale, as it is better armed and defended. The cachalots swim in large herds, and are found in almost all seas, and from the poles to the equator there is not a spot that does not contribute to their nourishment. The large-headed cachalots sometimes attain the enormous length of 80 feet; they are agile and courageous, while the whales, on the contrary, are timid, and seldom leave their customary resorts. There are seven sorts of cachalot whale: its principal distinction is, that it has the lower jaw furnished with a great number of teeth. While the upper has but three; it has the nose blunted, and its head alone forms nearly the half of its whole bulk. It has a small tongue, but a throat large enough to swallow an ox whole, and a shark 15 feet long has been found in the body of a cachalot. The cachalot furnishes less oil than the whale, but this deficiency is amply supplied by the spermaceti, a shining, semi-transparent matter, very light, inflammable, and easily dissolved in oil. This substance, when fresh, has but little smell and an agreeable taste. It is used in medicine, and candles are made from it, the whiteness of which is fully equal to those manufactured of wax.'

While we were thus conversing about the cachalot, and calculating its value, Ernest suddenly uttered a loud scream. 'A man! A savage!' said he, and he pointed out to us in the distance a sort of canoe dancing over the waves. The person who conducted it seemed to have perceived us, for he advanced, and then disappeared behind a projecting point, as if to communicate his discovery to his companions. I leave our sensations to the imagination of the reader. I had not the slightest doubt that we had fallen in with a band of savages, and we began to fortify our boat against their arrows, by making a bulwark of the stalks of maize and corn we had brought with us. We loaded our cannons, guns, and pistols, and, everything arranged, we stood ready behind our rampart, resolved to defend it as long as we were able. We dared not advance, for there was the savage; and Ernest, growing tired of the pantomime, observed that, if we used the speaking-trumpet, possibly our savage might understand some words of the half-dozen languages we were familiar with.

The advice appeared good. I took up the speaking-trumpet and bellowed out, with all my force, some words of Malay; but still the canoe remained immovable, as if its master had not comprehended us.

'Instead of Malay,' said Jack, 'suppose we try English.' So saying, he caught up the trumpet, and in his clear, loud tone pronounced some common sailor-phrases well known to all who have ever been on board ship. The device succeeded, and we saw the savage advancing toward us, holding a green branch in his hand. Nearer and nearer he came, and at last we recognized, in the painted savage, our own dear Fritz.

'Fritz! 'tis Fritz, 'tis Fritz; there is his kayak, and the walrus's head in front; it is Fritz disguised like a savage', exclaimed Jack.

We soon received our intrepid adventurer; he was naked to the waist and painted white and black, just like a Caribbee Indian. We embraced him tenderly, and tears of joy streamed down his mother's cheeks as she gazed on her first-born again.

64
Adventures of Fritz; Our Adopted Sister; her Courage and Industry

WHEN we had freed Fritz from our oft-repeated embraces, we commenced asking him all manner of questions; and, speaking all together, the poor boy was so confused he did not know what to do. I demanded an answer on two points only—whether his excursion had been pleasant, and why he had played this farce of dressing himself like a savage, and causing us such anxiety.

'As to the purpose of my excursion,' said he, with a joy he could scarcely conceal, 'I have attained it'; and the young man, as he said these words, pressed my hand, which he held in his. 'As for my costume, I mistook you for a tribe of Malays, or some other nation, and, in the fear that you were enemies, I endeavoured to disguise myself by painting the upper part of my body with powder, soaked in water. The two reports of the cannon that I heard convinced me more and more that you were enemies; the Malay words that you addressed to me confirmed me, and I should still have been endeavouring to deceive you, and you would still be in fear of me, if Jack had not bawled out those sailor-phrases in his unmistakable voice.'

We all began to laugh over the farce we had been enacting; and Fritz, drawing me aside, said, in an eager, joyous tone, 'I have succeeded, papa: the hand of God conducted me to the dwelling-place of the poor shipwrecked girl—for it was a woman that had written those lines. Three years had she lived on that smoking rock, all alone! destitute of every thing! Can you believe it, but the poor girl has conjured me not to betray her sex, excepting to you and my mother, for she is afraid of my brothers, although I assured her that none would welcome her more gladly. I have brought her with me: she is near by, on a little island, just beyond the Bay of Pearls; come and see her. Oh! do not say anything to my brothers. I want to enjoy their surprise when they find I have brought them back a sister, for I am sure she will allow them to call her so.'

I consented to the wish of my son, and, without saying anything to the rest of the family, I ordered them to hoist the sails, weigh anchor, and make ready to depart. Fritz, who had changed his dress and washed off his disguise, flew about, hastening his less eager brothers; then, jumping into his kayak, he piloted us through the shoals and reefs that were scattered along the coast. After an hour's sailing he turned off, and directed his course toward a shady island not far from the Bay of Pearls; we sailed close up to the shore, and fastened the pinnace to the trunk of a fallen tree. Fritz, however, was quicker than we, and he was on shore, and had entered a little wood in the middle of the island before we had yet landed. We followed him into the wood, and soon found ourselves in the presence of a hut, built like those of the Hottentots, with a fire burning before it on which some fish were cooking in a large shell. Fritz uttered a peculiar kind of halloo, and what was our surprise to see, descending from a large tree, a young and handsome sailor, who, turning his timid eyes on us, stood still, as if he dared not approach.

It was such a long time since we had seen a man—ten years!—society had become so strange a thing to us, that we remained stupefied; our hearts felt for the young stranger, but our tongues remained dumb.

The silence was broken by Fritz, who, taking the young sailor by the hand, advanced toward us. 'My father, my mother, and you, my brothers,' said he, in a voice broken by emotion, 'behold a friend—a brother—that I present you, a new companion in misfortune—Sir Edward Montrose, who, like ourselves, has been shipwrecked on the coast.'

'He is welcome among us', was the general cry; and, approaching the young sailor, whom I easily recognized as being a woman, and taking her by the hand I comforted and encouraged her, assuring the seeming man that among us he would always find food and sustenance; my wife and myself would be his parents, and my sons his brothers. My wife, moved by compassion, opened her arms, and the young sailor rushed into them, bursting into a flood of tears as he thanked us for our kindness. The most lively joy now reigned in our little circle, and his brothers poured question after question upon Fritz, who joyfully replied, 'I will tell you all afterward; let us attend now to our new brother.' Supper was served, and my wife brought out a bottle of her spiced hydromel to add to the feast. Everybody spoke at once, and my sons addressed their new companion with such vivacity as to embarrass the timid stranger: my wife saw his distress, and, as it was late, she gave the signal for retreat, taking the sailor with her on the pinnace, where she said she intended to provide a bed for him that would amply console him for the uncomfortable nights he had hitherto passed. We then separated, my wife and the stranger retiring to the boat, while my sons and I stopped to light and arrange our watch-fires.

The newcomer naturally became the subject of conversation.

'I should like to know,' said Francis, addressing himself to Fritz, 'what put it into your head to go to the succour of our new brother. How did you know there was a man shipwrecked on the coast.'

Fritz smiled without answering.

'Are you endowed with second sight, after the manner of the Scotch?' said Ernest.

'No,' added Jack; 'I think Sir Edward must have written him a letter by the carrier pigeons.'

'A good idea—you almost guessed right', said Fritz; and he then recounted to his brothers the whole history of the albatross; he spoke of his thoughts and actions, but he became so excited in his narration that he forgot himself and the secret he had to keep. A word escaped him, and he called the young sailor 'Emily'.

'Emily!—Emily!' repeated his brothers, who had begun to doubt the mystery, 'Emily!—Fritz has deceived us, and Sir Edward is a girl!—Our adopted brother turned into a sister.'

I leave to the imagination to picture the embarrassment of Fritz when he discovered his imprudence. In vain he endeavoured to bring back his words; it would not do, and the girl could no longer hide her sex by the hat and pantaloons.

This discovery changed the conversation. Fritz explained to his brothers the motives which had induced Emily to conceal her sex, for she was afraid to trust herself among four young men, whose character and manners she was utterly unacquainted with; but the boys declared that nothing pleased them better than to have a new sister, and that this change would not lower Emily at all in their esteem.

The next morning it was a comic sight to see the embarrassment and awkwardness with which my sons approached one whom they had the day before embraced as a companion and a brother. My poor boys were not acquainted with the usages of polite society and the ease it inspires, and they appeared to a great disadvantage by the side of the beautiful English girl. The name of sister was substituted for that of brother, but pronounced with reserve and embarrassment. As for Emily she was very much astonished at the discovery the young men had made, and she retreated, as if for protection, to the arms of my wife; but a moment after, recovering herself, she advanced, and, extending her hand to each one of the boys, gracefully demanded for the sister the friendship they had extended to the brother. This amiable frankness dissipated the embarrassment of my three sons; they assured the young girl of their fraternal regard, and begged that they might consider her as a sister. Gaiety was re-established and we sat down to breakfast which was composed of fruits, cold meat, and chocolate of our own making, which was a great treat to my new daughter, and recalled her native land to her mind. After breakfast I proposed to weigh anchor and return to the Bay of Pearls, where the cachalot stranded on the shore offered us a magnificent prey. Arrived there, we debated in what manner we could carry away the oily matter with which the head and dorsal bone of this animal is filled. Unfortunately we had no barrels in which we could gather the precious product. Emily rescued us from our dilemma, by mentioning a process she had seen employed in India, which was, to put the half-liquid matter in wet linen bags. The idea appeared excellent, and we immediately put it into practice. I gathered all the sacks I could find, and, dipping them in the seawater, stretched them open with pieces of branches. We were two hours engaged in these preparations. The tide was not yet high enough to allow the pinnace to approach the bank where the whale lay; but we took the canoe and the kayak and set off, leaving the two women under the safeguard of Turk, and taking with us Folb, Braun, and the jackal. The monster lay extended like a huge wall; our dogs ran up to it, and a moment after we heard some animals howling dreadfully. We hastened up and found our brave dogs valiantly contending with a troop of black wolves, who were devouring the whale. Two of their number were already stretched on the sand, two others were yet engaged with the dogs, and the rest had fled at our approach toward a little wood. We perceived some jackals among the troop of fugitives, and our surprise was great at seeing Jager, our jackal, run away with them, leaving us all astonished at his desertion.

Our dogs acquitted themselves bravely: four wolves lay stretched upon the sand, but the noble animals had paid dearly for their victory; the blood streamed from all parts of their bodies, and the ears of Folb especially were dreadfully torn. Jack dressed their wounds with some hydromel, while Fritz and Francis aided me in another work. The former, after having armed his feet with cramp-irons,* climbed like a cat up the back of the monster, and cut open the enormous head of the cachalot with a hatchet, and then with a ladle dipped the spermaceti out of the head, and emptied it into one of the sacks which I held ready, while Francis covered the outside with wet sand and mortar, forming a solid crust through which none of the grease could escape. Our sacks were soon full, for as fast as Fritz emptied the head the cavity was filled by a fresh supply from the back-bone. The operation was very fatiguing, and I was glad when it was over. We then cut a quantity of willows, and wove them into little pointed caps, with which we covered the sacks, in order to shield them from the sun and the birds of prey, who were fast assembling in great numbers.

We now thought of returning. The tide was high, but the load was too heavy for the boat; we therefore were obliged to leave them and return to the verdant little island, which we had named 'Good Rencounter', because there we had first found Emily. The appearance of the sacks ranged on the sand was very droll; they exactly resembled little Chinese with their pointed hats, and we could not help laughing heartily at the sight.

After having recounted our adventures, and shown our four fine black wolves, with their superb skins, we were invited by

our dear housekeepers to sit down to an excellent dinner,
enriched by a new dish—a sauce after the manner of the
Caribbees, made with the eggs of land-crabs, with which the
island abounded. I was undecided as to what means I should
adopt to transport the spermaceti to the island, for the pinnace
could not approach the bank near enough, without risk of
running aground, and our other boats were not large enough.
Everyone gave his advice; when it came to Emily's turn, she
observed, in her soft, silvery tone, 'If you are willing, my dear
papa'—for already she had accustomed herself to address me
by that endearing name—'if you are willing, while you and my
brothers are engaged in that disgusting tannery, I will promise
to bring over your sacks: and, if you will give me a piece of
wolf's skin,' added she, laughing, 'I will make a charm that I
am certain will bring back the runaway companion of my
brother Jack.'

This proposition was received with a shout of laughter from
my boys, who could not believe that an inexperienced girl
would be able to effect a thing which appeared to *them* so
difficult. They commenced a perfect battery of pleasantries on
their adopted sister, who received them with a very good grace,
and ran to seek my wife, who was preparing our supper. The
young girl had been a little hurt by the sarcasms directed
against her, and my wife consoled her as well as she could,
assuring her that the unlucky jokes proceeded not from a
malicious heart, but from ignorance of politeness. The dear
girl dried her tears, and, after having tenderly embraced her
adopted mother, endeavoured to make a muzzle for the jackal
out of a piece of wolf's skin.

The next morning, before my sons were awake, Emily
prepared for her expedition; she took a bladder of fresh water,
a basket of provisions, and, lightly descending the ladder of
the pinnace, she seated herself in Fritz's kayak, untied it, and
rowed off with a grace and ease that surprised me. I would have
called her back, but the little vixen gaily kissed her hand, and
soon was far on her way toward the bank of sand. She had
chosen just the right time; the tide was rising, and had just
commenced to wet the bottom of the sacks. The adventurous
girl jumped on shore, fastened all the sacks by cords to a rope
which she had with her, and tied the rope to the kayak, and,

again embarking, drew after her all the sacks, the contents of which, being light, floated like bladders on the water.

The capture of the jackal gave the most trouble to the industrious girl, for she was obliged to get out and fasten the boat with a large stone. From the point where I stood I had been able with my spy-glass to trace her every movement; but when I saw her disappear in a little wood near the shore I felt some slight alarm, which was soon dissipated, however, as she immediately returned, and, sitting down on the grass, began to eat the provision she had taken with her, scattering pieces of bread and meat all about, while she called 'Jager! Jager!' as loud as she was able. The poor animal, who was almost dead with hunger, not being accustomed to seek his food in the woods, soon approached near the young enchantress, who kept on throwing him pieces of soaked biscuit, each time shortening the distance between them; at last she offered him a vessel of water, and the poor beast ran right up to her: she threw a rope round his neck, fixed the muzzle on his mouth, and taking him on board the kayak, placed him on the second seat Fritz had constructed. In this position poor Jager, very much ashamed of his bad conduct, found himself placed; the little witch who had captured him then made a little cap of rushes, and put it on his head, throwing a shawl around his shoulders, which gave him exactly the appearance of another passenger.

During this time, my sons, who had been occupied with their wolf-skins, commenced to feel uneasy about the long absence of their sister, and they were just going to step into the pinnace to go in search of her, when suddenly she came in sight. The appearance of her new companion surprised them greatly.

'Where did our new sister find that new brother?' said one.

'Do men sprout up here like mushrooms?' said another.

'Perhaps a magician has aided her', added Francis.

Fritz alone remained silent; but he seemed greatly confused, and ran out into the water to see who it was accompanying his sister. Suddenly he burst into a loud laugh, and, clapping his hands and sputtering the water all over us, he cried out to Jack, 'Oh, 'tis he—your rascally jackal. See how he sits there—the scamp—with all the dignity of a prime minister.'

We burst into a laugh at the odd appearance he made, and then turned toward his mistress, who, jumping on shore, set the captive at liberty, and then, with a triumphant air, showed us the long line of sacks which followed her little skiff. We received her with every testimonial of joy and gratitude. Jack especially was very lavish of his thanks, the poor fellow was so glad to have recovered his old playmate.

It was now full noon; we sat down to table, and, after dinner, began our preparations for setting out for Felsenheim, where we desired to install our new companion. We packed up everything we had, including Emily's treasures, both those she had saved from shipwreck and those she had fabricated herself. Fritz had made her a box which held them all, and they really were very curious, consisting of clothes, ornaments, domestic utensils, and all sorts of articles which she had made in her exile, out of the scanty material she had at her disposal. There were fish-lines made of the twisted hair of the young girl's head, with fish-hooks attached, made of the mother-of-pearl; some needles made from fish-bones; piercers and bodkins, made of the beaks of birds; two beautiful needle-cases, one made of a pelican's feather, the other of the bone of a sea-calf. The skin of a young walrus sewed together served for a bottle, a lamp made of a shell, with a wick of cotton drawn from her handkerchief, over the lamp another shell served as boiler; a turtle-shell used to cook food in, by throwing in hot stones, some fish-bladders, shells of all sizes, serving for glasses, spoons, dishes, etc.; little sacks full of seeds gathered by the young solitary, a quantity of antiscorbutic* plants, as the cochelaria,* sorrel, celery, and cress, which grow among the rocks.

For wearing apparel, she had a hat made of the downy breast of the cormorant,* which was stretched over some feathers from the same bird, forming a complete shelter for the head and neck against the rays of the sun, a little waistcoat with sleeves, made from the belly-skin of the sea-calf, the skin of the forelegs serving as sleeves; some other garments made of bird-skin or walrus-skin; belts, stockings and shoes, all made of skin.

Emily's jewels were few in number, consisting of a gold comb and a string of fine pearls, which she had happened to have on when shipwrecked; she had also some boxes made of turtle-shells, which contained pieces of amber, and some pearls of a beautiful red tint, which she had extracted from some sort of shellfish, and, besides, some pencils made of feathers and hair, with which the poor girl amused herself by writing. I must not forget to mention a beautiful little purse made of sea-calf skin, and containing some elegant and rare shells, which she had gathered on the sea-shore. The whole of these objects were enclosed in a large case which Fritz had made, and placed on our vessel, already loaded with the bags of spermaceti, the wolf-skins, etc. The rest of the day was consumed in packing up; and, during the last repast we made on the island, Emily's skill, and the means which she had employed to pass away the time of her dreary exile, furnished interesting and spirited subjects for conversation.

The next day, when we were all ready to start, Emily brought us another proof of her patience and industry; she ran into a little plot of shrubbery, the branches of which dipped into the sea, and brought out a large bird, confined by a cord, which she presented to us as being a skilful fisherman; it was a cormorant, which the young girl had tamed, and taught, after the manner of the Chinese, to capture fish. Emily now bade adieu to the island that had received her, and the trees that had sheltered her, during her short sojourn. We could not leave the place without giving it a name, so we called the bay in which we anchored 'Happy Bay', in allusion to the joyful meeting we had had there. We now took the direction of the Bay of Pearls, where we were obliged to make a short stay before returning to Felsenheim, to which we were impatient to introduce our new companion.

65
Fritz's Story

FRITZ, seated in his kayak, served as pilot to assist us in penetrating safely through the rocks and shoals into the bay, where at last we all arrived in safety. Everything was found just as we had left it—the table and benches yet standing, our fireplace undestroyed, and, what was more, the air was purified; and, the oysters having all been dried up by the sun, they had lost their unpleasant odour. The dead bodies of the lions and the wild boar were but heaps of whitened bones, the birds of prey having completely stripped them of every particle of flesh.

All appeared tranquil, and we thought it safe to stop long enough to extract the pearls from their shells; this operation, which was certainly not very agreeable, did not long detain Emily, who ran away to find my wife, whom she demurely asked whether she would like a plate of nice fish for dinner. The good woman smiled incredulously, and said that she knew no means by which we could procure fish enough for seven persons in half an hour.

'Well, well,' said the young lady, 'let me alone, and I will promise to bring you a nice mess before dinner-time.'

She took her cormorant under her arm, and, jumping into the kayak, in two strokes of the oars was twenty paces from the shore; she then passed a large copper ring round the neck of the cormorant, so that he could not swallow the fish he caught. Thus prepared, she placed him on the edge of the boat, and remained perfectly still. The fishing soon commenced, and it was a droll sight to see the feathered fisherman, his neck stretched out, his eye fixed steadily on the water, and every now and then giving a plunge and reappearing with a fine fish—a trout, a silverfish, or a salmon—which he carried to his young mistress. After he had taken enough, she took the ring off his neck, gave the faithful bird some small fish as a reward, and hastened home with her successful capture to her adopted mother.

When our pearls were all extracted we gathered them up in a linen sack, counting 400, among which were some extra-ordinary large ones. There was nothing for supper; my sons, therefore, took their guns and game-bags, intending to go and shoot some birds in the Wood of Truffles. Emily wished to accompany them, and when I observed that it was impossible she should be familiar with the use of arms, she, smiling, assured me it would be a pity indeed if she, the daughter of a colonel and a hunter, did not know how to fire a gun, and, besides, I had nothing to fear, as she would not leave her brothers. I let her go, not placing much confidence in her shooting; but a snipe* killed on the wing by the young huntress elicited unbounded applause from my sons, who, when they returned home, lauded her performance to the skies.

We wished to go direct to Felsenheim immediately; but an unexpected discovery detained us a longer time than we had intended. I had noticed among the stones which strewed the shore, a sort of rock which appeared to me could be easily converted into lime. It was a discovery too precious to be neglected, and I resolved to establish a lime-kiln without delay on the beach. It did not take us long to make one to suit our purpose; but the calcination of the stones occupied us much longer, and we were obliged to sit up a great part of the night. During this time we made some barrels of pieces of pine bark, circled with strong withes of willow; a round piece of bark served for the bottom, and another for the cover. To enliven our labour, and to abridge the length of the evening, I persuaded Fritz to give us a more complete account than he had hitherto done of the manner in which he had found our new sister, and the details of his voyage. It was the best way in which to employ our remaining time, and the curiosity of my sons was so excited that they formed a circle about Fritz, who thus commenced his narration.

'You all remember', said he to his brothers, 'the manner in which I left you, after having given my father a letter which contained an account of my intended excursion. The sea was calm; but I had scarcely passed the Bay of Pearls, when suddenly a violent wind came up, and gradually increased to a perfect hurricane; the waves rose high in the air, the rain, the thunder and lightning, all confounded in horrible confusion. My little barque was not strong enough to resist the waves, and all that I could do was to abandon myself to their direction, without allowing myself to be frightened at the violence with which they hurried me along. I put my trust in God, hoping that he would extend his hand over me to save me, as he had often done before. My hope was not to be deceived. After several hours of dread, the wind fell, the air calmed, and my canoe again found its equilibrium upon the surface of the waters. I was far from all the places that we were acquainted with; the tempest had thrown me on a coast entirely new to my eyes; the conformation of the rocks, the gigantic cliffs which seemed to lose themselves in the clouds, the vegetation, the animals I perceived on the coast, the birds which flew about me, all announced a new world. My first care was to look carefully around and see whether some light smoke did not rise from behind the rocks; for, as you know, the Smoking Rock was my only thought—it was the point of my expedition, and I felt within me a voice which said my search would not be in vain. I could perceive nothing as yet; but, full of hope, I rowed along the coast. Night came on, and I passed it in the kayak, after having made a miserable supper on pemmican.

'The next morning I continued my journey, and the farther I advanced the more the coast appeared to change its aspect. I encountered, from time to time, majestic rivers, which flowed silently on and mingled with the sea. The mouth of one of them resembled an immense bay, and I decided to ascend it some little distance; its banks were covered with large trees, willows, and vines, so thickly woven together that they resembled a huge mat, covered with birds, monkeys, and even squirrels: among the aquatic birds, were some who would fall down on the water as if struck by a ball, but no sooner did they touch the surface than they would suddenly rise, and stretch out toward me their long necks terminated by a flat head and pointed beak, forming an exact resemblance of the serpent; it is called, I think, the Amhingu, or snake-neck bird,* which lives in the water, but hatches and makes nests in trees. Toward the middle of the day the heat became so insupportable that it was impossible to resist the desire of seeking some shade under the trees. I turned my boat and ascended the difficult current of a broad and noble river, and went on shore with the intention of shooting some bird; but scarcely had I fired, when an enormous mass rushed from the willows, and, catching up my bird, I fled to the kayak as quickly as possible. I then perceived, on the surface of the water, an enormous hippopotamus and her young ones, who, frightened by the report of my gun, were endeavouring to gain the opposite bank. I descended the river, and having regained the sea, took refuge from the heat of the sun under a rock which overhung the waves. After being slightly refreshed, I pursued my route, and sailed on a long time without being able to land; the rivers and shores were both defended by guards I had little desire to come in contact with, for I recognized elephants, lions, panthers—in one word, a complete reunion of all the ferocious animals of creation. I also saw antelopes and troops of gazelles; but these timid animals seemed only to have been placed there as food for beasts that range the forests. After travelling several leagues farther, the appearance of the coast suddenly changed, and, as if the ferocious animals had had their district marked out to them, I ceased to perceive any. The shore appeared peaceable, but desolate; the breeze which murmured among the vines and the song of some inoffensive birds were the only noises which broke the calm stillness, and I felt reassured, and resolved to land and procure a repast. I accordingly fastened my kayak as strongly as possible, and jumped lightly to the shore; and, being hungry, I lighted my fire, and began to prepare a juicy dinner from a fat goose which I had shot while landing, and a dozen of oysters. While I was thus occupied in appeasing my hunger, I saw advancing toward me, from a little wood which skirted the river, a sort of being, which, by its movements, height, and formation, I at first took to be human. The fire did not frighten him. He walked upright, holding a stick in his hand, and advanced toward me without the least hesitation. At this sight I felt an emotion of mingled joy and fear, for I thought I saw a man; but this illusion was of short duration, and I soon recognized in the strange being an orang-outang monkey.* I would willingly have let him approach, but I perceived that he was followed by a whole troop of the same species; and, as I was afraid of such a number, I fired my pistol, and the entire cavalcade, screaming with terror, disappeared in the woods. Night now approached, and I resolved to spend it on shore. I made no fire for fear that it would attract the orang-utans, and I fired my gun two or three times in order to frighten them more effectually. I had noticed under the roof of the cavern, which I had chosen for the purpose of passing the night in, a sort of hideous bird, which, by its form and manners, might well pass for the harpies* of the fable. It was an enormous bat, called, I believe, the "Vampire Bat",* and which suck the blood of persons whom they may chance to find asleep. I commenced by firing some shots to drive away these unwelcome guests, and three or four of the horrid monsters flew out, uttering piercing cries, and you may believe that I did not sleep very well with such unpleasant neighbours near, who kept up a continual rustling all night.

'I rose long before daylight, and felt glad to get away from the cavern, which I named "Vampire Cave". The country through which I now sailed was of an aspect entirely different from any I had ever yet seen. There were beautiful green plains, dotted over with clumps of towering palms; little lakes surrounded with osiers, upon the borders of which sported herds of elephants; thick tufts of cactus of all sorts, loaded with flowers and fruits, which the enormous rhinoceros seemed to devour without paying any attention to the thorns; beautiful clumps of the mimosa, the high tops of which the towering giraffe devoured with as much facility as a goat would a small shrub. Never had the works of the creation appeared so grand, so imposing to my mind, and I admired the wisdom of the divine Author of all things, who had willed that so many different beings, so many grand and terrible animals, should find in the desert nourishment for each day, and this thought sustained my courage, and appeared a warrant of the success of my enterprise. "You will not, oh my God", cried I, "you will not allow a human being to perish for want of succour, when your beneficent hand extends itself over all the habitants of the desert", and I rowed on with more strength and courage, and my eyes searched with more confidence for the sight of the Smoking Rock.

'On I sailed, and, once more, seduced by the picturesque appearance of a river which lost itself in a tranquil bay, I resolved to ascend it. The water slipped gently under the prow of my little kayak—nothing appeared to indicate any danger; there were no serpents on the bank, no horrid beasts in the forests, and I floated tranquilly on, enjoying the fresh breeze and the cool shade of the overhanging trees, when suddenly there appeared before me a long throat, armed with rows of strong, sharp teeth; it was distended to its full capacity, as if it would take in at one mouthful myself, the kayak, and the oars. I instantly comprehended the extent of my danger, and, seizing one of the oars, I drove it with all my strength direct into the yawning mouth of the monster, who disappeared in an instant, leaving a long trace of blood behind him, showing that the wound I had made was of some importance. I did not remain long on this river; two other monsters of the same nature as the first rose up to the surface of the water. They were crocodile-alligators, the most terrible kind among these animals, but whose tremendous voracity is happily balanced by a natural laziness, which retains them always near the spot where they were born. The alligator rarely goes in search of his prey, generally patiently waiting its approach; all its science consists in hiding under the water, and rising to the surface just in front of its destined victim. I had escaped from one danger only to fall into another; for, at a little distance from the River of Alligators, while coasting along a little wood, I observed that the trees were loaded with the rarest and most beautiful of birds, among which were lyras,* paroquets, humming-birds, and birds of paradise—in one word, a complete assemblage of all that array of beautiful plumage which decorates the forest of the New World, and I could not resist the desire of attacking them. I landed, attached my kayak to the bank, and walked up to the wood, holding my eagle, unhooded, in my hand. I cast him off, and he returned with a superb paroquet, whose flame-coloured feathers sparkled in the sun's rays. While I was occupied in examining him, I heard, behind me, a light rustling on the sand, which I thought was merely caused by a little land turtle, or some such animal, and I turned carelessly around. It was well I did so; for, not twelve paces from me there was a splendid royal tiger, with open mouth, crouched down as if about to spring upon me. I stood as if struck with stupor; a mist came over my eyes, and scarcely could I raise my gun, so much had horror paralysed my strength, when suddenly my brave eagle, comprehending my danger, flew boldly at the advancing tiger, and began to pick at his eyes. This timely succour saved me; for it enabled me to collect my senses, and, levelling my gun, I discharged its contents into the right flank of my enemy, and then two pistol-balls, lodged in the throat, completed my victory. The tiger lay dead; but, alas! my victory had cost me dear, for my poor eagle fell at the same time with his conquered enemy, who had seized him in his claws and had torn him in pieces. I picked him up, weeping bitterly over my loss, and carried him to the kayak, hoping some day to have him stuffed and placed in our museum.

'I quitted the shore with a sorrowing mind, and again I prayed to my heavenly Father that he would give me strength to continue my voyage. I doubled a little cape, and, suddenly, from the summit of the grey rocks which bordered the coast, I perceived a light cloud of smoke rising in the air. Oh! what vivid emotions of joy I experienced—all my presentiments were realized—there was the Smoking Rock, and I felt convinced that I should now have an opportunity of saving a fellow-creature. I turned my canoe in the direction of the long-sought-for signal. The irregularities of the rocks along the coast were the only difficulties I had to encounter, and it appeared to me that I should never get through them. At last I landed, and, with infinite difficulty, scrambled up the rocks until I arrived at a platform, on which I perceived a human creature. After a space of ten years, this was the first strange face I had seen, and you may judge of my emotions by comparing them with those you have just experienced. At the noise which I made in approaching, the individual, who was arranging the fire, rose, perceived me, uttered a cry of surprise and joy, then, joining his hands, stood still, as if waiting for me to speak. Notwithstanding the midshipman's dress* she wore, her exclamation, and the delicate contour of her features, convinced me that I was in the presence of a female. I stopped about ten paces from her, and, calling to my memory all I knew of English, I said, in a subdued tone, "I am the liberator whom God has sent you. I have received the message of the albatross". I must have pronounced these words very badly, as Emily did not at first comprehend them; I repeated them, however, and after a few moments we understood each other well enough to make a mutual interchange of our feelings. Gestures, looks, accents all filled up the blank that words had left vacant. I spoke to my new sister about the castle of Felsenheim, the bay of Falcon's Nest, our shipwreck, and ten years' sojourn on the coast, where we lived in almost European luxury; on her part, she recounted to me the history of her childhood, her ship-wreck, and existence on the Island of the Smoking Rock, making a fine story for my papa to write out in the long winter evenings. We had thus suddenly become brother and sister, the sympathy of misfortune supplying the want of the ties of blood. Emily graciously invited me to supper, after which we passed the remainder of the night, I in my kayak, she in the branches of a tree where she always slept from the fear of wild beasts.

The next morning we again met. Emily had already prepared breakfast, which consisted of fruit and broiled fish. The repast being over, the sea looked so calm that I thought we had better start; so, after packing up all her curiosities, and putting them on board the kayak, we took our seats and set off. We sailed on a long time; but an accident happened to my little barque, and I was obliged to put in at the little island which you have called "Happy", in memory of our meeting; it was there I left my new-found sister, who, doubtful of her reception in a strange family, begged me to go on and ask permission of my father to bring her among them. I consented; and my canoe having been repaired, I took the well-known road home: it was then that I encountered you, and, from fear that you were pirates, I disguised myself, and played you such a trick.'

'Oh! I am so sorry it is done,' cried Jack, as Fritz finished his story; 'but you must now tell us the history of our sister.'

Fritz was about to commence a new narration, of greater interest than the former, but I stopped him, and advised him to take a little rest before he talked any more.

66
Return to Felsenheim; the Winter

THE story of Fritz had detained us longer than I had anticipated, as, upon looking at my watch, I discovered that it was midnight. The audience were not at all sleepy, however: but, as we had to execute labours on the morrow, which would require strength and agility, I thought that if they sat up all night they would be overwearied the next day; I therefore deemed it necessary to cut short the narration, deferring its completion till a more convenient time. This decision was received with a very bad grace; but it was positive, so each one sought his accustomed resting-place either on shore or in the pinnace.

The next morning, when all the family was assembled for breakfast, the enterprise and courage of Fritz became the subject of conversation; this naturally brought on the story of last night, and I was obliged to consent that Emily's history should open the day. I wanted the dear girl herself to tell it; but she was so timid, though at the same time so lively, busied in her domestic occupations, that I could do nothing with her. Fritz was therefore entreated to act as her proxy, and resume his recital.

'As soon as I was able to understand my new sister,' said he, 'I asked her by what course of events she had been thrown on the desert coast where I now found her.

'She told me that she was born in India, of English parents, and that her father, after having served as major in a British regiment, obtained the command of an important place in the English Colony in India. The commandant, Montrose—for that was the name of Emily's father—had the misfortune to lose his wife only three years after his marriage; and, profoundly afflicted by this loss, all his affections centred in their only child. He took charge of her education, and he devoted all the time he could spare from his official duties in developing the precious qualities which nature had endowed his dear daughter with. Not content with providing her with every means for mental improvement, he endeavoured to make her a strong, healthy woman, capable of facing and resisting danger. Such was Emily's education up to the age of 17; she managed a fowling-piece as well as a needle, and rode as gracefully and firmly as the best cavalry officer, and shone resplendent in her father's brilliant saloons.

'The commander, Montrose, having been appointed colonel, was ordered to return with part of his regiment to England. This circumstance forced him to separate himself from his daughter, as naval discipline did not allow women on board a line-of-battle ship in time of war. It was arranged, however, that she should sail, the same day that he did, in another ship, the captain of which was an old friend of her father's, and who would take every care of his daughter. The old soldier wept bitterly at parting with his dear child; he foresaw all the dangers of the long and tedious voyage, and it was not without a great deal of self-command that he resolved to entrust his beloved Emily to the treacherous waves of the ocean. The voyage at its commencement was prosperous and agreeable, but before many days a terrible tempest arose. The ship was thrown off her course, and a furious wind drove her down upon our rocky coast; two shallops* were launched upon the angry waves, and a chance of safety offered to the shipwrecked. Emily found a place in the smallest—the captain was in the other. The storm continuing, the boats were soon separated, and the one that contained Emily was broken in pieces, and the poor girl alone, of all the crew, was fortunate enough to escape death. The waves carried her, half-fainting, to the foot of the rock where I discovered her. She crawled under the shade of a projecting rock, and, sinking on the sand, slept for twenty-four hours. There she passed several days, abandoned to dark despair with no nourishment but some birds' eggs which she found on the rocks. At the end of that time, the sun reappearing and the sea growing calm, the poor castaway thought of the crew in the large shallop; and, in the hope that they might find her, she resolved to establish signals of distress. As she wore a midshipman's uniform on board ship, by order of her father, she had a box in her pocket containing a flint, knife, and other articles. She picked up some pieces of wood which the sea had thrown on the sand, carried them to the summit of the rock, and there kindled a fire which she never allowed to go out. You can easily imagine how drearily passed the first days of Emily's exile; she had to contend against all the horrors of hunger and the desert. How thankful she felt for the semi-masculine education that her father had given her: it had endowed her with courage and resolution far beyond her sex. She comprehended the whole extent of her situation, and, turning to Heaven, she placed her trust in God and hoped on. She built a hut, fished, hunted, tamed birds— among others a cormorant, which she taught to catch fish—in one word, she lived alone, with no earthly succour, for three long, dreary years.'

Fritz stopped; his eyes fell upon the heroine of his story, who could hardly conceal her embarrassment.

'My child,' said I, 'you are but another proof that God never withholds his aid from those who desire it. That which you have done for three years a poor Swiss family has done for ten, and heavenly aid has never been withheld from them.'

I allowed some little time for commentaries on Emily's history; but, as I had resolved that the day should be an active one, I soon gave the signal for work. The manufacture of lime had succeeded. I submitted some pieces to the proof of water, and found it excellent.

Toward evening the pinnace was laden with all that we could carry away, and we talked seriously of returning to Felsenheim. The poetic description we had given concerning the salt grotto, and our aerial palace at Falcon's Nest, had rendered Emily exceedingly curious to judge for herself concerning all these wonders. The next day we weighed anchor just as dawn was breaking; the sail of the pinnace fluttered gaily in the fresh breeze, and Fritz's kayak, containing himself and Francis went before us as pilots. When we hove in sight of Prospect Hill, I proposed to stop and take a look at the farmhouse; but Fritz and his brother asked permission of me to go on home, so that they could have all things prepared for us. I consented, and they set out, while we landed at Prospect Hill. All was in order at the farmhouse. Emily, who for three years had not seen a human habitation could not restrain a cry of admiration. My wife ostentatiously showed her the colonies of feathered inhabitants which she had established, and which had prospered beyond our most sanguine hopes.

We again embarked in the pinnace, and from Prospect Hill we sailed to Shark Island, where we secured, in passing, a fine quantity of the soft wool of the Angora rabbit. From Shark Island we directed our course toward Felsenheim, and we could just distinguish it when a salute of ten guns greeted our ears. This produced a very good effect, doctor Ernest only regretting that the salute was not composed of an odd number of guns. 'An even number', said he, 'is entirely contrary to general usage.'

We returned the polite salute of our two artillerymen by a salvo of eleven guns, the execution of which was undertaken and performed by Jack and Ernest in a style that would have done honour to a practised cannoneer. Soon after we saw Fritz and Francis coming toward us in their canoe; they received us at the entrance of the bay, and followed us to the shore. They landed before us, and the moment Emily's foot touched the sand, a hurrah resounded through the air, and Fritz, springing forward, presented her his hand, like a gallant cavalier, and led her up to the portico of the grotto. There a new spectacle awaited us: a table was spread in the middle of the gallery, and loaded with all the fruits that the country produced. Bananas, figs, guavas, oranges, rose up in perfumed heaps upon flat calabashes. All the vases of our fabrication, coconut cups and ostrich eggs mounted on turned wooden pedestals, urns of painted porcelain, all were filled with hydromel and milk; while a large dish of fried fish, and a huge roast turkey, stuffed with truffles, formed the solid part of the repast. A double festoon of flowers surrounded the canopy above the table, sustaining a large medallion on which was inscribed, 'Welcome fair Emily Montrose'. It was a complete holiday, and as pompous a reception as our means would allow. Emily sat down to table between my wife and myself; Ernest and Jack also took their places, while the two caterers of the feast, each with a napkin on his arm, did the honours of the table. The most poetic and bombastic toasts were successively drank, and Emily's name echoed from every side.

We passed from the table to the interior of the grotto, and our young companion had the apartment next ours for her use. She could not restrain her admiration at the effects our industry had accomplished: she was astonished that a man and four children could have effected so much. We conducted her to the kitchen-garden, the just pride and special domain of my good Elizabeth; we showed her our orchard, our dove-cot—not a corner of Felsenheim passed unnoticed. The chateau in the tree at Falcon's Nest next received a visit; it had fallen into decay, from neglect, and we passed a whole week in fitting it up. We then set out for Waldegg to gather our rice and other grains, for the season was advancing, and some violent showers already warned us to hasten our preparations for the coming winter. Emily gave proof, during these labours, of an intelligence and good will which rendered her assistance very valuable; and she inspired everybody with such zeal and industry, that when the winter set in we were all prepared for it. Ten years had accustomed us to the terrible winters, and we calmly listened to the wind and storm as it raged furiously without. We had reserved for the winter several sedentary occupations, in which our new companion proved her skill and industry; she excelled in weaving and plaiting straw, osiers, etc.; and, under her direction we made some light straw hats for summer, some elegant baskets, and conveniently-arranged game-bags. My wife was delighted with her adopted daughter, and Ernest found a companion, whose fine education rendered her a conversable and intelligent woman. In fact Emily had become to my wife and myself a fifth child, and a beloved sister for my sons.

67
Conclusion

IT is with a thousand different sensations that I write the word *conclusion*. It recalls to my mind all that has passed. God is good! God is great! is the reigning sentiment in my heart. I have so many favours for which to thank Providence, that I hope the reader will pardon me for the disorder in which I finish my story.

It was toward the end of the rainy season, the wind had lost its violence, and a patch of blue sky could now and then be seen; our pigeons had quitted the dove-cot, and we ourselves ventured to open the door of the grotto, and taste the fresh air.

Our first cares were for our gardens, which had suffered injury; we repaired the damage as well as we were able, and then set out for our more distant possessions. Fritz and Jack proposed to make an excursion to Shark Island, to inspect our fort and colony there. I consented, and they set off in the kayak.

My sons, on their arrival, having examined the interior of the fort, and assured themselves that nothing of importance was damaged, began to look round and see if anything appeared on the horizon, but all was blank. Wishing to see whether the cannons were in good order, they began firing away, as if they had all the powder in the world at their command. But what was their astonishment and emotion when, a moment after, they heard distinctly three reports of a cannon in the distance. They could not be mistaken, for a faint light toward the east preceded each report. After a short consultation as to what should be done, the two brothers resolved to hasten home and recount their adventure to us. To jump into the canoe and put off was the work of an instant; the boat appeared scarcely to touch the surface, so rapidly did they impel her over the waves.

We had heard the reports of the cannons they had fired, and we could not imagine why they were hurrying back so fast.

I called out, as loud as I could, 'Halloo, there! What is the matter?' On they came, and, jumping on shore, fell into my arms, faintly articulating, 'Oh papa, papa, did you not hear them?'

'Hear what?' said I. 'We have heard nothing but the noise your waste of powder made.'

'You have not heard three other reports in the distance?'

'No.'

'Why, we heard them plainly and distinctly.'

'It was the echo,' said Ernest.

This remark nettled Jack a little, and he replied rather sharply, 'No, Mr Doctor, it wasn't the echo; I think I have fired cannons enough in my lifetime to know whether that was an echo or not. We distinctly heard three reports of a cannon, and we are certain that some ship is sailing in this part of the world.'

There was something so decided and truthful in the young man's voice that it was impossible to disbelieve the news he had brought. The discovery of a ship was a weighty matter in the history of our existence, and all felt the necessity of calm deliberation in regard to an event the consequences of which might be so important.

'If there is really a ship on our coasts,' said I, 'who knows whether it is manned by Europeans or by Malay pirates—who knows whether we ought to rejoice, or be sorry at its presence, and that, instead of preparing for deliverance, we should make preparations for defence?'

My first resolution was to organize a system of defence, and provide for our safety. We watched alternately under the gallery of the grotto, so that we could be ready in case of surprise; but the night passed quietly away, and in the morning the rain commenced, and continued so violently during two long days that it was impossible for us to go out.

On the third day the sun reappeared. Fritz and Jack, full of impatience, resolved to return to Shark Island, and try a new signal. I consented; but, instead of the kayak, we took the canoe, and I went with them. My wife, Emily, Ernest, and Francis remained in the grotto. On arriving at the fort we hoisted our flag, while Jack, ever impatient, loaded a cannon and fired it; but scarcely had the report died away in the distance when we distinctly heard a louder answering report in the direction of Cape Disappointment.

Jack could not contain himself for joy. 'Men, men,' cried he, dancing about us; 'men, papa; are you sure of it now?' And his enthusiasm communicating itself to us, we hoisted another and a larger flag on our flag-staff. Six other reports followed the first one we had heard.

Overpowered with emotion, we hastened to our boat, and were soon in the presence of the family. They had not heard the seven reports, but they had seen our two flags flying, and they were eagerly waiting for circumstantial news.

'Quick, tell us,' cried they, all at one time, 'are they Europeans?—English?—is it a merchant vessel?—a corvette?'

We could not answer half these questions; we could only positively announce the presence of a ship on our coast. My children were half-wild with joy; and Emily especially, giving loose rein to her lively imagination, assured me that it was certainly her father come in search of her, and that God himself had brought him to this spot.

I ordered that everything in the grotto should be put in a place of safety. My three youngest sons, my wife, and Emily set off for Falcon's Nest with our cattle, and I embarked in the kayak with Fritz, to reconnoitre. This separation pained us exceedingly; my good Elizabeth, whom age had rendered less confident, could not restrain her tears, and she enjoined us to be particularly prudent during our expedition.

It was near midday when we set out; we coasted along without discovering anything, and the illusion of the moment began to dissipate. On more calm reflection, however, the certainty that we had heard the seven reports of the cannon kept up our courage, when suddenly, on doubling a little promontory which had hitherto concealed it from us, we beheld a fine European ship majestically reposing at anchor, with a long-boat at the side and an English flag floating at the mast-head.

I seek in vain to find words that will express the sentiments which filled our souls. We elevated our hands and eyes toward heaven, and thus returned our thanks to God for his great beneficence. If I had permitted it, Fritz would have thrown himself into the sea and swam off to the ship; but I was afraid that, notwithstanding the English flag, the vessel before us might be a Malay corsair,* which had assumed false colours in order to deceive other vessels. We remained at a distance, not liking to venture farther without being more certain what they were. We could see all that was passing on board the vessel. Two tents had been raised on the shore, tables were laid for dinner, quarters of meat were roasting before blazing fires, men were running to and fro, and the whole scene had the appearance of an organized encampment. Two sentinels were on the deck of the vessel, and when they perceived us they spoke to the captain, who stood near, and who turned his spy-glass toward us.

'They are Europeans,' cried Fritz; 'you can easily judge from the face of the captain. Malays certainly would be more dusky than that.'

Fritz's remark was true; but yet I did not like to go too far. We remained in the bay, manœuvring our canoe with all the dexterity of which we were capable. We sang a Swiss mountain song, and when we had finished I cried out through my speaking-trumpet these three words, *Englishmen, good men!* But no answer was returned: our song, our kayak, and, more than all, our costume, I expect, marked us for savages, and we saw the captain making signs to us to approach, and holding up knives, scissors, and glass beads, of which the savages of the New World are generally so desirous. This mistake made us laugh; but we did not approach, as we wished to present ourselves before them in better trim. We contented ourselves with exclaiming, once more, *Englishmen*, and then darted off as fast as our boat could carry us. The joy that we felt, redoubled our strength, and we instinctively knew that the morrow would be a new era in our existence, and that our ties with mankind would again be renewed.

We landed near Falcon's Nest, where our dear ones were anxiously awaiting us. Our prudence was approved; Emily alone thought that we should have gone and discovered who the strangers were. My wife, on the contrary, praised us especially for not presenting ourselves before people in such a machine as a miserable kayak.

'Truly,' said she, laughingly, 'it would give too small an idea of the importance of our establishment. We must take the best of our boats when we go, or else the captain of the ship will think we are nothing but poor shipwrecked creatures.'

This little display of vanity on the part of my wife made me smile, and it was decided that the pinnace should be prepared to carry us to the English ship. Everyone but myself and wife were half-mad with joy; they ran hither and thither, acting like crazy people. As for myself, I scarcely wished to renounce my patriarchal mode of life and my possessions, which had cost me so much labour, and had become so dear to me; and neither my wife nor myself would ever again consent to a sea voyage; but all this was but a dream: we as yet knew nothing about the ship or its character.

We passed a whole day in preparing the pinnace, and loading it with presents for the captain, as we wished him to see that those whom he had taken for savages were beings far advanced in the arts of civilization. We set off at sunrise; the weather was magnificent, and we sailed gallantly along, Fritz preceding us as pilot. My wife and Emily were dressed as sailors. Ernest, Jack, and Francis managed the boat, while I attended to the rudder. As a precaution, we loaded our cannons and guns, and took with us all the defensive arms that we could find. We counted on the friendly disposition of the English; but if they deceived us, we were disposed to sell our lives dearly.

When we could clearly distinguish the ship, a sensation of vivid joy was experienced by us all: my sons were dumb with pleasure and eagerness.

'Hoist the English flag,' cried I, in the voice of a Stentor;* and a second after a flag similar to the one on the ship fluttered from our mast-head.

If we were filled with extraordinary emotions on seeing a European ship, the English were not less astonished to see a little boat with flowing sails coming toward them. Guns were now fired from the ship and answered from our pinnace, and, joining Fritz in his canoe, we approached the English ship to welcome the captain to our shores.

The captain received us with that frankness and cordiality that always distinguishes sailors; and he conducted us to the cabin, where a flask of Cape wine cemented the alliance between us.

I recounted to the captain, as briefly as possible, the history of our shipwreck and our sojourn of ten years on this coast. I spoke to him of Emily, and asked him if he had ever heard of

Sir Edward Montrose. The captain not only knew him, but it was a part of his instructions to explore these latitudes, where, three years before, the ship Dorcas, which had on board the daughter of Commander Montrose had been supposed to be wrecked, and to try to discover whether any remains of the young girl could be found. In consequence he manifested the greatest desire to see her, and assure her that her father was alive. He informed us that a tempest of four days' duration had thrown him off the course, which he followed for Sydney and New Holland, and had driven him on this coast, where he had renewed his wood and water. 'It was then,' added he, 'that we heard the reports of cannon, which we answered; on the morrow new discharges convinced us that we were not alone on the coast, and we resolved to wait until, by some means or other, we discovered who were our companions in misfortune. But we find an organized colony and a maritime power, whose alliance I solicit in the name of the united kingdoms of Great Britain.'

This last sally made us laugh, and we cordially pressed the hand which Captain Littleton extended to us.

The rest of the family were waiting some distance off in the pinnace. We took leave of the captain, who, ordering his boat to be manned, arrived almost as soon we did on board our vessel. We received him with every demonstration of joy and friendship, and Emily was half wild with happiness at the sight of a fellow countryman, and one who brought intelligence of her father.

The captain brought with him an English family, that the fatigues of the passage had rendered ill; it was that of Mr Wolston, a distinguished machinist,* and consisted of himself, wife, and two daughters. My wife offered Mrs Wolston her assistance, and promised her that her family should find every comfort and convenience at Felsenheim, if they would return with us. They gladly consented, and we set out with them, taking leave of the captain, who did not like to pass the night away from his ship.

My readers can form an idea of the astonishment which was evinced by the Wolston family on seeing all our establishments. We ostentatiously pointed out to them Felsenheim with its rocky vault, the giant tree of Falcon's Nest, Prospect Hill, and all the marvels which were comprised in our domains. A frugal repast in the evening united both families under the gallery of the grotto, and my wife prepared, in the interior, apartments and beds to receive the newcomers.

The next morning Mr Wolston came up to me, and, tenderly stretching out his hand spoke as follows:—'Sir,' said he, 'I cannot express all the admiration that I feel on regarding the wonders with which you are surrounded. The hand of God has been with you, and here you live happily, far away from the strife of the world, alone with your family among the works of creation. I came from England to seek repose: where will I find it better than here? And I will esteem myself the happiest of men if you will allow me to establish myself in a corner of your domains.'

This proposition of Mr Wolston filled me with joy, and I immediately assured him that I would willingly share with him the half of my patriarchal empire.

Mr Wolston hastened to communicate to his wife the success of his application, and the morning was consecrated to the joy and pleasure that this news caused. Considerations of a painful nature occupied my mind: the ship which now presented itself was the first one we had seen for ten years, and probably as long a period might elapse before another appeared, should we let Captain Littleton and his ship leave us without any addition to his crew. These questions affected the dearest interests of the family. My wife did not wish to return to Europe; I was myself too much attached to my new life to leave it, and we were both at an age when hazards and dangers have no attraction, and ambition is resolved into a desire for repose. But our children were young, their life was but just commencing, and I did not think it right to deprive them of the advantages which civilization and a contact with the world presented. On the other hand, Emily, since she had heard that her father was in England, did not conceal her desire to return; and although we regretted losing this amiable girl, yet it was impossible to detain her. At last I decided to call my children together, and ascertain their sentiments. I spoke to them of civilized Europe, of the resources of every kind which society offered to its members, and I asked them if they preferred to depart with Captain Littleton or to be condemned to pass the remainder of their lives upon this coast.

Jack and Ernest declared that they would rather remain. Ernest, the philosopher, had no need of the world to interrupt his studies; and Jack, the hunter, found the domain of Falcon's Nest large enough for his excursions. Fritz was silent, but I saw by his countenance that he had decided to go: I encouraged him to speak; he confessed that he had a great desire to return to Europe, and his younger brother, Francis, declared that he would willingly accompany him.

At last, the family of the old pastor was to be dismembered: two of our sons were about to leave us, and perhaps we should never again see them. My good Elizabeth submitted to the sad necessity; she was a mother, but she studied the advantage of her children.

Mr Wolston also dismembered his family: he kept but one of his daughters; the other went on to New Holland. These family arrangements were very painful, and when they were finished I hastened to inform the captain of the Unicorn. He readily consented to take our three passengers.

'I resign three persons,' said he, 'Mr and Mrs Wolston and one of their daughters; I take three more, and my complement will be complete.'

The Unicorn remained eight days at anchor, and we employed them in preparing the cargo which was to be the fortune of our voyagers on arriving in Europe. All the riches that we had amassed—pearls, ivory, spices, furs, and all our rare productions, were carefully packed and put on board the ship, which we also furnished with meat and fruits.

On the eve of their departure, after having exhausted myself in a last conversation, in which I advised my sons always to carry out the principles in which they had been instructed, and so to live in this world that we might be united in the next, I gave Fritz this narration of our shipwreck and establishment on the desert coast, enjoining him expressly to have it published as soon after his arrival as he possibly could; and this desire on my part, exempt from all vanity of authorship, had for its only object and hope that it might be useful to children as a lesson of morality, patience, courage, perseverance, and of Christian submission to the will of God. Perhaps some day a father may take courage from the manner in which we supported our tribulations; perhaps some young person will see, in the course of this narrative, the value of a varied education and the importance of becoming acquainted with first principles.

I have not written this as a learned man would have done, and all my results may not have been carried out according to the correct theory; but we were in an extraordinary position, and were obliged to depend on our own resources. We placed our entire trust in the mercy of God, and he ever watched over and protected us.

We none of us slept much during the night preceding the day of separation. At the dawn of day the cannon of the ship announced the order to go on board. We conducted our children to the shore; there they received our last embraces and benedictions, and went on board; the anchor was weighed, the sails unfurled, the flag run up to the mast-head, and a rapid wind separated us from our children.

I will not attempt to paint the grief of my dear Elizabeth—it was the grief of a mother, silent and profound; and when we lost sight of the ship which contained our children she burst into a flood of tears. Jack and Ernest also wept bitterly. Curbing my grief and heartfelt sorrow, I led my wife to our now desolate and deserted home.

I write these last pages, which the ship's boat is waiting to take on board. My sons will receive these lines, with which I send my last blessing. May God ever be with you. Adieu, Europe! Adieu, dear Switzerland! Never shall I see you again! May your inhabitants be always happy, pious, and free!

THE END

EXPLANATORY NOTES

16 *firkins*: small casks.

17 *fowling-pieces*: long-barrelled guns for shooting game-birds.

   auger: a tool for boring holes.

19 *crow*: a crowbar.

   Archimedes' lever: Archimedes, a philosopher of ancient Syracuse, said that with a lever long enough and a place to stand upon, he could move the world.

22 *carbines*: firearms shorter than a musket and often used by horsemen.

24 *palm trees . . . coconuts*: traditional accoutrements to tropic islands, perhaps here first established as necessary to the castaway experience. See note to p. 40.

25 *isosceles triangle*: a triangle having two equal sides.

   penguins and flamingos: the penguin, a seabird native to Antarctic regions, is not capable of flight, and is here anomalously joined with the tropical flamingo, an exotic if idiosyncratic touch typical of much wildlife to follow. (cf., however, pp.165-66).

26 *tenter-hooks*: hooked nails with sharp points upon which objects may be hung; can also be used to stretch fabric, as here.

29 *fox . . . stork*: reference to the fable of the fox and the stork who entertain each other at dinner. The stork serves milk in a pitcher too deep for the fox, who returns the favour by serving milk on a flat stone from which the stork cannot drink.

30 *agouti*: a large rodent resembling the guinea-pig.

   M. de Courtills: unknown to us.

31 *patates*: Fr. *patate*—sweet potato.

33 *Cain*: the eldest son of Adam and Eve and a tiller of the ground (Genesis 4: 1–25). In a jealous fit, he kills his brother, Abel, and as punishment is banished to lead a nomad's life.

34 *vespers*: evening prayers or devotions.

35 *kids*: here, small goats (not children).

36 *Robinson Crusoe*: his adventures, the creation of Daniel Defoe, provide the model for all subsequent castaway stories, and are

several times referred to below—not always, however, with complete accuracy. Thus, there are no references in Defoe's novel to coconuts or other indigenous tropic fruits, his hero subsisting chiefly on goat meat and (imported) European grains.

40 *coconut*: more erroneous information. The coconut is found in nature surrounded by a thick fibrous husk, not the 'thin skin' described.

42 *gourd-tree kind*: the bottle-tree gourd resembles a large cucumber.

46 *crowns*: coins bearing the symbol of a crown, testifying to the royal impress on matters of mintage.

48 *from thy loins*: cf. Genesis 35: 11.

   young patriarch: Jacob, whose loins were referred to above, was a patriarch of the Jewish people.

53 *hydromel*: the same as mead, a fermented drink made from honey and water.

56 *hyena*: this will be a long-deferred encounter. See p. 425-26.

57 *milk of almonds*: a beverage of sweet blanched almonds and water.

62 *jackals*: wild dogs indigenous to Asia and Africa.

63 *Caribbees*: Caribs, a people indigenous to the Caribbean, so often accused of cannibalism during the early years of European exploration in the New World as to become synonymous with cannibals.

69 *sailyard*: a pole extending from and swinging upon the mast, used to hold a sail: a boom.

70 *Vandals*: a Germanic tribe which invaded Western Europe in the fourth and fifth centuries, whose name became a byword for barbarism and pillage.

   roasting-jack: machine for turning meat on a spit.

   Westphalia hams: Westphalia is a province of Germany known for its succulent hams.

71 *maize*: the corn of the Americas: Indian corn.

77 *Mr Currier*: a currier dresses leather as a trade.

78 *Cain . . . Abel*: see note to p. 33 concerning this tragic pair of siblings.

79 *The same which Robinson Crusoe . . . island*: Defoe's hero does indeed eat turtle eggs (and turtle), though they are not a main staple of his diet.

80 *Canary wine*: light, sweet wine from the Canary Islands.

85 *Condor*: large, vulture-like bird of South America (see p. 368). Not, however, to be confused as here with the ostrich, which Mme de Montolieu added to the menagerie (see pp. 359 ff.)

86 *brache*: measurement equal to the length of an extended arm.

94 *shagreen*: untanned leather from shark skin.

97 *Laplanders*: inhabitants of a polar region adjacent to Scandinavia.

packthread: thick thread used for rough stitching.

101 *Tartar*: a member of the various tribes which overran Asia Minor in the Middle Ages.

103 *as the wolf . . . with the . . . lamb*: reference to Aesop's fable in which a wolf resorts to sophistry so as to justify his eating a lost lamb.

104 *bustards*: large, ground-running birds related to cranes.

105 *Nimrod*: a biblical figure (Genesis 10: 8–12), the great-grandson of Noah, noted for his hunting abilities; prototype of all hunters.

106 *margay*: a tropical American tiger-cat, now nearly extinct.

107 *mango tree*: i.e. the mangrove, a tropical tree that grows in mud flats at low-water mark with its roots showing well above ground. The mango is quite another, fruit-bearing tree.

108 *fig-trees*: large members of the mulberry family, indigenous to Asia Minor.

Antilles: islands of the West Indies.

packing-needles: large needles, used with packthread.

112 *triangles . . . lines*: using here the Pythagorean theorem: the square of the hypotenuse of a right-angled triangle is equal to the sum of the squares of the other two sides.

117 *park-paling*: a fence made of vertical stakes.

119 *cavaliers*: i.e. knights.

123 *tares*: a vetch, synonymous (in Elizabethan, hence biblical, usage) with weeds. The father's allegory is heavily derived from Christ's parable of the talents (Matthew 25).

137 *karata*: Karatas, or silk-grass, is a plant native to South America and the West Indies, valuable for its fibre.

Linnaeus: Carl von Linné (1707–78), Swedish botanist who devised the Linnaean system of classification, still used today and a major creation of the Enlightenment.

138 *aloe*: a genus of plants used for medicinal purposes.

139 *cochineal*: insect of South America used to make brilliant scarlet dye.

141 *Bruce*: James Bruce (1730–94), Scottish naturalist and explorer.

143 *parachute*: though developed by 1785 as a safety device associated with balloon ascension, 'parachute' makes no sense here. 'Basket', or 'car', would be more appropriate.

145 *ortolans*: buntings; like other small song-birds, once regarded as delicacies.

149 *without fear and without reproach*: a phrase applied to Seigneur de Pierre Terrail Bayard (1470–1524), a famous French soldier-hero nicknamed *le bon chevalier*, and considered a type of exemplary behaviour.

151 *New Holland*: name given to the western half of Australia by the Dutch navigator Abel Tasman (1603?–59), in 1644.

  Captain Cook: James Cook (1728–79), English naval captain and explorer associated with the opening of the Pacific islands to commerce.

157 *farrier's instruments*: a farrier is a combined blacksmith and veterinarian.

  tortoise: generally applied to the terrestrial genus; 'turtle' would be more accurate here.

159 *quintals*: one quintal equals 100 kilos.

162 *manioc . . . cassave*: manioc is technically the Brazilian name for the cassave plant, but can also refer to the flour derived from its roots.

164 *bulkhead*: an upright partition dividing the hold of a ship into water-tight compartments.

  Herculean: the mythic Greek hero who sought immortality by performing twelve presumably impossible tasks for King Eurystheus, Hercules lends his name to all difficult labours, save that of childbirth.

165 *Lilliputians*: the residents of Lilliput in Swift's *Gulliver's Travels* are 6 inches tall.

167 *pollard*: a type of wheat.

171 *bulkhead*: see note for p. 164. There is no part of a ship's hold known by this term.

173 *mortar*: heavy, cup-shaped vessel in which ingredients are pounded with a pestle.

  train: a fuse.

173 *pitched*: smeared with pitch (tar) to seal.

   cracker: a fire-cracker, but here something much larger.

179 *ananas*: pineapple.

   maize: see note to p. 71.

181 *Patagonians*: Indians native to the southernmost tip of South America. The weapon described is called a *bolas*.

183 *bustard*: see note to p. 104.

187 *Mancenilla*: manchineel, a tree of West Indian origins, which has a poisonous sap and an acrid fruit resembling an apple.

   Yguana: the iguana, a large arboreal lizard common to the West Indies and South America, sometimes reaching 5 feet in length.

189 *eyes of the sun*: this poetic phrase is perhaps a misprint for the more sensible 'rays of the sun'.

   guava: acidic fruit used for jelly, apple-like only in shape.

190 *oak . . . green*: the Live Oak of the southern regions of North America, an evergreen.

192 *erect and bristling*: the description is that of the American ruffed grouse. cf. p. 277.

194 *Myrica cerifera*: the wax-tree, indigenous to the southern USA, whose berries can be used for the purposes described.

   chaffinch: a common songbird of Great Britain. See p. 208, for a specific identification of the exotic species referred to.

196 *cephalate*: a term usually applied to a large-headed mollusc, but obviously inappropriate here.

   kakerles: cockroaches.

197 *marmoset*: a type of monkey. What is meant is 'marmot', a burrowing rodent of the squirrel family.

198 *caoutcnouc tree*: the rubber tree of Brazil.

199 *sago palm*: a Malayan feather palm, the sap of which yields a starchy foodstuff, sago.

202 *quincunx*: an arrangement of fives, four plants forming a rectangle or square, with the fifth centered in the middle, a design associated with the extreme formalism of seventeenth-century gardens.

206 *four-pounders*: small cannons that take a four-pound ball.

207 *land of Canaan*: the land promised to the children of Israel, and figurative of heaven.

212 *buffaloes*: probably intended to mean the water buffalo of India, as per early illustrations, but the American bison is equally a candidate, as per the reference to 'Americans' on p.215.

218 *farinaceous*: cereal-like, starchy.

220 *ligneous*: woody.

226 *Mr Huber of Geneva*: M. Huber Lullin published a dissertation on the habits of bees, despite the fact he was blind.

231 *karata*: karatas, cf. note to p.137.

algava: algarroba? Probably a type of mesquite, the wood of which is highly flammable.

233 *onagra*: a wild ass. 'Onagra' denotes the female of the species, 'onager' the male. In the 1826 text, the latter is often used as an alternative to the former, a gender blurring that we have silently amended throughout, given the original emphasis and situation.

240 *heath fowl*: the ruffed grouse. See note to p. 192.

242 *flax-plant*: Flax (*Linum usitatissimum*) is the common source of a textile fibre much used in the western world before the introduction of cotton. The variety found in New Zealand, however, has fibres in its leaves (as here) not its stems.

244 *beetles*: heavy instruments for beating the fibres loose from leaves (see pp. 249, 250).

250 *turnery*: the process or art of shaping objects on a lathe.

spinning-wheel: a machine used to twist (spin) fibres into thread or yarn.

251 *Robinson Crusoe*: that Crusoe found 'a spacious cavern that merely required arrangement' contradicts the statement on p. 248 that he had 'cut himself a habitation out of the solid rock'. In point of fact, both statements are true. Crusoe's primary cave is one that he dug himself, but later in the narrative he finds a natural cavern, which he uses as a storeroom but not for his habitation. The situation in *The Swiss Family Robinson* conflates the two caves in *Robinson Crusoe*, as we shall see.

252 *calcareous stone*: i.e. limestone.

253 *mephitic*: poisonous.

255 *squib*: a type of firework that explodes after a period of incandescence.

tapers: candles.

Bucephalus: the war-horse of Alexander the Great, tamed by him as a boy, hence the reference.

256 *French horn*: one of the most difficult of the brass instruments to master, with a long, slender, coiled tube leading to the mouthpiece.

grotto of Antiparos: Antiparos is a small, rocky island, one of the Greek Cyclodes in the Aegean Sea. On the south coast lies a subterranean grotto that has enormous stalactites and houses a tiny chapel of St John.

259 *pilasters*: a pilaster is a square or rectangular pillar, often part of a supporting wall, emerging from it in half-relief.

263 *sea-dogs*: the common harbour seal.

266 *windlass*: a winch.

Neptune: the sea-god in Roman mythology, identified with the Greek god Poseidon, traditionally armed with a trident, or three-pronged spear.

267 *tunny fish*: the tuna.

the rind of a tree: the bark.

272 *loom*: a machine used to weave thread or yarn into fabric.

274 *acacia*: a shrub of the mimosa family, that yields (as below) a gummy sap.

gum-mastic: used in making varnish.

275 *Ceylon*: island nation off southern tip of India; the modern Sri Lanka.

277 *Canada heath fowl*: once again, the grouse (cf. pp. 192, 240, etc.).

pine strawberry: as described.

278 *jetty black*: native to Australia, the black swan was first discovered by Willem de Vlaming, a Dutch navigator, on what is now known as the Swan River. See also p. 433.

279 *Beast with a Bill*: the duck-billed platypus, a marsupial indigenous to eastern Australia and Tasmania.

Arcadia: the region near Rome associated with the idyllic landscape of Virgil's pastoral poetry.

282 *Milo*: Milo or Milon of Crotona, a Greek athlete of the late sixth century BC, noted for his feats of strength.

283 *a wash of size*: size is a viscid substance used to stiffen fabric or seal porous surfaces.

285 *flageolet*: a wind instrument with six stops, sometimes keyed.

286 *ortolans*: see note to p. 145.

286 *Pelew islands*: the Pelew (or Palau) Islands are part of the Caroline Chain in the western Pacific, east of the Philippines.

287 *root of anise*: the anise is an umbelliferous plant found in Egypt, the oil of which has medicinal properties.

290 *monkey root*: ginseng, a sweet-tasting, finger-sized root, often assuming anthropomorphic shapes, used in China for medicinal purposes, early exported from America to the Orient.

beccafico: another edible songbird.

291 *resin*: a vegetable product coming from certain conifers, used in making varnishes and turpentine.

295 *pigeons of Molucca*: Molucca was formerly called the Spice Islands, part of Eastern Indonesia.

296 *isinglass*: semi-transparent mica, often used as a substitute for glass.

298 *familiar spirit*: a demon at the beck and call of dealers in magical powers.

300 *mages*: i.e. magi, members of the Persian priestly caste, credited with powers of sorcery and astronomy.

magic lantern: an optical instrument that magnifies images projected off a tiny mirror against a screen.

301 *grey moss*: i.e. Spanish moss.

302 *Barbary*: Arab countries on the coast of Africa.

osiers: a species of willow with tough, pliant branches, useful for baskets and other wicker work.

303 *jackal*: this is the first mention of Jack's new pet, the origin of which seems to have dropped out between translations.

304 *Hottentots*: inhabitants of South-West Africa, noted for their short stature.

gom-gom: or gum-gum, more properly an iron bowl struck by a wooden stick; a word of Malay origin, and an instrument quite different from the one described.

305 *canals*: i.e. conduits.

naiads: river nymphs.

308 *turning-lathe*: we learned earlier that the father was forced to do without a lathe: see p. 250.

309 *Babel*: a city named in the Bible (Genesis: 10: 10; 11: 9) as having suffered because its polyglot citizens could not communicate with one another.

310 *sea-lion*: a large-eared seal, or otary, so called because of its bellicose roar.

Admiral Anson: Baron George Anson (1697–1762) circumnavigated the globe in just four years and captured a Spanish galleon near the Philippines while *en route* back to England (by way of Macao), a prize that made him a wealthy (if not universally popular) man.

312 *smoke-jack*: i.e. a roasting-jack (see note to p. 70).

polypi: the tiny, tentacled anthozoan organisms that build and inhabit coral reefs.

313 *dewlaps*: usually a pendulous fold of skin under the throat of bovine animals, but here the 'whalebone', or baleen, by means of which the Greenland whale filters out its food and women gave unrealistic shape to their bodies with corsets.

314 *Jonah*: the biblical prophet (Jonah 2; or John 1: 42; 21: 15–17) who as punishment for disobeying a divine command is swallowed by 'a great fish' (traditionally read as a whale) in which he lives for three days and nights until vomited forth on dry land. Jonah's ordeal is a type of rebirth, and is at the heart of the castaway tradition, starting with *Robinson Crusoe*.

318 *steam-boat*: perfected by Robert Fulton in 1807, the paddle-wheel steam-boat is an up-to-date inclusion by Mme de Montolieu, but is anachronistic given the context. The year would be about 1800.

320 *mammoth*: a large species of elephant of the Pleistocene epoch, rendered extinct by the glacial age.

321 *antediluvian*: literally, 'before the flood', but here intended to mean prehistoric.

322 *Neptune*: See note to p. 266.

324 *August*: in Locke's translation, January. We have modified the date for the sake of uniform chronology.

Robinson Crusoe: Defoe's hero actually notches his calendar on a cross, a symbol of penitence, but in subsequent versions edited for (Protestant) children, a penitential stick or rod is substituted.

328 *Milo*: see note to p. 282.

331 *palanquin*: a stretcher-like conveyance, consisting of a basket with two poles projecting fore and aft, used in east Asian countries for carrying persons of importance.

332 *singular animal*: it is perhaps needless to point out that the following episode, de Montolieu's most memorable contribution to

the story, is a creation of a perfervidly romantic imagination. The boa is a gentle creature, incapable of ingesting a jackass whole.

335 *videttes*: mounted sentries.

338 *pigs*: hogs can kill and eat rattlesnakes because their thick layers of fat make them immune to venom.

339 *Hooded Snake*: the cobra.

341 *Captain Stedman*: Captain John Gabriel Stedman (1744–97) travelled to Surinam to help placate social tensions between white masters and black slaves. See his *Narrative of the Five Years Expedition against the Revolted Negroes of Surinam* (1796).

344 *gold mine*: Crusoe discovers no such mine, but occasionally discourses on the uselessness to one in his condition of the gold coins salvaged from the wreck. However, he keeps the coins.

347 *bittern*: a bird related to the heron, but smaller, and known for the peculiar 'boom' it utters during the mating season.

348 *cabiai*: or capybara: a rodent, near-cousin to the guinea-pig, which inhabits the rivers of South America.

351 *ondatra*: since the ondatra is essentially a muskrat, there is duplication here.

manchineel-tree: see note to p. 187.

353 *Tujacus*: We must take the Father's word on this species, about which we can find nothing more.

Otaheitan pigs: i.e. Tahitian pigs; for Cook, see note to p. 151.

peccaries: a type of wild swine indigenous to the American South-West and South America.

355 *ravensara*: a small Madagascar tree of the laurel family. See also pp. 369, 388.

356 *Sydney, Port Jackson, and New Holland*: Sydney is the capital of New South Wales, Australia (New Holland); Port Jackson is a harbour town of New South Wales.

357 *Kamtschatkans*: i.e. Kamchatkans, natives of Kamchatka, a Siberian Peninsula on the Bering Sea.

358 *Arabia Petrea*: a section of north-west Arabia, noted for its rocky terrain.

'*Ad angusta, per angusta*': 'to difficulties, through difficulties' (Locke's note).

358 *Nile . . . Nubia*: the Nile is one of the longest rivers in the world, the longest in Africa, passing through Egypt. Nubia is a desert region of north-east Africa which includes the Nile valley.

359 *Tartars*: see note to p. 101. Tartars were noted for their horsemanship.

Bedouins: nomadic tribesmen of the Arabian desert, fierce warriors and skilled horsemen.

Camelopards: giraffes (built like camels, spotted like leopards).

360 *euphorbia*: a genus of plant having an acrid milky juice. See also p. 375.

362 *25 to 30 eggs*: a dozen would be a more likely number.

364 *shower of frogs*: a learned reference beyond our reach.

Cape of Good Hope: the southern tip of South Africa.

366 *Tibetan*: 'Thibetian' in the Locke translation. Tibet is a remote and mountainous region in southern China.

368 *birds of prey*: turkey buzzards.

condor: see note to p. 85.

369 *bucaniers*: not to be confused with the word for 'pirate', though the etymology is related. A *boucan* is a framework of wood upon which meat was smoked by Brazilian natives, or 'bucaniers'.

Greenlanders: inhabitants of Greenland, a large continental island within the Arctic Circle, near Iceland.

ravensara: see note to p. 355.

370 *the sons of Jacob*: about to die, the biblical patriarch calls his sons to him, and prophesies to each his future (Genesis 49: 1–33).

amianth: a variety of asbestos, a natural mineral made up of white fibres that can be woven into fire-proof fabric.

talc: generic Arabian term for mica-like minerals; cf. 'isinglass', p. 296 .

371 *Angora rabbits*: noted for their long, fine white fur, sometimes 7 inches in length, and pink eyes.

372 *Levaillant's Voyages*: François Le Vaillant (1753–1824), French explorer and ornithologist.

374 '*cuckoo-indicator*': the cuckoo is a European bird whose name echoes the mating call of the male; the reference here is to the African honey-guide, genus *Indicator*, used by bee-hunters as a guide to hives.

374 'Hop o' my Thumb': the folk-tale about a small boy of that name
who is abandoned with his brothers in the woods. They are taken
in by a family of ogres, the father of whom is tricked by the boy
hero into eating his own daughters instead of him and his
brothers.

375 euphorbia: see note to p. 360.

377 hoodwinked: hooded, so as to lose the power of sight.

380 palisade: a fence made of stakes, or pales, placed tightly together
and pointed at the top, for defence.

Vanilla bean: the source of the familiar flavouring.

382 fikes: fyke, a long bag-net or fish trap, kept open by a series of
hoops.

383 tunny fish: see note to p. 267.

asafetida: this gum, with its strong, garlicky smell, was used in
cooking and as a medicinal specific, hung in bags about the neck
to ward off disease.

384 magic lantern: see note to p. 300.

386 brevet: an official document granting certain privileges.

388 hydromel: see note to p. 53.

ravensara: see note to p. 355.

389 William Tell: the well-known exploits of the legendary Swiss
patriot belong more to the spirit of emerging nationalism in the
late eighteenth and early nineteenth century than to any verifi-
able events.

390 ondatra: see note to p. 351.

391 transparent: 'translucent' might be a more appropriate word.

392 Sèvres: town in France famed for its fine porcelain.

kayak: 'cajack' throughout the Locke translation. As described,
the boat is a hunting-craft made by stretching a skin made of
hides over a light framework. Its occupant is enclosed around
the waist within the cockpit to make the hull watertight.

395 sea-calves: seals.

398 cameos: the word is most properly used in reference to a precious
stone, similarly carved, and not to this less expensive version.

buccins: whelks.

ostracism: whatever the etymology, Athenians used potsherds and
clay tiles in determining banishment.

398 *ambrosia*: the fabled food of the gods.

401 *Tolay of Buffon*: The name Buffon is the common appellation for
the famous naturalist, Georges-Louis Leclerc, comte de Buffon
(1707–88). A tolay has the appearance and habits of rabbit and
is larger than the ondatra (muskrat).

402 *hippogriff*: a winged horse.

403 *Zurich*: the most populous canton of Switzerland, and near the
Wyss homeland in Berne.

404 *pneumatic machine* : an air pump.

408 '*The ox . . . turneth*': cf. 1 Corinthians 9: 9; Deuteronomy 25: 4;
Timothy 5: 18.

412 *morse*: a walrus, or *sea-horse*, more properly. The *sea-cow* is the
dugong or manatee, though the two are often, as here, conflated,
as a later reference to the walrus (p. 413) suggests.

413 *dugon*: the dugong, a herbivorous mammal of the Indian ocean,
similar to the walrus in appearance, but lacking the prominent
tusks.

417 *Prince Fortunatus*: fabled prince who possessed many magic
articles, including an enchanted cap that would transport him
instantly to any place he wished to visit.

418 *clove-tree*: a tree indigenous to the Moluccas. It bears buds which
when dried are used as the familiar spice.

Dutch of Ternate: Ternate is one of the Moluccas or Spice Islands,
now part of eastern Indonesia, where the Dutch held sway for
over 250 years.

sturgeons: or paddlefish, are indigenous to the Atlantic's North
Temperate zone and are valued as a source of caviare, made from
the roe (eggs) found in females of the species.

Vienna . . . Danube: Vienna is the capital of Austria, and a port on
the Danube, a river in central and south- eastern Europe flowing
to the Black Sea.

419 *Peyceres*: a naturalist unknown to us as well.

leech: a blood-sucking aquatic worm. But it is probably the
lamprey eel, a blood-sucking parasite devastating to game fish,
which is referred to here.

421 *buffaloes*: the reference is here, in the de Montolieu sequel,
clearly to the American bison, whatever was Wyss's original intention. (Cf. note to p. 212 .)

423 *Turk Pigeon*: probably a large-plumed pigeon indigenous to the East Indies, noted for its turkey-like displays.

425 *hyena*: the following account, like that of the boa constrictor, takes liberties with observable nature. There are three distinct species, conflated here as the striped hyena. Though they live alone, hyenas do hunt in bands, are chiefly carrion eaters, seldom attack anything bigger than a sheep or goat, and are notorious cowards.

426 *ounce*: the lynx.

427 *Sydney, Port Jackson*: see note to p. 356.

traveller pigeon: usually referred to in America as the passenger pigeon, a small, dove-sized bird with a long, cuneal tail, capable of flying great distances. The 'celebrated naturalist' mentioned below (p. 428) is probably Alexander Wilson, who estimated the numbers of this bird at over 2,000 million. It is now extinct, due to the ravages described below.

428 *telegraphs*: in 1820, the word would signify a semaphore device, not the invention of S. F. B. Morse.

433 *black swans*: see note to p. 278.

heron royal: perhaps the great white heron of Florida.

tapir: an animal resembling a pig, except for a long, flexible snout.

434 '*Numidian Girls*': the 'demoiselle', a crane of Asia and Africa, known for its grace of form.

the Bird of Paradise: male birds of this species are noted for their long, colourful plumes. More accurately classed *Paradiseidae*, but the classification given was common *c.* 1800.

436 *Pine cones*: 'Pine-apples' in Locke.

doura: or *durra*, Indian (or true) millet. Black millet is one of several grain families known generally as 'millet'.

438 *caravanserai*: 'caravansary' in Locke, a type of inn accommodating caravans of the Middle East.

Nemesis: in classical mythology, the goddess of divine retribution.

Valete, valete: 'farewell, farewell' (Locke's note).

441 *Kamschatdale*: i.e. Kamchatka. See note to p. 357.

Talipot palm: a palm with especially large leaves, of perfect size for a fan.

442 *gherkin*: 'a sort of cucumber' (Locke's note).

444 *a celebrated writer*: Bernadin de Saint Pierre (1737–1814), French author who stressed man's enduring nobility in a natural environment, a Rousseauean doctrine he illustrated in his popular romance, *Paul et Virginie* (1788), a castaway story that takes place on the Île de France. His *Studies of Nature* is cited below.

Isle of France: i.e. Île de France, an island of volcanic origin situated in the Indian Ocean due east of Madagascar. Seized from the French by the British in 1810, its name was changed to Mauritius.

Indian Brahmin: a member of the highest or priestly caste among the Hindu.

. . . for all his wants: 'from St Pierre's *Studies of Nature*' (Locke's note).

445 *Admiral Anson*: see note to p. 310.

Sultan Cock: a small domestic fowl originating in Turkey, also called a sultana bird. For Buffon, see note to p. 401.

446 *pintados*: 'pintadoes' in Locke; the guinea fowl.

Antilles: see note to p. 108.

cayman: reptiles of Central and South America related to alligators, but characterized by a much narrower snout.

452 *merops, or bee-eaters*: bright-coloured birds of the family *Meopidae* found throughout the subtropics, including Australia.

454 *elysium*: the abode of the blessed after death in Greek mythology, but, generally, any ideally happy place, here defined as an impregnable garden, in keeping with earlier allusions to Paradise, Eden, Arcadia, etc.

tunnels: i.e. conduits.

455 *great squirrel of Canada*: probably the fox squirrel.

loriots: orioles.

456 *many articles . . . in abundance*: 'Vide vol. I, Part I' (Locke's note, but read on): I fail to find any earlier mention of this handy equipment.

458 *resembled the turkey*. The fowls used to replenish the gene pool of the family's poultry yard seem closely related to the fabled wahoo bird.

461 *narwhal*: 'narval' in Locke; a small grey whale of the Arctic. The male's front tooth develops into a straight tusk extending up to half the length of the animal.

babiroussa: 'babirossa' in Locke; a wild hog from eastern Asia.

463 *Salargane*: more correctly, salangane, perhaps a misprint.

464 *a river of that name in Florida*. The St Johns river was made famous in literature by William Bartram's account of it in his *Travels* (1791).

albatross: or frigate-bird, having long, narrow wings and a 12-foot wing-span. Made famous in the romantic age by its ominous role in Coleridge's 'Ancient Mariner'.

sea-otter: a marine otter found along the shores of the North Pacific.

468 *apalogue*: apologue: a fable or allegory with a moral message.

469 *toga virilis*: the name of the robe assumed by Roman youth at the age of 17 (Locke's note).

470 *Sea-horses*: Walruses. See note to p. 412.

nautilus: a mollusc with a smooth, coiled shell that acts as a float. A *gondola* is the famed water taxi of Venice.

471 *Gothic cathedrals*: masterpieces of medieval architecture, usually gigantic structures with decorative facial features and pointed arches, often associated by romantic writers with forms found in nature.

472 *the Doric, the Ionic, the Corinthian*: the three orders of Greek architecture, progressing from the earliest, simplest style to the latest and most complex.

pretium affectionis: 'Literally, the worth of desire' (Locke's note).

473 *vidette*: see note to p. 335.

charivari: 'uproar, clatter—Meadows's Dictionary' (Locke's note).

474 *tubercles*: protruberances or nodules; tubers, cf. p. 136.

truffles: underground fungi found in Europe, where they are considered a rare delicacy.

476 *Nankin cotton*: a yellow variety of cotton used to make the cloth called nankeen, after the place in China where it was made.

479 *poetry*: in the original, perhaps, but lost in translation.

480 *Buffon*: see note to p. 401.

481 *Hercules . . . Nemean forest*. For the first of his twelve labours, Hercules killed the fierce Nemean lion. See note to p. 164.

482 *cachalot*: a sperm whale, which grows to about 62 feet in length and has an enormous head with a squarish profile; valued for its oil and most especially for spermaceti, a waxy white substance used for making candles and cosmetics.

488 *cramp-irons*: a piece of metal bent at both ends, used to hold pieces of masonry together; not to be confused with *crampons*, but used for a similar purpose here.

491 *antiscorbutic*: foods preventing scurvy, a disease formerly common among sailors, caused by an inadequate diet.

cochelaria: perhaps the cockle, or corn-cockle, whose medicinal properties are not known to us.

cormorant: an aquatic diving bird, with dark feathers, a hooked bill, and a long, distensible neck.

493 *snipe*: a small game-bird, distinguished by a plump body and relatively long bill.

495 *Amhingu, or snake-neck bird*: the anhinga, or water turkey, found in swampy regions of the New World, and distinguishable by its unusually long neck.

496 *orang-outang monkey*: more often orang-utan, an anthropoid ape distinguished by its orange-red hair. The name comes from the Malay 'man-of-the-woods', perhaps too literally presented here.

harpies: fabulous winged monsters with a woman's face and birds' talons.

Vampire Bat: dear to lovers of horror fiction and to enemies of the American tropics, the so-called Vampire Bat named here is in fact the giant fruit-eating bat of Brazil, which for years was falsely accused of the blood-sucking proclivities of its much smaller cousin, *Desmodus rufus*, which feeds chiefly on horses.

498 *lyras*: presumably the lyre bird of Australia, named for its lyre-shaped tail.

499 *midshipman's dress*: a midshipman is a young man in training to become a naval officer; a young lady of Emily's gentle birth would hardly don the costume of an ordinary sailor.

502 *shallops*: sloops.

508 *corsair*: a pirate ship.

509 *Stentor*: a Greek warrior in the Trojan War, famed for his powerful voice.

510 *machinist*: an engineer.